PRAISE FOR

Troublemakers

"Leslie Berlin is a master historian of Silicon Valley, and the publication of this book is a landmark event. Kaleidoscopic, ambitious, and brilliant, the book draws on a dazzling cast of characters to chart the rise of the five industries that have come to define technology today and, collectively, to remake the world."
—**Eric Schmidt, former CEO of Google and Executive Chairman of Alphabet, Inc.**

"Apple's late co-founder Steve Jobs liked to hang out with an older generation of Silicon Valley entrepreneurs because, he claimed, 'you can't really understand what is going on now unless you understand what came before.' For the rest of us, Leslie Berlin's sweeping new history of the Valley is the next best thing. . . . essential reading . . . the book does a masterful job explaining profound discoveries like recombinant DNA and microprocessors, while also keeping up a brisk narrative that's aided by off-color accounts of naked hot tub meetings and pot-addled programmers."
—**Fortune**

"[A] deeply researched and dramatic narrative of Silicon Valley's early years. . . . meticulously told stories permit the reader to gain a nuanced understanding of the emergence of the broader technology ecosystem that has enabled Silicon Valley to thrive. . . . compelling history."
—*The New York Times*

"Stress, sleep loss and operatic emotion . . . a fetching portrait of the less chronicled years of Silicon Valley."
—**Michael Moritz, *The Wall Street Journal***

"This kaleidoscopic history alternates among seven 'troublemakers'—entrepreneurs, investors, and managers who helped drive the tech revolution. . . . Berlin sketches their lives in vivid detail, showing the vital contributions of these 'audacious' leaders."
—*The New Yorker*

"[Berlin's] charge is to deliver the narrative, a task that she does with great style and storytelling skill. She relies on a trove of unpublished primary sources, including interviews, that few would be able to synthesize so deftly. . . . There is much to learn from Berlin's account, particularly that Silicon Valley has long provided the backdrop where technology, elite education, institutional capital, and entrepreneurship collide with incredible force."
—*The Christian Science Monitor*

"Engrossing. . . . Troublemakers offers a corrective to the regnant great man theory of technological progress of which the virtuosic Mr. Jobs is exhibit A. In narrating these innovations, Berlin shows the village that brought them forth."
—*San Francisco Chronicle*

"While it is a hefty tome, the book is compelling as it maps out the 'building' of the Valley, the challenges its early tech pioneers faced, as well highlighting those who reached dizzying success only to suffer as the dotcom bubble burst."
—*Financial Times*

"Leslie Berlin has done it again. Following on her richly informative biography of Intel cofounder Robert Noyce, *The Man Behind the Micro-chip*, Berlin now brings us a definitive account of Silicon Valley's 'break-through years' in the 1970s. *Troublemakers* recounts the fascinating careers of six little-known but enormously impactful players who shaped the Valley's unique high-tech ecosystem. As entertaining as it is authoritative, *Troublemakers* is required reading for anyone seeking to understand how the tech revolution took root in the San Francisco Bay Area and eventually transformed the entire planet's way of life."
—**David M. Kennedy, Donald J. McLachlan Professor of History, Emeritus, at Stanford University and winner of the Pulitzer Prize**

"If you're old enough to remember a life without your iPhone clutched in one hand at all times, then you may have found yourself wondering why that iPhone has rapidly become something you can't live without (you were living without it before, I promise.) . . . [*Troublemakers*] digs deep into exactly how we arrived in the technology-obsessed/dependent landscape we currently occupy today."
—*Bustle*

Troublemakers

SILICON VALLEY'S COMING OF AGE

■■■

Leslie Berlin

Simon & Schuster Paperbacks

New York   London   Toronto   Sydney   New Delhi

Simon & Schuster Paperbacks
An Imprint of Simon & Schuster, Inc.
1230 Avenue of the Americas
New York, NY 10020

First Simon & Schuster trade paperback edition November 2018

SIMON & SCHUSTER PAPERBACKS and colophon are registered trademarks
of Simon & Schuster, Inc.

For information about special discounts for bulk purchases,
please contact Simon & Schuster Special Sales at 1-866-506-1949
or business@simonandschuster.com.

The Simon & Schuster Speakers Bureau can bring authors to your
live event. For more information or to book an event contact the
Simon & Schuster Speakers Bureau at 1-866-248-3049 or visit our
website at www.simonspeakers.com.

Interior design by Lewelin Polanco

Manufactured in the United States of America

10 9 8 7 6 5 4 3 2 1

Library of Congress Cataloging-in-Publication Data is available.

ISBN 978-1-4516-5150-8
ISBN 978-1-4516-5151-5 (pbk)
ISBN 978-1-4516-5152-2 (ebook)

Here's to the crazy ones. The misfits. The rebels. The troublemakers. The round pegs in the square holes. The ones who see things differently. They're not fond of rules. And they have no respect for the status quo. You can quote them, disagree with them, glorify or vilify them. About the only thing you can't do is ignore them. Because they change things. They push the human race forward. And while some may see them as the crazy ones, we see genius. Because the people who are crazy enough to think they can change the world, are the ones who do.

—APPLE COMPUTER ADVERTISEMENT (1997)

Indiana Jones: I'm going after that truck.
Sallah: How?
Indiana Jones: I don't know. I'm making this up as I go.

—LAWRENCE KASDAN,
THE RAIDERS OF THE LOST ARK (1981)

Contents

Introduction

████████████ A BIT LIKE LOVE

Tens of millions of people have watched the video of Steve Jobs's 2005 commencement address at Stanford University. Executives and journalists quote from it. An exhibit at London's Victoria and Albert Museum featured excerpts. Every year, a neighbor festoons a pair of trees along Palo Alto's bicycle boulevard with photos of Jobs and quotes, many from the commencement address. The speech, given on a sunny June morning and lasting just over fourteen minutes, is indeed remarkable; in it, Jobs talks about his adoption, his cancer diagnosis, his work, his family, and his inspirations.

About six minutes into his address, Jobs tells the story of being fired from Apple when he was thirty years old. "The focus of my entire adult life was gone, and it was devastating," he says. The next lines are easy to overlook but vital for anyone who wants to understand how Silicon Valley works: "I felt that I had let the previous generation of entrepreneurs down—that I had dropped the baton as it was being passed to me. I met with David Packard and Bob Noyce and tried to apologize for screwing up so badly."[1]

Throughout his career, Jobs spent time with older entrepreneurs such as Intel's Robert Noyce and Andy Grove, as well as with Regis McKenna, a former employee of National Semiconductor who founded Silicon Valley's leading public relations and marketing firm. In 2003, I asked Jobs why he spent so much time with the semiconductor pioneers. He said he "wanted to smell that second wonderful era of the valley, the semiconductor companies leading into the computer." He added, "You can't really understand what is going on now unless you understand what came before."[2]

Jobs did not mention it in his Stanford speech, but after he returned to Apple, he passed the baton forward to a new generation of entrepreneurs. Google cofounders Sergey Brin and Larry Page went to Jobs for advice when they were starting their company. Facebook's Mark Zuckerberg considered Jobs a mentor.[3] I once saw Zuckerberg eating dinner in the same hidden corner of a low-key Palo Alto Mexican restaurant where, years before, I had seen Jobs. Zuckerberg was in the same seat, at the same table, sitting, as Jobs had, alone, his back to the window.

The period covered in this book marks the first time that the generational handoff happened on a grand scale, as pioneers of the semiconductor industry passed the baton to younger up-and-comers developing innovations that would one day occupy the center of our lives.[*] The result was staggering. Between 1969 and 1976, the narrow peninsula south of San Francisco was the site of the most significant and diverse burst of technological innovation of the past 150 years. In the space of thirty-five miles and seven years, innovators developed the microprocessor, the personal computer, and recombinant DNA. Entrepreneurs founded Apple, Atari, Genentech, and the pioneering venture capital firms Sequoia Capital and Kleiner Perkins Caufield & Byers. Five major industries were born: personal computing, video games, advanced semiconductor logic, modern venture capital, and biotechnology. In this same remarkable period, a laboratory at the Stanford Research Institute received the first transmission of data from the Arpanet, the precursor to the Internet. Stanford University pioneered a model of turning faculty research into commercial products that has since earned nearly $2 billion for the university.[4] The independent software industry was born.[†] Xerox opened the Palo Alto Research Center

---

[*] Former employees of the Valley's first wave of innovative tech companies founded in the 1930s and 1940s (companies such as Hewlett-Packard and Varian Associates) typically had little direct involvement with the rise of the semiconductor companies that followed.

[†] In 1969, IBM, facing antitrust action, unbundled its hardware, software, and services businesses, after which the revenues of independent software vendors exploded from $20 million to $400 million.

(PARC), which would pioneer computing's graphical user interface, icons, Ethernet, and laser printer. Upstart CEOs with an eye to the future lay the foundation for the now-tight political alliance between Silicon Valley and Washington, DC.

These years marked Silicon Valley's coming of age, the critical period when the Valley was transformed from a relatively obscure regional economy with the microchip industry at its heart into an economic engine whose rate of job growth has been double that of the country as a whole for the past five years.[5] In the process, Silicon Valley has spawned countless imitators and numerous industries that together have created the modern world. *Troublemakers* charts these changes through the lives of the people who made them happen. The book opens in 1969, when the transformation is just beginning in the shadow of the Vietnam War. The story ends in 1983, after the world has begun to experience the effects of these revolutionary innovations. By that date, a new generation of innovators and industries was on the rise.

In 1969, the name *Silicon Valley* did not exist; it would be two years before the region renowned for its plum and apricot orchards gained its current moniker. Starter homes in Palo Alto that now sell for nearly $2 million cost $60,000, but a computer with less processing power than today's $250 programmable thermostat cost hundreds of thousands of dollars.[6] The term "biotechnology" had not been coined, and the prospect of combining genes to create a hybrid life-form sounded like science fiction. Computers occupied rooms, not desktops—and certainly not pockets or wristbands. Silicon Valley companies built products for engineers, not consumers. To the extent that the average American thought at all about computers, it was only to curse them as the culprits blamed by banks when a checkbook did not balance properly.

Today we consider Silicon Valley the hub of the information economy. In 1969, however, the Valley was a manufacturing center; 60 percent of people working in the local electronics industry worked production jobs.[7] Defense-based companies such as Lockheed and GTE Sylvania were pillars of the local high-tech economy, and the sophisticated global electronics supply chain was only beginning to take shape. Federal law barred pension funds from investing in high-risk/high-reward young companies. Entrepreneurs were suspect—weirdos who could not cut it climbing the ladder in a good corporate environment.[8]

Yet within a dozen years, products from Silicon Valley companies such as Apple, Atari, and their competitors were reshaping how people worked and played. Insulin had been synthesized using recombinant DNA, and the Supreme Court had declared that genetically engineered life-forms could be covered by patents.[9] Software companies were going public. Pension funds had invested hundreds of millions of dollars with venture capitalists backing the riskiest early-stage investments. The US economy was beginning to reshape itself around information and services, rather than manufacturing. No longer did politicians claim that what was good for General Motors and other large companies was good for America. Entrepreneurs were the new American business heroes.

The 1960s set much of the tone for Silicon Valley, as engineers and scientists built the first microchips, and entrepreneurs launched startup companies that offered generous stock options and flat organizational structures. But the dozen years that followed—the years that unfold in these pages—laid the foundation for the world we know today.

Today, when five of the six most valuable companies on the planet are high-tech firms, three based in Silicon Valley;[10] when the high-tech sector accounts for 9 percent of US employment, 17 percent of gross domestic product, and 60 percent of US exports;[11] when 400 hours of video are uploaded to a single platform (YouTube) every minute,[12] and more than 200 billion emails are sent every day;[13] when the video game industry is larger than the movie business, and the biotech industry generates $325 billion in revenues in the United States alone;[14] when US manufacturing jobs continue to disappear, many of them lost to automation;[15] when the electronics manufacturing and equipment industry spends almost $60 million on lobbying in a single year;[16] when a sitting US president guest-edits an issue of *Wired* magazine devoted to the future of innovation, and his successor rides to victory through savvy use of online social networks also blamed for disseminating misinformation; at a time when our relationship with technology is so intimate that 46 percent of Americans say they cannot live without their smartphones and one-third of adults would rather give up sex than their phones[17]—it makes sense to ask how we got here.

Understanding our modern world means understanding Silicon Valley's breakthrough era, and a close examination of those years makes it clear that Silicon Valley was not built by a few isolated geniuses. Anyone who has worked in Silicon Valley (or anywhere else, for that matter) knows that although the spotlight often has room for only one person, those just outside its illuminated circle are at least as responsible for the success that has given the star a moment of glory. At a party I attended a number of years ago, someone who had served as chief operating officer of a major Silicon Valley company with a superstar CEO sang a little song about this phenomenon. The only lyrics were "I did all the work. He got all the credit." Innovation is a team sport, suffused with similar passion and drama. "When a really difficult thing is being worked on and you get synergy from the small team in just the right way, you can't describe it. It's like love. It is love," says Alan Kay, a visionary computer scientist who worked at Xerox PARC, Apple, and Atari. "You're trying to nurture this thing that is not alive into being alive."[18]

Silicon Valley's success today and for decades has rested on personal connections and collaboration that transcend companies, industries, and generations. Men and women working in Silicon Valley during the breakthrough years brought technology from Pentagon offices and university laboratories to the rest of us. The history is far richer and more complex than the isolated tales of famous entrepreneurs—yet that history has never before been told. Looking at multiple contemporaneous developments reveals the overlapping experiences and individuals that made them possible.

This history is a tale of upstarts who defied the norms of their professions and peers, not just once but again and again. The entrepreneurs were not the only mavericks in Silicon Valley. Suit-wearing attorneys, venture capitalists, and angel investors risked their careers to back barefoot hippies. Public relations experts invented a new type of all-American media darling—the high-tech entrepreneur—in an effort to make complex technology more appealing to consumers. Executives banded together across industries to influence politicians. Scientists and financiers dared to play with DNA, the very stuff of life.

Troublemakers tells these stories, which feature some of the most famous names in Silicon Valley history, while also profiling seven other individuals in depth. Bob Taylor kick-started the precursor to the Internet, the Arpanet, and masterminded the personal computer. Mike Markkula served as Apple's first chairman, with an ownership stake equal to that of Steve Wozniak and Steve Jobs. Sandra Kurtzig, an early software entrepreneur, was the first woman to take a technology company public. Bob Swanson cofounded Genentech. Al Alcorn designed the first wildly successful video game, Atari's *Pong*. Fawn Alvarez rose from an assembler on a factory line to the executive suite. Niels Reimers changed how university innovations reach the public; in the process, he helped launch the biotech industry.

I chose these seven individuals not only for what they have done but also for who they are. Their work deserves more attention. Their fascinating stories speak directly to the forces that built Silicon Valley and continue to resonate in our lives today.

The troublemakers did not always have much in common, but they shared two traits: they were persistent, and they were audacious. None of the changes they pushed came easily. Inventing the future is a fraught business. At times, the previous generation—the one passing the baton— did not know what to do in the face of new technologies. When there were no blueprints to trace with a finger and no maps to follow, the troublemakers remained undaunted. They made it up as they went along.

ARRIVAL
1969–1971

As 1968 turned to 1969, the United States was in turmoil. In the previous few months, assassins' bullets had felled Martin Luther King, Jr., and Robert F. Kennedy. Police had clubbed and teargassed protestors on the sidewalks outside the Democratic National Convention in Chicago; inside, rampant dissension threatened the future of the party.* The Vietnam War was claiming the lives of 1,200 American soldiers every month. Eighty percent of the dead were between the ages of eighteen and twenty-five. Life opined, "Wherever we look, something's wrong."[1]

But in the young electronics industry that had taken root near Stanford a decade earlier, late 1968 was a time of unbridled optimism. Technology developed on the San Francisco Peninsula had helped send Americans beyond the planet's atmosphere—and then to beam back to Earth photos taken by the astronauts. Established and important companies such as Lockheed, IBM, and Sylvania had built research or manufacturing operations in the area. The homegrown electronics giant Hewlett-Packard, now nearly three decades old, employed thousands of people. Eight microchip companies had

---

* The convention marked a turning point for the Democrats. The party won seven of nine elections between 1933 and 1968 and then lost all but three of the ten between 1968 and 2008.

been born on the narrow forty-mile stretch of peninsula in just the past four years. Electronics companies thrived.[2] One magazine called the moment "the Age of Electro-Aquarius."[3]

Meanwhile, the population of the Peninsula was exploding. The number of residents had tripled in two decades, from roughly 300,000 in 1950 to more than 1 million.[4] It was as if a new person had moved into Silicon Valley every fifteen minutes for twenty years. The new arrivals were mostly young and often well educated. They tended to come from other parts of the state or the country, from Southern California or Boston or Chicago or Utah or farm towns in the Midwest. They came on their own or with families, car trunks hardly closing over jumbles of clothes and books and household goods and photos from home tucked neatly among the sheets and towels. They came because they were starting college or wanted to join the hippie scene or because they needed a job, and they knew that the electronics manufacturers were willing to train unskilled labor. Together these new arrivals, collaborating with those who had come before, would unleash an era of unprecedented innovation, a flowering of creativity and technical virtuosity that remains unmatched to this day.

By 1969, the newcomers had transformed the land around them. In a bucolic farming region fragrant with orchards and known as the Valley of the Heart's Delight, the new arrivals built suburbs and shopping malls. They pressed local governments to open schools and parks. Early every spring they hosed down the orchard trees that remained to prevent messy fruit from dropping on manicured yards and still fresh sidewalks.

At first the changes were hardly noticeable. "We watched the electronics factories spring up like mushrooms from Redwood City to San Jose, and we did not in the least realize or understand the magnitude of the transformation," recalled Stanford professor and novelist Wallace Stegner. "Then one spring we drove through and the endless froth of blossoms was no more than local patches. One summer we found that there were no longer any orchards where we could pick our own apricots at a pittance a pailful."[5]

By 1969, the newcomers had created a new business culture, much of it centered around silicon microchips. Silicon came to the Valley in 1956, when Nobel Laureate William Shockley launched a company to build transistors. The very next year, eight of Shockley's top young scientists and engineers left to launch Fairchild Semiconductor, the first successful silicon company in Silicon

Valley.[6] *In the ensuing dozen years, Fairchild gained renown for the quality and innovative record of its researchers, engineers, and sales and marketing teams. The company was also a huge financial success. In 1965, it was the fastest-growing stock on the New York Stock Exchange, its share price rocketing fifty points in a single month.*[7] *Fairchild's success inspired others to want to start their own companies, and the first of many waves of entrepreneurial frenzy gripped the valley.*

Much was being lost on the San Francisco Peninsula. Much was just beginning. A bucolic paradise was being plowed under, and it seemed that anything was possible.

Prometheus in the Pentagon

BOB TAYLOR

It started with a crash. On October 29, 1969, a $700,000 Sigma 7 computer at UCLA sent a command to a slightly leaner SDS 940 machine at the Stanford Research Institute.

The UCLA machine looked formidable. A half-dozen or more refrigerator-sized components lined the perimeter of a special room dedicated to their use, the entire operation controlled by an expert sitting at a typewriter-looking console in the center of the room.

But when the UCLA machine sent its command—LOGIN—up the California coast, the Stanford computer crashed before the word, typed letter by letter, even got past *G*.

After a bit of reprogramming, the message was sent again and received, and the first computer network, called the Arpanet, was online.[1]

This transmission, often hailed as the "birthday of the Internet," has been celebrated in conferences, books, speeches, and news reports. Plaques have been erected in its honor. One man who worked on the network at UCLA has since recast the failed initial login in biblical terms: "And so the very first message ever sent over the Internet was 'Lo!' as in 'Lo and behold!' Quite a prophetic message indeed."[2]

Of course, no such message was intended. That LOGIN was the computing equivalent of Alexander Graham Bell's "Mr. Watson, come here": a practical effort to determine if a message had been received. Whatever importance the LOGIN transmission has achieved by now, back in 1969, the sent message was not a momentous achievement. At UCLA and Stanford, there was a bit of applause and a lot of relief but no announcements from press offices, no reporters waiting to hear if the connection would work. A simple notation in a UCLA log book, "2230, Talked to SRI host to host," served as recognition. At the universities—and at the Pentagon, where the Department of Defense had funded this new network; and in Cambridge, Massachusetts, where a small company called Bolt Beranek and Newman was building the equipment and writing much of the software—most people involved with the network scarcely paused to note the transmission.

And Bob Taylor, the man who had jump-started the network, paid no attention at all.

■ ■ ■ ■

Three years earlier, in 1966, three thousand miles from Silicon Valley in Washington, DC, Taylor had been walking back from lunch with his new employee, twenty-three-year-old Barry Wessler. To be more exact: Taylor was walking, and Wessler was sneaking in the occasional half jog to keep up. Taylor—thirty-four, slight, a cloud of pipe smoke obscuring a face that reminded more than one person of John F. Kennedy—did nothing at a sedate pace. Every morning Taylor squealed his rare BMW 503 into the giant Pentagon lot after driving as fast as possible from his home in suburban Maryland. He grabbed his heavy, hard-leather briefcase, clenched his pipe between his teeth, and strode through the halls of the Pentagon to his office in the D-ring, stepping neatly around the soldiers on adult-sized tricycles who rode up and down the endless corridors and the ramps between floors, delivering mail. Or he would skip the office and head to the airport for a trip to Boston or Pittsburgh or Palo Alto, so he could check on the research he was funding with an annual budget of $15 million to advance computer technology.

Taylor and Wessler worked at the Advanced Research Projects Agency (ARPA), which oversaw research initiatives for the Department of Defense. (ARPA is today called DARPA.) Taylor ran ARPA's Information Processing Techniques Office. The $15 million he had to work with was only a tiny fraction of ARPA's $250 million annual budget, but he could direct it all toward computing.[3]

The Department of Defense believed that advanced computing would give the United States an edge over the USSR, and Taylor, too, was an ardent

believer in the promise of computers. He thought that no other technological field was "able to so strongly affect all other areas of human endeavor."
Moreover, he was confident that "in no other domain is there a greater
technological distance between the United States and the rest of the world."
Computers, he believed, would help the United States win the Cold War.[4]

Walking with his protégé, Taylor stopped for no apparent reason.
"Let's take a little detour, Barry." Taylor tended to bark, not because he
was angry but because he was impatient. He pivoted on his heel and began
walking in the other direction, Wessler close behind.

A few minutes later they were beelining through the E-ring, where the
muckety-mucks of the Pentagon had offices behind thick wooden doors
guarded by secretaries. With a jovial "Hi. I'm just going in to talk to him,"
Taylor swept past the secretary of an assistant secretary of defense. Before
the desk attendant could even push back her chair, Taylor had opened the
office door, slipped inside, and taken a seat.

Wessler sat down next to him, stunned. Only minutes before, Taylor had been complaining about the byzantine rules of communication
within the Pentagon, laying out how if one man wanted to speak with
another of higher rank or status, protocol dictated that the lower man's
secretary call the higher one's secretary to set up a phone call, and then,
at the agreed-upon time, the lower-level man's secretary had to call the
higher man's secretary and then get the lower-level man on the line before
the higher-level man's secretary would bring her boss on. "A lot of work
just to talk to a guy," he had groused.

Barging into the office of a man several reporting levels above him,
Taylor had just broken every rule.

Meanwhile, the assistant secretary, surprised at his desk, had risen to
his feet, face blazing. His secretary began apologizing.

Taylor began talking about the projects his office was funding: what he
was excited about, where he had concerns. The next thing Wessler knew,
the assistant secretary was walking back around to his seat, dismissing his
secretary, listening, and nodding his head.[5]

■ ■ ■ ■

The primary function of bureaucracy, Taylor believed, was to block communication, and nothing irritated him more than obstacles that prevented
people from sharing ideas and getting feedback. Taylor, like any good

Texan, loved football—he always regretted having been too small to play in college—and for his entire career he anticipated any encounter with a bureaucratic system as a new play to strategize. Should he barrel through, as he did in the assistant secretary's office? Or should he finesse the situation to his advantage? Both finesse and assertiveness could get him what he wanted.

It was Taylor's good fortune that the director of ARPA, Charles Herzfeld, felt the same way about the Pentagon's bureaucracy. Herzfeld, a University of Chicago–trained physicist from Vienna, was, in one memorable description, "an old Washington hand who knew high-level B.S. when he heard it and wasn't afraid to call it by name."[6] An informal conversation between the two men—one that came about when Taylor raced up to Herzfeld's office in the E-ring and rapped on the door—led to the October 29 transmission that brought the Arpanet online.

When Herzfeld invited him in, Taylor saw no need for a preamble.

"I want to build a network."

Tell me more, Herzfeld responded.

■ ■ ■ ■

Bob Taylor had taken an unconventional path to become what he liked to call a "computerist." His first encounter with a computer had been a nonencounter. Taylor was finishing a master's thesis in physiological psychology at the University of Texas. His research, to determine how the brain locates the origin of a sound, had yielded pages of data that he needed to analyze. His adviser sent him to the university's computer center to run the data through a statistical tool on the school's prized machine.[7]

The computer center was cavernous. A wall separated the room-sized computer from its users. In front of the wall, a student in a white lab coat sat at a high desk. Taylor could not see much of the machine, but he could hear its dull humming behind the young sentry.

Yes, the student said, Taylor could use the computer. He went on to explain how. First, Taylor needed to take an empty stack of cards that were roughly the size of dollar bills to one of the center's keypunch machines. He then needed to type in his data by punching a specific pattern of holes, up to 72 digits per card, that corresponded to his data set. The computer would read the cards—

Taylor interrupted, not sure he had heard correctly. "You mean I have to sit down and punch holes in these cards to get my data in, and then I have to take the cards over to the computer, and I give the cards to a guy who runs them through the computer, and I go away and come back and get the results on a long printout of paper?"

The student said that Taylor had it right. It could take several hours or even days, depending on demand, for the printout to be completed.[8]

Taylor thought that system was ridiculous. "I'm not going to do that," he said.

He left the building, irritated. There's got to be a better way, he thought. It would be just as fast to input by hand every calculation into the lab's Monroe calculator, a heavy, typewriter-sized beast with more than a hundred round clacking keys. And that was exactly what he did.

About a year after graduating, Taylor took a job at the Glenn L. Martin Company (now Martin Marietta) as an engineer to work on the system design for the Pershing missile. It was a desk job with a twist. He analyzed and integrated the computerized test and checkout processes, and he also occasionally donned a special suit and ran parts of the missile through trials in a cold-test lab so frigid that he was warned to blink frequently so his eyeballs did not freeze.[9]

During a break at Martin, Taylor read a paper that changed the course of his career. Just over seven pages long, it appeared in a new technical journal, *IRE Transactions on Human Factors in Electronics*. The paper addressed what its author, J.C.R. "Lick" Licklider, called "Man-Computer Symbiosis." Like Taylor, Licklider was a quantitatively oriented psychologist whose work was more about the physics of perception than the nuances of private feelings.

The paper was less a report than a vision: "The hope is that, in not too many years, human brains and computing machines will be coupled together very tightly, and that the resulting partnership will think as no human brain has ever thought and process data in a way not approached by the information-handling machines we know today." The computer, in Licklider's imagining, could become something more than a tool for calculation; if responsive and easy enough to use, it could help a person to think and create.[10]

"When I read [the article], I just lit up," Taylor recalls. "I said, 'Yes. That's the answer to my key punch dilemma. That's worth working on.'"[11]

For years, he had been hearing that a computer was like a giant brain, but he had never before understood the comparison. The overgrown calculator in the University of Texas computer center and its cousin machines around the world were nothing at all like the human brain. Licklider's interactive computers, however, were brainlike: tools that extended humans' capacities to think and create.

Two years after reading Licklider's article, Taylor met the author. Taylor was then working at NASA, allocating money for computing research, and Licklider was running the Information Processing Techniques Office at ARPA, a job Taylor would himself assume a few years later. The two men became friends, so much so that Licklider's wife, Louise, would later tell Taylor, "You are a very special son to us, and we love you."[12] Licklider and Taylor traveled together for work and enjoyed many evenings on the road in small, unassuming bars. Sometimes they would talk about their jobs, sharing ideas about the researchers who were pushing to attain Licklider's vision, but just as often they shared personal stories. One memorable time, Taylor, who sang and played guitar, took the stage with a folk band in Greece, and Licklider insisted on staying until early morning to hear every piece his colleague played.

■ ■ ■ ■

Taylor joined ARPA's Information Processing Techniques Office in 1965 as assistant director to Licklider's successor, Ivan Sutherland, another leading proponent of interactive computing. Taylor and Sutherland worked closely together—they were the only people in the office, aside from a secretary—and Taylor enjoyed traveling to the many campuses where ARPA had funded the development of new computer science programs. Before 1960, there were no computer science departments. People interested in computing clustered in engineering, math, or physics departments. But the ARPA money allocated by Licklider, Sutherland, and Taylor had changed all that. Now schools such as MIT, Carnegie Mellon, the University of Utah, the University of Illinois, the University of California at Berkeley, and Stanford had fledgling computer science departments. Many of these departments were built around a new type of computer whose development ARPA had also helped fund. The machines, called time-sharing computers, could switch from running one person's program to running the next person's so quickly that each user, sitting alone at one of many typewriterlike terminals, often had the illusion that the computer was working exclusively on his or her program. Time-sharing computers did not require punch cards and displayed some results immediately.

Taylor, who racked up more than 500,000 miles on United Airlines alone during his four years at ARPA, developed a bit of a routine for each visit to the universities' computer science programs.[13] He would rent a cheap car, head to campus, and, after following up with the research team to offer assistance, would find the graduate students who cared about computing. Graduate students, Taylor felt, did the most interesting work in the field.

To Taylor's surprise, many students told him that the time-sharing computers had helped them find friends and colleagues. A person could log on to the computer, look in a directory to see what had changed since his or her last login, and then, if a new program or clever bit of code had been added, it was easy to tell who was responsible and invite the coder for a beer. Licklider had envisioned computers that helped people to think and create by connecting one person to one machine. Now Taylor saw the potential for computers to do even more: to connect one person to another. A time-sharing computer, in one sense, was the hub of a wheel with people at the end of each spoke.

Taylor believed in the power of communities built around a common core. A clinical psychologist (of the sort Taylor did not want to be confused with) might point out that Taylor had been raised in precisely this type of community.[*] In 1932, when Taylor was twenty-eight days old, a Methodist minister and his wife had adopted him. Taylor describes his father as a quiet intellectual ("not the pulpit-pounding preacher you might imagine") and claims that he "introduced southwest Texas Methodism to existentialism in the late 1940s." Jobs were scarce in Texas in the middle of the Depression, and Taylor's earliest memory, from when he was three, is leaving church with his parents after a prayer meeting and

---

[*] Taylor says that he did not go on for a PhD in psychology in part because because he would have had to qualify in areas such as child psychology and other "soft" disciplines that he thought were "crap."

finding the back seat of the family car filled with food—his father's pay for the month.

By the time Taylor was twelve, he had lived in six towns, most of them tiny. Yet wherever he lived, he was immersed in a world of church socials, church potlucks, church services, and church camp. He spent several summers at a camp in the Texas hill country set aside for Methodist ministers and their families. It is easy to imagine that from his peripatetic-but-rooted childhood he imbibed a lesson that would help him understand the potential of a network connecting people around the world: that community comes not from geography but from shared beliefs and interests.*

■ ■ ■ ■

Taylor was particularly interested in connecting the *right* people with one another. For a minister's son who had once wanted to be a minister himself, he could be merciless in his judgments. To him, people were either geniuses or duds. He was interested in the geniuses. Once he found someone worth finding, he considered it his calling to help that person meet and work with others of the same caliber.

To this end, one of Taylor's early moves at ARPA was to revamp the mandatory meetings of the lab leaders, or "principal investigators," funded by the Information Processing Techniques Office. Taylor turned the principal investigators' meetings into destination events. Under Licklider and Sutherland, the meetings had been relatively short—a few hours, maybe— and folded into a larger meeting such as the Fall or Spring Joint Computer Conference, which the principal investigators would be attending anyway. Now Taylor scheduled them for appealing times and places (Alta, Utah, during ski season; New Orleans during Mardi Gras; Hawaii in the dead of winter) and planned the formal sessions so as not to conflict with the best hours on the slopes or watching the parades in Fat City. Taylor wanted the

---

* Taylor claims his childhood influenced him only insofar as his parents always told him he was special because they had chosen him to be their son, while most parents just had to take whoever showed up. That notion, he admits, instilled in him a certain unshakable self-confidence. Apple cofounder Steve Jobs, also adopted, said that his parents had told him the same thing.

principal investigators to interact again and again, not only around conference tables but also on a chairlift or at the beach.

At the formal sessions, Taylor asked each principal investigator to give a talk, urging him to prepare the kind of presentation he would like to hear. After the talks, he says, "I got them to argue with one another, which was very healthy, I think, and helpful to me because I would get insights about strengths and weaknesses."[14]

In the meetings, Taylor set himself up as an outsider, more conductor than musician. For him, it was not a difficult role to assume. Nearly everyone else had a PhD in physics, engineering, or math. (Computer science was still such a young field that the first generation of PhDs was only beginning to graduate.)[15] Taylor, meanwhile, had a master's degree in psychology. Almost everyone else in the room came from one of the coasts or had been educated there. Not Taylor. In a Texas twang, he talked up his summer jobs herding cattle—he had once climbed onto the back of a 2,000-pound Brahma bull, just to see what would happen—or working in the oil fields.

Years later, he would discover that some of the principal investigators had been unhappy to have him in charge. He was the only head of the Information Processing Techniques Office not to hold a doctorate and the only one to have taken the job without a significant record of contributions to the field.

Taylor nonetheless says that for the entire time he was at ARPA, he had no sense of inferiority, even though he knew that he had not been the top choice to succeed Ivan Sutherland as director of the office.[16] "There was not a one of them who I thought was smarter than me," he says. ("I thought Lick was smarter than me," he adds.) Taylor had graduated from high school at sixteen, had an IQ so high (154) it was written up in his small-town newspaper when he was a child, and finished college with a double major in psychology and math, as well as substantial course work in religion, English, and philosophy.[17] He did not possess the principal investigators' depth of computer science knowledge, but the breadth of Taylor's mind was wide and his chutzpah, limitless. He was unafraid to fire question after question in meetings, an assertive technique that nonetheless often helped the experts clarify their own thinking. (As Barry Wessler put it, "Bob would not necessarily have a firm technological grasp on the details, but you would come in because you had a problem, and you'd walk out with a solution.") Taylor's different background also meant he had nothing invested in any technical or social orthodoxy and could maintain a useful distance from the principal investigators. He had not gone to school with any of them. He was not their colleague or student, nor would he train students who might want to work in their labs.

As Taylor listened and probed at the principal investigator meetings and on university campuses, an idea began to pull at the edges of his mind. Almost all of the principal investigators worked on time-sharing computers; why not come up with some way to connect the computers to one another? If connecting a few terminals to a time-sharing computer had brought together diverse users on the individual campuses, imagine what could happen if each of those time-sharing machines, with its dozens of users, was connected to other time-sharing machines. The community would grow and grow, until it became what Taylor would come to call a networked "supercommunity" or "metacommunity."[18] A network, he wrote in 1968, could "overcome the geographical barriers which prohibit the cooperative working together of people of common interests—be they students, scientists, soldiers, statesmen, or all of these."[19]

Taylor resolved to build this network one day when he was back at the Pentagon and happened to turn his head to look at the entrance to the small room next door. He spent a lot of time in that room, which housed three computer terminals, each one linked directly to a different time-sharing computer: one at MIT, one at Berkeley, a third at the System Development Corporation in Santa Monica. The three computers could not talk to one another.[20] But Taylor, who participated in all three computing worlds, knew that the ideas and problems being addressed in each place were similar. "You don't have to be half smart to see this thing ought to be designed such that you have just one terminal and you can go wherever you want to go," he said years later, recalling his thoughts when he had looked at the array of machines and imagined a network connecting them.[21]

Taylor said that he went to ARPA wanting to do "one big thing," and in this computer network, he saw his opportunity. Networks had come up before in a "wouldn't it be nice" kind of way, and Licklider had outlined his thoughts on what he called an "intergalactic network" in 1963 in a memo to principal investigators. By the fall of 1966, when Taylor turned to look at the three terminals, he and Licklider had been friends for years, but Taylor says he had never connected his ideas to Licklider's imagined "intergalactic network." Nor did he and Licklider talk about a computer network until after the Arpanet project was under way, he said, though Licklider supported the idea once he heard it. Taylor says, "It could be that our discussions subconsciously prompted me to say maybe we can connect all these computers. I just know that it wasn't explicit."[22]

Taylor moved the network idea from "wouldn't it be nice" to "we can do this—now." He understood that there were enough people in computer science—hundreds—who would value being networked to each other and who also could figure out how to build the network. To demonstrate that a network was technically doable, he funded a small experiment to send a few bits back and forth, rather slowly and across telephone lines, between a TX-2 computer at MIT's Lincoln Laboratory and an incompatible Q-32 computer at System Development Corporation. The experiment also made clear that long-distance communication between two different computer systems, though feasible, would require a huge amount of work to be made quick, reliable, and scalable.

There was one more reason why the time to build a network was now, Taylor knew: ARPA would pay for it. If the project had to be funded piecemeal, with different government agencies or universities each financing a portion of its development, the work would be slowed down by endless bickering around fair shares and conflicting reporting procedures and requirements.

Taylor likely marshaled all these arguments in the fall of 1966 as he sat forward on his chair in Charles Herzfeld's office, explaining why ARPA should fund a computer network. (Neither Taylor nor Herzfeld has an exact memory of the conversation, nor is there a written record.) Taylor's primary interest was in connecting ARPA-funded researchers and thereby amplifying the power of their brains. He told Herzfeld that a computer network would enable principal investigators at one site to use programs and data at other sites, thereby cutting down on wasteful duplication. Moreover, because the network could connect different types of machines, there could also be a direct payoff for the Department of Defense: a network would free up people in the field or at the Pentagon to buy the best computer for a given job, rather than being locked into a single system for everyone. And, Taylor continued, it was worth building this network only if it could be fast. Otherwise, "the illusion of being a local user, near the computer, [would] be destroyed."[23] Contrary to popular beliefs about the origins of the Internet, there is one important argument that Taylor did *not* make for the network: he never claimed that the network would be good for national security.[24] Later, Herzfeld and his successors, petitioning Congress to continue funding for the Arpanet, would argue that the network could enable communications to continue even in the event of a nuclear attack. But that was not a motivation for the network's creation.

Herzfeld had been quiet for almost half an hour while Taylor made his case. Both men knew that the Information Processing Techniques Office budget was committed for the fiscal year. But it was not unusual to fund a new project using money earmarked for something else and then to write the new project into the next year's budget as maintenance of an ongoing effort. Congress seemed much more inclined to fund projects already under way than to take risks on new ones.

Taylor was raising his eyebrows, as he often did to indicate that it was someone else's turn to talk.

Herzfeld nodded. This was not the first time he had heard of a network. Years earlier, he had listened as Licklider had shared his vision of a giant computer network. It had been just a dream at that stage (as its "intergalactic" name made clear), but now, according to Taylor, the network could be real.[25]

"Do it." Herzfeld said. He told Taylor he could spend at least $500,000.[26]

■ ■ ■ ■

Convincing other people was much more difficult. The network project was a departure for both the Information Processing Techniques Office and ARPA. Under standard procedure, a researcher, usually at a university, would initiate a project and submit a proposal for funding. Often the "proposals" were conversations, since most of the researchers were known to Taylor. Indeed, many formal proposals were written months after the projects had been funded.

The network project, by contrast, was initiated by ARPA—more specifically, by Bob Taylor, who was determined to see the network succeed.

Taylor's first move was to hire Al Blue, an expert in ARPA's legal, contractual, and financial protocols, to document the network program. Blue, who had worked at ARPA for years, could not believe the "strange dichotomy" between the computing world surrounding the Information Processing Techniques Office and the reality experienced by the rest of ARPA. While Taylor's office was funding state-of-the-art time-sharing machines and now a new computer network, the rest of ARPA ran on the punch-card batch-processing computers of the sort that Taylor had refused to work with when finishing his thesis. For certain reports, Blue would leave his Information Processing Techniques Office with its sleek terminals and high-level talk about coding and networking to drop off his hand-annotated papers for an operator at Cameron Station to keypunch onto cards for the computer. With his new perspective from Taylor's office, Blue now saw the approach he had been using for years as "primitive."[27]

It is a measure of how much the network project meant to Taylor that, to ensure its launch, he was willing to increase his bureaucratic and paperwork responsibilities to the point that he needed to hire new staff. In addition to Blue, he needed a program manager. The network, only one of seventeen projects Taylor was funding, was going to send messages among incompatible computing systems at a time when there were no agreed-upon standards for operating systems, programming languages, or word sizes.[28]

Taylor had neither the technical expertise nor the time to take day-to-day responsibility for the network's construction. He needed somebody whose technical instincts he could trust and, moreover, someone whom the principal investigators could respect. He knew whom he wanted for the job: Lawrence G. Roberts, a lead researcher at MIT's Lincoln Lab. Roberts held a PhD in engineering from MIT and had made major contributions in the field of computer graphics before he worked on the small proof-of-concept networking project that had brought his talents to Taylor's attention. He was a tenacious and productive scientist, and already, at twenty-nine, a leading figure in the field.

There was only one problem. Roberts did not want to run the networking project. He did not want to leave his own research. He did not like the idea of reporting to Bob Taylor, not because of anything personal about the man, but because in the universe in which Larry Roberts was educated and now lived, it just wasn't right for a guy with an MIT PhD in engineering to report to a guy with a master's degree in psychology from the University of Texas.

So when Taylor asked Roberts to come to ARPA to take charge of a new computer network, Roberts said no. He refused again a few months later and again after that.

But Taylor wanted the network to happen, and he wanted Roberts to make it happen, and Taylor was stubborn. He found a way to get Roberts—blackmail.[29]

Taylor asked ARPA director Charles Herzfeld to call MIT's Lincoln Lab and remind the head of the lab that 51 percent of its funding came from ARPA. Then he asked Herzfeld to suggest that the head of the lab let Roberts know that it would be in his best interest, and the best interest of the lab, to take the ARPA job overseeing the network.

Not too many days later, Roberts was called into the office belonging to the head of Lincoln Lab and told, "It'd probably be a nice thing for all of us if you'd consider this."[30]

By December, Roberts was in Washington, DC. The next month, he reported to his new job as program manager for the network.[31]

Privately, however, Roberts did not think of himself as the program manager. Instead, he was the office director in waiting, who would assume control as soon as he had a chance to learn the paperwork and reporting niceties of the job. Taylor was a placeholder for the succession that should have happened, a succession that Taylor's predecessor Sutherland had wanted: from Licklider, to Sutherland, to Roberts.[32]

It is a credit to both Taylor and Roberts that their differences never surfaced publicly. When the Roberts family first moved down to DC from Boston, they spent the Christmas holidays with Taylor's family. Within a short time, such socializing ended, but Taylor and Roberts always maintained a cordiality in the office, even after Roberts was named chief scientist and began reporting to the deputy director of ARPA, rather than to Taylor; even after, in some people's minds, Roberts seemed to want people to think that he, not Taylor, ran the Information Processing Techniques Office.

Taylor let Roberts say what he wanted. What mattered was getting the network built, and Roberts was the man to do it. Hiring him, Taylor later said, "was probably the hardest and most important [element] of my Arpanet involvement."[33]

Roberts, five years younger than Taylor, was slight and tall, with a cleft chin, intense dark eyes, and a hairline that appeared to have been beaten into retreat by his blazing mind. Roberts ran at only two speeds: off and full throttle—and no one ever saw the first. "When you were working with Larry, you came into the office and you hit the ground running, and you did not quit running until the whistle blew, and then you ran some more. He was on the go every minute. When Larry ate a meal it was like shoveling coal into the steamboiler," Blue recalls.[34] He proved an inspired choice to run the network.

In March 1967, Roberts and Taylor, jointly leading a meeting of ARPA's principal investigators in Ann Arbor, Michigan, told the researchers that ARPA was going to build a computer network and they were all expected to connect to it. The principal investigators were not enthusiastic. They were busy running their labs and doing their own work. They saw no real reason to add this network to their responsibilities. Researchers with more powerful computers worried that those with less computing power would use the network to commandeer precious computing cycles. "If I could not get some ARPA-funded participants involved in a commitment to a purpose higher than 'Who is going to steal the next ten percent of my memory cycles?,' there would be no network," Taylor later wrote. Roberts agrees: "They wanted to buy their own machines and hide in the corner."

In the face of this lack of interest, Taylor and Roberts increased the pressure. They explained that ARPA would not fund new computers for principal investigators until the existing computers were networked and, in Roberts's words, "you have used up all of the resources of the network." Taylor and Roberts also said that if the principal investigators wanted to make sure the network did not become the resource drain they feared, they should help design it. [35]

The tenor of the meeting began to change as many of the principal investigators, despite their initial reaction, became intrigued with the concept of a network. Though trepidatious, they could not resist the appeal of the problems involved. As Robert Kahn, an engineer who joined the project later, put it, "The effort to actually create the ARPANET design was actually a pretty intense intellectual activity . . . like the first missile that you send up, or the first rocket into space; you don't know if it's going to work or survive."[36] The opportunity to build something from scratch, using cutting-edge ideas and designing every part of the system, appealed to the researchers. Roberts led a technical discussion that grew quite animated, while Taylor kept pace, making sure that the fine-sounding ideas could be translated into practical reality. "Somebody would say, 'Well, that is going to mean that you will have to buy a bunch of something-or-others,'" recalled MIT's Wes Clark, who was at the meeting, "And Bob Taylor would say, 'Well, let's see, I guess we can just take that out of something-or-other program funds.' And he would turn to Al Blue [the expert in ARPA compliance] and say, 'Can we do that, Al?' And Al would think for a half a second and say, 'Yes, we can do that.'"[37]

The technical discussions in Ann Arbor ran so long that they continued while Taylor drove a number of participants to the airport. During the car ride, Wes Clark proposed an idea: rather than networking computers together directly, the group should build an underlying backbone network of routers to streamline connections between computers and help make the network more reliable.[38] His contribution was critical.

Taylor kick-started the network. He got it funded. He helped to cajole researchers to contribute to it. No wonder Paul Baran, the inventor of packet switching, called the Arpanet Taylor's "baby."[39] But after the Ann Arbor meeting, once it was clear the network was going to be built, Taylor stepped back and Roberts stepped forward.

Roberts spent hours sketching out possible network configurations and meeting with principal investigators in subgroups that addressed topics ranging from software protocols to hardware design, to appropriate bandwidth allocations, to packet storage and routing. By summer 1967, a number of key issues had been resolved. The network would be packet-switched and run on a subnet system of the sort Clark proposed. The first four nodes would be at UCLA; Stanford Research Institute; the University of California, Santa Barbara; and the University of Utah. Consensus usually dictated such decisions, but it was Roberts who cast the deciding technical vote if necessary.

By mid-1967, Taylor's primary role in the network project had shifted to keeping his bosses at ARPA informed and excited so that funds would flow unimpeded. He spent much of the Christmas holidays at Licklider's house, coauthoring an article to introduce the notion of networking to the business world. "In a few years, men will be able to communicate more effectively through a machine than face to face," Taylor and Lick- lider wrote in "The Computer as a Communication Device," published in *Science and Technology* the following spring.[40] The coauthors explained how networks would enable people to "interact with the richness of liv- ing information" in a new way, not as passive readers or learners but as active participants. The article ended with a prophetic statement about networking: "For the society, the impact will be good or bad, depending mainly on the question: Will 'to be on line' be a privilege or a right? If only a favored segment of the population gets a chance to enjoy the advantage of 'intelligence amplification,' the network may exaggerate the disconti- nuity in the spectrum of intellectual opportunity." Taylor estimates that this article was seen by "essentially no one." The magazine ("for the tech- nical men in management") failed soon after the article was published.[*]

Taylor says that he intervened in the network project one other time after the Ann Arbor meeting. In July 1968, the Information Processing Techniques Office issued a request for bids to build the network. This huge job required writing complex software, designing and building specialized routers called interface message processors (IMPs), and any number of other tasks. Twelve companies submitted bids.[41] Taylor says that a sub- committee of the networking group selected a large military defense con- tractor, Raytheon, to do the job. Taylor thought Raytheon was a terrible choice. The winning company would have to work with the academic re- searchers. Raytheon's command-and-control culture would not mesh well with the "you can't make me do it" academic culture. "It was just oil and water; it would not mix," he says.[42]

Taylor overruled the subcommittee and awarded the contract instead to Bolt Beranek and Newman, a small firm where Licklider had once

---

[*] The article also described an artificial-intelligence agent called OLIVER that would act as an über-personal assistant in the age of networked machines ("Lick's idea," Taylor says). The piece ended with a vision of nerdutopia in which all unemploy- ment has disappeared from the planet because "the entire population of the world is caught up in an infinite crescendo of on-line interactive debugging."

worked and that was staffed mostly by MIT grads. Taylor believed the cultural fit would be better.[43] The issue was not technical, but about human interactions, and the person with the final say was not an engineer but a psychologist.[44]

■ ■ ■ ■

Taylor had many more responsibilities than the network, which by this time was being called the Arpanet. He was involved in the launch of a new program in computer graphics at the University of Utah, a program whose graduates would come to include the founders or key minds behind Pixar-Disney Animations, Adobe, Silicon Graphics, Xerox PARC, and Atari.[45] Forever concerned about getting like-minded people together, he also initiated a conference for graduate students working with his principal investigators. Many of those students would go on to do pioneering and critical work in computer science.

Taylor continued to travel to visit principal investigators, all of whom had ARPA-funded projects they were working on in addition to the network. A demonstration by one of those researchers, a craggy-faced forty-two-year-old engineer named Doug Engelbart at the Stanford Research Institute, impressed Taylor. Taylor had quietly been supporting Engelbart's rather out-there research since 1961, when Taylor, then at NASA, had funded Engelbart to build a new interactive computing system that included, among its many innovative features, the world's first computer mouse.[46] (Taylor later said, "Remember when NASA was advertising Tang [the bright-orange-colored drink consumed by astronauts] as its big contribution to the civilized world? Well, there was a better example, but they didn't know about it.") Taylor's respect for Engelbart only grew when, in the first meeting about the Arpanet, the researcher, alone among the principal investigators, expressed immediate support. Engelbart volunteered Stanford Research Institute to archive documents related to the project and make them available as paper copies or, later, via the network itself.

When he had moved to ARPA, Taylor had continued to fund Engelbart's work, directing some $500,000 (nearly $3.6 million in 2016 dollars) to the researcher. He was forever pushing Engelbart, who was by nature soft-spoken and self-effacing. At a dinner in about 1966, Taylor turned to the researcher abruptly. "The trouble with you, Doug, is that you don't think big enough," he said. "What would you really want to do?" When Engelbart, flustered, admitted that if he had enough money, he would buy a million-dollar time-sharing computer that would enable him to test and evolve his ideas more quickly, Taylor said, "Well, let's write a proposal." And they did.[47]

In October 1967, Taylor went to the Stanford Research Institute to attend an invitation-only review of Engelbart's new computer system, called NLS (for oNLine System). Engelbart assembled the few visitors at a number of terminals, and, wielding his newfangled mouse in one hand and a specialized chord keyboard he had developed in the other, he demonstrated the closest approximation of Licklider and Taylor's vision for computing that Taylor had seen. Here was Engelbart, showing how he could edit documents by pointing and clicking on a screen, in 1967, when most machines could only show information by spitting out fanfold paper. Here was Engelbart toggling between multiple windows. Here was Engelbart urging the participants to play with their terminals and see how they could pull up new information by highlighting certain words: the world's first demonstration of hyperlinks. Here was Engelbart giving what amounted to an interactive rudimentary PowerPoint presentation. Anything he typed on his computer would appear on the other terminals, and if anyone at those terminals moved a mouse to track a spot on his individual screen, the tracking was visible on all the other terminals.

Taylor had been promoting interactive computing for years, and here was Engelbart doing it. The presentation was not a complete surprise, since Engelbart had kept Taylor abreast of his progress, but it was impressive nonetheless.

There was even more to come. Engelbart explained that everything he was doing could work on the Arpanet, once the network technology was sophisticated enough. A researcher in California would be able to point to a spot on his interactive terminal, and thousands of miles away, at MIT for example, someone else on a networked computer would be able to see that cursor move.

Taylor told Licklider about the demonstration, and the two men included a detailed description of it in their 1968 "The Computer as a

Communication Device" article. They told their readers, "If our extrapolation from Doug Engelbart's meeting proves correct, you will spend much more time in computer-facilitated teleconferences and much less en route to meetings."

More significantly, Taylor insisted that Engelbart find a way to show more people what this system could do. "I was working with the rest of the computer world, as well as with Engelbart," he recalls. "Engelbart's project was way off in weird land, compared to all my other projects, and I wanted the other project people to see it."

"This was the future of computing, as far as I was concerned," he explains.[48] While the field of human-computer interaction was attracting more interest every year, only a handful of people—Engelbart's lab team and the invited participants in the project review—had seen a physical demonstration of its possibilities, and then only in a low-key way. Reorienting the trajectory of computer science would require something much bigger and much more dramatic.

■ ■ ■ ■

You need to show more people what you are doing, Taylor told Engelbart again at the end of the invitation-only demo. Engelbart admitted he had considered it, largely because he was afraid that "if we didn't get people's attention, ARPA would have a hard time maintaining its support for us."[49] Taylor and Engelbart batted around ideas of venues that might make sense for a public demonstration and settled on the Fall Joint Computer Conference. The conference, one of the young industry's two most important, would be held in San Francisco in December 1968. More than a thousand people, many of them researchers, would attend.

Engelbart liked the proposed venue but still had objections. A demonstration would be expensive. Since he could not move his system, which ran on the giant time-sharing computer Taylor had helped him write the grant for, the only way to do a demonstration would be to set up a huge display in San Francisco and somehow connect it to the Stanford Research Institute machine room some thirty-five miles south.

Taylor would not be deterred; he said his office would pay for the demonstration.[50] In making that offer, he was departing from his brief at

ARPA. Instead of backing advanced research, he was supporting evangelism.* But he pushed ahead, convinced that a public demonstration of Engelbart's work would advance the field. He helped in other ways, as well. When Engelbart's lead engineer, Bill English, determined that the only projector that could give the right effect belonged to NASA, Taylor vouched for the lab.

Engelbart's December 9, 1968, demonstration at the Fall Joint Computer Conference has come to be known as the Mother of All Demos. Even the world's most advanced computer scientists had never encountered anything like his system.

What a show it was. The San Francisco auditorium seated two thousand, and Engelbart's presentation had been given the full ninety-minute slot typically split among a panel of presenters. Engelbart sat on the right side of the stage, facing the audience, his headset resting on the silvering hair at his left temple, his hands hovering above an unusual workstation that housed a standard keyboard, a special five-key chord key set, and the mouse that Taylor had seen at the invitation-only event. Above and behind Engelbart hung a twenty-two-foot-high screen that alternately projected the image of his face and the display output of his workstation's monitor—and sometimes both at the same time, side by side or one overlaid on the other.

When the lights went down, Engelbart began speaking. His voice, amplified through the loudspeaker from the microphone on his headset, was confident, if a bit halting. "If in your office you as an intellectual worker were supplied with a computer display backed up by a computer that was alive for you all day and was instantly responsible . . ." Prompted by Bill English, who was stage-managing the demonstration and speaking straight

---

* Taylor recalls, "The regional contracting officer, who looked after DOD [Department of Defense] money in this region, saw these expenses going across his desk for these strange things, and he got in touch with Engelbart about this: 'What's all this about?' Engelbart explained it to him, and this contracting officer . . . said, 'This thing sure sounds crazy to me, and I'll tell you what. If it doesn't work, I'm going to deny I ever heard about it.' I didn't know that story until the demo was already finished, and Engelbart told me what this guy had said."

into Engelbart's headset, Engelbart corrected himself: "*responsive* to every action, how much value could you derive from that?" He allowed himself a small smile. "I hope you'll go along with this rather unusual setting. If every one of us does our job well, it will all go very interesting." He then whispered to himself: "I think."

For the next ninety minutes, Engelbart showed off the same technology that had captivated Taylor. But here, so much bigger, so much louder, with the audience's murmurs of surprise and approval punctuating the demonstration, Engelbart's system to "augment human intellect" seemed even more incredible. Throughout the presentation, his computer buzzed and beeped, which only contributed to the overwhelming sense that this was some sort of science fiction extravaganza from a futuristic research pod.

Engelbart pulled up an imaginary shopping list and showed how he could reorder and reorganize it with the click of a mouse. He opened a link to a map. About halfway through the presentation, he explained that he wanted to connect to his lab team at the Stanford Research Institute in Menlo Park. He had been remotely using the computer at the lab, an impressive feat in itself, and now he wanted to bring in video images.

"Come in, Menlo Park," he said. The entire audience could hear him inhale and hold his breath. Engelbart knew that connecting with Menlo Park required the perfect synchronization of two custom-built modem lines and two video microwave links relayed to San Francisco from a truck parked on a hilltop midway between the city and the laboratory.

Only after a new image flickered onto the screen—a young man's well-manicured right hand, grasping a mouse, hove into view—did Engelbart exhale and resume his talk: "Okay, there's Don Andrew's hand in Menlo Park."

And he was off again. He introduced the mouse ("I don't know why we call it a mouse. Sometimes I apologize. It started that way, and we never did change it."). He showed the hardware that was driving the system. He demonstrated how someone in Menlo Park could see the same document that Engelbart had on his screen and how, if the man in Menlo Park moved his mouse, the cursor (Engelbart called it a "tracking spot" or "bug") moved on Engelbart's own screen projected for the auditorium.

Engelbart also gave a nod to the Arpanet. Noting that his lab would be the second node on the network and the keeper of the network's library, he explained that the network designers "plan to be able to transmit across the country [fast enough] that I could be running a system in Cambridge over the network and getting this same kind of response."

As the presentation neared its end, Engelbart began thanking people. He mentioned his seventeen-person team. He dedicated the show to his wife and children, who were in attendance. He singled out Bill English, who had prepared and stage-managed the entire "performance," as Engelbart called it.

Engelbart mentioned only one other person by name. He thanked Bob Taylor "for backing me all these years in this wild dream of doing this sort of thing."

The demonstration ended. The clapping was loud and sustained, and after a few seconds, the audience rose to its feet for a standing ovation. Watching Engelbart "dealing lightning with both hands," in one memorable description, counted for many as a near-religious experience.[51] Taylor says with satisfaction that "Nobody'd seen anything like it."

■ ■ ■ ■

In the spring of 1969, Taylor received good news about the computer network. The contractor Bolt Beranek and Newman, the firm Taylor had chosen over Raytheon, had used the IMP router network to transmit data between two computers acting as nodes. Taylor saw this feat as solid evidence that the network was going to work—perhaps not instantly (he knew bugs would need to be worked through) but eventually. He had accomplished his goal of doing one big thing during his ARPA tenure.

And now, he felt, he was ready to move on. He had been at ARPA for five years, longer than average.[52] He had an excellent successor in Roberts.

He was ready to leave for another reason, as well. As the war in Vietnam escalated, he no longer wanted to be associated with the Department of Defense. In the first stages of the conflict, he, like many of his peers, had been moderately supportive: the South Vietnamese needed aid, he thought. So in 1967, when he was asked to go to Vietnam to help reconcile conflicting reports from computers at various bases, he had reported to an office in Virginia, where he was given a cholera shot and an ID card listing his rank as a general in the US Army, a rank he enjoyed both for its absurdity (he had last fired a gun in high school) and for its status.[53] He was given two copies of the card; one, he was told, was to give to the enemy in the event of his capture.

A few weeks later, Taylor was bound for Vietnam with three friends from the office of the Joint Chiefs of Staff. The trip got off to an auspicious start, with an overnight in Hawaii. When the plane stopped on the tarmac, a driver pulled up in a pristine car, jogged to the stairs that had been pushed up to the plane, and unrolled a red carpet from the bottom step right to the rear door of the automobile.

One of Taylor's traveling companions nudged him. That's for you.

"What is this for?" Taylor asked.

"You're a general. These people are assigned to you, to look after you while you're here."

Taylor laughed. "I can't do this. This is just silly." He asked one of the men from the Joint Chiefs office to send the car away. The carpet was rolled up, the car drove off, and the four friends rented a cheap clunker so they could, in Taylor's words, "do crazy things" that he declined to specify.[54]

Taylor's early days in Vietnam seemed to confirm his belief in the value of the war. South Vietnam was breathtaking. The women were lovely—one, in fact, kept trying to convince Taylor's traveling companion to arrange for her to sleep with the handsome young general.[55] After days visiting the bases, Taylor would return to his hotel room and shake his head at how once beautiful Saigon was now buried beneath sandbags and military vehicles.

He soon grew disillusioned. People who had been in the country for a while told him that the South Vietnamese government ran on patronage and nepotism; any victory by the South Vietnamese and their US allies would only install a corrupt bureaucracy. Taylor began to question the value of the war: Why fight for years and years in what is essentially a civil war, lose many Vietnamese and American lives, and then, when it's over, leave behind a government that will not help the South Vietnamese people? The trip to Saigon marked the beginning of the end of Taylor's time in the Department of Defense.

Opposition to the US involvement in Vietnam also led a few principal investigators to wonder if they should continue to accept ARPA money. Nearly all did. There was a general sense among researchers that Taylor's office was more part of the research community than part of the

Department of Defense monolith. "How can that bunch of guys be really interested in murdering civilians?" the researchers asked themselves.[56] To be sure, by the end of the 1960s, every program funded by Taylor's office needed to include a relevance statement addressing how the program would benefit the military, but people in the research community understood that Taylor's office was "inventing stories" on this front. The statements were "fiction in many cases," according to Al Blue.[57]

Taylor nonetheless felt that as long as he was working for ARPA, he was supporting what was happening in Vietnam. He knew that his office had paid to develop a prototype time-sharing system for the National Military Command System Support Center and that once that system was in place it would be, as the Arpanet program plan put it, "a natural recipient for an interactive computer network."[58] In mid-1968, Taylor asked Oregon Senator Wayne Morse to help him find a job outside of the Department of Defense. Morse was one of only two senators to oppose the Gulf of Tonkin Resolution, which authorized military action in Vietnam without a formal declaration of war.[59]

At about the same time, Dave Evans, a principal investigator who had established the University of Utah's graphics center, offered Taylor a job coordinating a number of computing research projects from around the university.[60] The job did not fire Taylor's imagination, but it was in his general field of interest, offered by a friend, and located well outside of the Pentagon. (And, Taylor adds, it would give him a chance to work on his skiing.) The University of Utah had an outstanding computer science program, Taylor knew. Moreover, the school was slated to come online as the fourth Arpanet node in a few months.

The West itself was also a big draw. The light reminded Taylor of Texas. The air was clear. And it seemed that there was an openness in the West, a kind of acceptance based on what you were doing now, rather than on where you came from. It was no accident, he thought, that the first four Arpanet nodes were going to be in California and Utah—the westerners who worked there just seemed more open to new ideas, "less hidebound," in Taylor's phrase, than their colleagues at East Coast institutions such as MIT.

When the L-O-G message was sent from UCLA to Engelbart's lab at the Stanford Research Institute in October 1969, Taylor was making his way from Washington, DC, to Salt Lake City. He was taking with him across the country his wife and three children, the family station wagon and his prized Corvette, boxes of household goods, and his Rolodex listing the researchers who he felt were doing the best computer science work in the country.

He also took with him one of the ID cards identifying him as a general. A military base not too far from the university had one of the few nearby bars, and Taylor wanted to be able to buy drinks at the canteen.

Taylor had been told that the second ID card was to give to the enemy. He left that one on his desk at the Pentagon. He had decided that the administration backing the unjust war in Vietnam was, by definition, his enemy.

Bob Taylor headed west.

Nerd Paradise

██████████████ AL ALCORN

On May 15, 1969, around the time Bob Taylor accepted the offer to move to Utah, Al Alcorn, an engineering student at the University of California, Berkeley, was working at his bench at the back of Hubbard Radio and Television Repair. He had taught himself to fix televisions and taken this part-time job to pay for his studies. On this particular afternoon, he could hear far-off shouts and chanting and the banging of makeshift cymbals. The sound was not unusual here on Telegraph Avenue, a few blocks south of a campus regularly rocked with anti–Vietnam War protests. He returned to his work.

A minute later, he looked up again. Something *was* unusual. The street people camped on the sidewalk outside his building were not playing their instruments or singing. Cars were not rumbling past. Shopping carts were not creaking by. Aside from the distant cacophony, there was no noise at all.

What was going on? Alcorn left his workbench and peered up the almost empty Telegraph Ave. He was surprised to see a huge group in the distance (he would later learn that it numbered around two thousand people), heading toward him.

He turned to look south and only then understood why his street was abandoned.

At that end of Telegraph, silent and shoulder-to-shoulder, stood 159 members of the Berkeley Police Department, Alameda County Sheriff's Office, and California Highway Patrol. They wore full riot gear. An eight-foot-high chain-link fence loomed behind them.

Behind the fence was People's Park, an expanse of green that only a month earlier had been a jumbled mess of broken concrete, rotting trash, desiccated weeds, and abandoned cars. Some of the hundreds of activists who had transformed the mess into a genuine park called it "a cultural, political, freak out and rap center for the Western world."[1] The University of California, which owned the land, considered the creation of the park an illegal trespass. Governor Ronald Reagan, who called the university a "haven for communist sympathizers, protesters, and sex deviants,"[2] believed the park was a "calculated political act" intended to "bring down capitalism."[3] Black Panther Eldridge Cleaver, who had once led a crowd of five thousand Berkeleyites in a jeering chorus of "Fuck Ronald Reagan," had recently challenged the governor, a regent of the university, to a duel.[4]

Al Alcorn liked People's Park and had made it the subject of a personal photography project. He had captured shots of people clearing the land and moving in slides, sandboxes, benches, trees, shrubs, and flowers. Later, he had photographed the men with porkpie hats and John Lennon spectacles, the barefoot women in flowing dresses, and the huge pots of vegetarian stew—free to anyone who was hungry—bubbling over a new fire pit.

Earlier in the week, the university had shut down the park, clearing it of seventy-odd people who had slept there and then surrounding it with the high fence. The thousands of protesters now descending on the park with their chants and cymbals, Alcorn realized, must be coming to try to reopen it.

Nothing good would come of this. Alcorn hurried back inside. He told the shop owner that they should move to the back of the store.

Who did what next on Telegraph Avenue is not clear, even decades later. The protesters reached the police. Someone threw a brick or a rock. A car's windows shattered. Someone opened a fire hydrant. Police called for backup.

And then—explosions.

At the back of the TV repair shop, Alcorn froze. He knew more than most people about explosions. He and a few friends had spent much of their free time in high school blowing things up. He recalls making TNT, plastic explosives, nitroglycerine—and most memorably, during Christmas break of their sophomore year of high school, a developing tank's worth of rackarock that the teens had ignited in an old cemetery south of San Francisco. (The outcome was a three-foot crater where the developing tank had been and an afternoon at the police station.)[5] Alcorn's reputation as an explosives aficionado had led his brothers at Alpha Sigma Phi fraternity to charge him with exploding a vat of chocolate pudding in the middle of the university's Channing Circle.[*]

Standing with Mrs. Hubbard, the shop owner, Alcorn knew what he had heard explode on Telegraph Avenue: tear gas canisters. He took Mrs. Hubbard to the second story of the building—tear gas, he knew, would stay low to the ground—and stepped out onto a mezzanine that overlooked the street. Most of the marchers were running away. After a few minutes, the tear gas dissipated.

Alcorn thought the worst might be over. "I think we can get out now," he told Mrs. Hubbard. They hurried down the stairs and out the door, careful to lock it behind them.

Hustling away from the park and Telegraph Avenue with Mrs. Hubbard, Alcorn surprised himself with a strange thought: maybe he should get his camera and go back. He had spent weeks photographing the park. Why not continue documenting the story?

He hesitated. He was a careful adventurer. He had tried LSD—but only a few times. He had stopped after deciding that LSD "makes you think of the big picture," but "it doesn't really change you essentially," so why bother?[6] He wore his hair long, but he planned to cut it when he graduated. And for all the time he had spent in People's Park, he had always been careful not to eat the communal stew. He did not want to get sick.[†]

---

[*] Alcorn says that the fraternity was defunct just two years after he pledged; with the Vietnam War and protest movements in full swing, few students were interested in joining a fraternity.

[†] This sort of considered caution was a hallmark of the early protest movements. Mario Savio, the initially reluctant leader of Berkeley's Free Speech Movement, took off his shoes before climbing onto a police car to deliver a speech to the student protesters who had sat down in front of it.

So Alcorn weighed the pros and cons of a return to People's Park. He might get hurt. But he was a big guy. Two hundred pounds and five feet ten, he had been recruited to play football at Berkeley after a high school career that had included playing against a young running back named O. J. Simpson.[7] Moreover, he was fast. When in shape, he could cover fifty yards in six seconds.

He decided to go back.

■ ■ ■ ■

Alcorn had been mistaken to think that the worst of the confrontation had passed. He heard the scene before he saw it. Sirens blared. People screamed. The smell of a smoldering vehicle—an overturned police car, he soon saw—wafted toward him.

As Alcorn focused his telephoto lens, trying to get a close-up, he heard the pop of a gun. He turned to see a man about his age fall, hands over his stomach. He swung about to see who had fired a shot. The officer still had the gun in his hand and was looking around.

Alcorn ran like a scalded cat.

When the chaos ended on May 15, one man, James Rector, had been fatally shot as he stood on a roof watching the scene below.[*] Another man was blinded by buckshot. An officer received a light knife wound, and twenty other police were also injured. Sixty-three protestors and bystanders were hurt badly enough that they risked possible arrest at the hospital. Hundreds more were likely injured but not treated.[8]

Governor Reagan, reactivating the state of emergency that Berkeley had technically been under since a series of student protests in February, imposed martial law and sent in 2,500 National Guard troops. Major

---

[*] The autopsy listed Rector's cause of death as "shock and hemorrhage due to multiple shotgun wounds with perforation of the aorta." An official report to Governor Reagan was careful to state that Rector was not a student but "on probation following conviction on charges of burglary and possession of marijuana" and, further, that inside his vehicle police had found "a Remington .22 caliber, semi-automatic rifle in a disassembled state; and a telephone induction coil, a piece of electronic equipment used for tape-recording telephone calls or for wire-tapping." A footnote noted that he had enlisted in the Air Force in 1963.

General Glenn C. Ames would later complain that "hippie-type females," in their own version of chemical warfare, had given his troops brownies and juices laced with LSD.[9] A curfew was imposed. Thousands of Berkeley residents continued to march in the streets, defying the bullhorn-magnified warnings against mass gatherings that rang through the days. False bomb threats were called in. The faculty voted to cancel classes. Buildings were evacuated. Graffiti and hand-lettered signs were everywhere: "Mothers and Children Against Troop Occupation." "Protect Your Park." In the final two weeks of May, police arrested nine hundred people in Berkeley.[10]

Alcorn took a striking image that captures the state of tortured, suspended animation that pervaded Berkeley in the spring of 1969. A tall young man—sideburned, blue-jeaned, hands on hips and abdominal muscles jutting above the shirt tied around his waist—faces off against a much smaller officer in fatigues and a heavy helmet. The officer's bayonet-tipped rifle slices a diagonal between the two men.[11] Another clash seemed inevitable.

On May 20, five days after the violence on Telegraph Avenue, some three thousand people, many wearing black armbands, marched against what many had come to call "Reagan's occupation." About seven hundred ended up at Sproul Plaza. No one seemed to notice that the soldiers and police stationed at either end of the plaza were letting people enter but not leave.

Troops on the ground pulled gas masks over their faces and looked up. A National Guard helicopter moved into the clear skies over the plaza.

Wisps, then puffs, then clouds of white smoke whispered forth from the chopper's belly and descended over the crowd. It was CS—a potent, nausea-inducing gas—spreading over the plaza.

Panic. Vomiting. Coughing. Fainting. The gas was almost impossible to outrun.

■ ■ ■ ■

Alcorn could not drop out of Berkeley. If he dropped out, he might be drafted. No way. He had classmates, teammates, and neighbors from high school who had come home from Vietnam injured or not come home at all. Alcorn had marched against the war and had opposed it in his own home, even though his father, a merchant marine, was shipping napalm into the war zone. When his father passed along an M16 automatic rifle he had somehow gotten from a returning soldier, Alcorn used a poster of President Lyndon B. Johnson for target practice.

Needing to stay enrolled and wanting out of Berkeley, Alcorn took a step any careful adventurer might admire: he signed up for a work-study program that would let him work off campus for six months but still maintain his student deferment. He would also earn some money.

His mother's boss knew someone at the pioneering audio and video firm Ampex, headquartered across the San Francisco Bay from Berkeley, in Redwood City. The town's motto, printed on an arch over the main street, was "Climate Best by Government Test."[12] Ampex had been started in 1944 by a Russian émigré, Alexander M. Poniatoff, who had named it by adding an -ex ("for excellence") to his own initials. The company had introduced the first practical audiotape recorder in 1948 and then, eight years later, the first practical video-recording technology. The boldly lettered AMPEX sign was a landmark to drivers along the five-year-old Bayshore/101 Freeway that connected San Francisco to the Peninsula.

The company's reach extended deep into American culture. In 1948, Bing Crosby had agreed to continue broadcasting his popular radio show only if he could use the new Ampex recorder to tape-delay his broadcasts. In 1959, Ampex technology had recorded the famous "kitchen debate" between Vice President Richard Nixon and Soviet premier Nikita Khrushchev at the opening of the American National Exhibition in Moscow. A few years later, Ampex video recorders had captured images of Earth sent by astronauts circling the moon aboard Apollo 8.[13]

His mother's connection got Alcorn an interview at Ampex. Soon he had an offer to cross the bay and work for six months as an engineer in a satellite office in Sunnyvale, in the heart of what would soon be known as Silicon Valley.

Alcorn, assigned to a remote building on Kifer Road, dropped into an orderly world of numbers and T-squares. While students at Berkeley protested nearly every day, the engineers at Ampex could not even bring themselves to complain too loudly about the coffee that was so godawful rumor held it was brewed in a giant vat and stirred with a canoe paddle.[14] The watchword at Berkeley was to trust no one over the age of thirty, but at Ampex, the younger engineers, even those who had grown out their hair and beards, admired the older men in their ties and pressed shirts. In many ways, the office resembled an old-fashioned European guild, with the senior engineers teaching the junior.[15]

Alcorn felt at home at Ampex, which he called his own "nerd paradise."[16]

Alcorn's group, some two dozen engineers and about 150 production workers and administrators, worked on a new product that could photograph documents and store the images on two-inch magnetic tape. Later, someone needing to see a document could call it up on a televisionlike monitor or print it out.[17] Today any mobile phone with a camera can do this job in an instant, but in 1969, the task required a copier-sized machine to capture the images, a refrigerator-sized machine with tape reels near the top for storage, and a desk-sized device with a screen for viewing. Ampex called the system Videofile.

When Alcorn arrived at Ampex, a number of organizations were considering a purchase of the Videofile system, even though the hardware cost roughly $1 million and the necessary microwave infrastructure cost another million. Both New Scotland Yard and the Los Angeles County Sheriff's Department thought the Videofile system might store fingerprints and mug shots. Southern Pacific Railroad wanted it to organize waybills and other paperwork. Several insurance companies hoped the system might help with saving and reviewing documentation for claims, complaints, and reimbursements.

In the end, only a few systems were installed, and the Videofile project was shut down. But Videofile was a fruitful failure; it helped to launch two major companies in brand-new industries. Larry Ellison, a cofounder of the database giant Oracle, worked on the Videofile system in the early 1970s. And Alcorn was about to meet the men who would found the world's first wildly successful video game company, Atari.*

■ ■ ■ ■

Near the desk Alcorn had been assigned at Ampex, two men shared a small office. They made an odd pairing, Alcorn thought. Ted Dabney

* Ampex, in general, is one of the great overlooked companies in Silicon Valley history. The audio pioneer Ray Dolby worked there, as well.

seemed like a standard-issue Ampex engineer. He had learned electronics in the marines. He spoke softly and often wore the abstracted look of a man who spent many hours in his own head. He was thirty-two, "a real grownup," Alcorn thought.[18]

The fellow who shared Dabney's office, Nolan Bushnell, was in his twenties, loud and brash, the kind of man who would later, without irony, describe himself as "the poet who interprets the gods for the masses, the gods being technology."[19] He was a trickster who while an undergraduate in college had secured a graduate student office by squatting in one and then telling the student who came to claim it, "I'm not sure we are supposed to share."[20] He had recently graduated from the University of Utah, where he had taken classes in the computer graphics program that Bob Taylor had funded at ARPA.

Even Alcorn could tell that Bushnell "wasn't the greatest engineer," but Bushnell appeared to feel no shame about his lack of technical prowess.[21] He liked to say that graduating last in his engineering class was proof of his efficiency: "I got the degree but didn't do more than I needed to."[22] Bushnell was always drawing attention to himself, which was not difficult, given his six-foot, four-inch frame and shaggy head of curly hair. Rather than just putting up with the lousy coffee, he brought in his own percolator. He started a stock-buying club that to Alcorn's unpracticed eye looked like a fancy version of gambling, with guys in the office pooling their money to bet on the market. Alcorn did not know it, but Bushnell had come to Ampex not because he wanted to work at a top-notch engineering company but because the job allowed him to live in California and earn more money than anyone else in his engineering class. He had worked his way through college as a carnival barker, and it showed.

The mismatched officemates worked well together, despite their differences. Dabney was an excellent practical engineer and Bushnell an inexperienced but quick study. The men shared a love of the Japanese game Go—so much so that Dabney carved a board with the Videofile logo on the back, so that he and Bushnell could hang it on the office wall for easy access. Both Dabney and Bushnell had young daughters, and the families spent time together on the weekends.

■ ■ ■ ■

After six months at Ampex, Alcorn completed a semester at Berkeley and then returned to his job. One of his first stops was Dabney and Bushnell's office.

Only Dabney was there.

Where's Nolan? Alcorn wanted to know.

Dabney lowered his voice. Nolan had left.

Left? Who would leave Ampex?

Dabney explained that Bushnell was trying to build something unusual. Dabney was helping him build it, and it was going well. So well that Dabney, too, was thinking of leaving Ampex.

What were they building? Alcorn asked.

A game you could play on a TV screen.

They're crazy, Alcorn thought.

Eight Quarters in Her Pocket

FAWN ALVAREZ

Less than five miles from Ampex's Sunnyvale offices, twelve-year-old Fawn Alvarez was in Cupertino, California. She was folding papers and stuffing them into envelopes, sealing each one with a swipe of a dampened sponge. She had come in to make a bit of money—$1.65 an hour, minimum wage—in an ongoing effort to grow her stack of 45 rpm records. The Beatles, Supremes, and Peter, Paul and Mary were battling for the top spot on the charts as 1969 drew to a close.

Fawn was working in a strip mall that also housed a dentist's office and Myberg's Deli. The small space was the one and only office of her mother's new employer, ROLM Corporation. The company, which had begun life two months earlier in an abandoned shed once used for drying prunes, had plans to build a computer for the military. The word "computer" meant nothing to Fawn. It conjured up no image at all.

Fawn was a fifth-generation Californian on her father's side, the family's roots stretching back to the time when California had belonged to Mexico. Her mother, Vineta Alvarez, had moved the family from Los Angeles to the Santa Clara Valley six years before, in 1963, when Fawn was six. Vineta, then twenty-six years old, had four daughters and an ex-husband who had stayed in Los Angeles. She had moved the family "to the country," as her daughters put it, because her maternal grandparents lived there and promised that there were plenty of jobs.

The grandparents worked in a Del Monte plant, canning food. They came from Shawnee, Oklahoma, during the Dust Bowl—just pointed the pickup truck, its bed stuffed with everything the family owned, due west on Route 66 and began driving. Sunnyvale, the town where they stopped in about 1933, was not the land of orange groves and palm trees that they had imagined. But there were jobs, and the ground didn't look like it would dry up and blow away any time soon, so they stayed. They bought a house close enough to the cannery that if the wind was right, they could stand in the front yard and smell peaches.

Shortly after coming to the Peninsula, Fawn's mother had found an assembly job at Lockheed, the aerospace giant and defense contractor. With 17,000 employees, Lockheed was the largest employer in the Valley. Vineta Alvarez told her daughters that there were so many employees that the city of Sunnyvale had asked the company to stagger its quitting times so it would not disrupt the evening commute for everyone else.

Vineta Alvarez had not attended school beyond the tenth grade, and her assembly experience consisted of two years working as a riveter in a stroller factory in Los Angeles. But Lockheed, always looking for employees, had a two-week in-house training program. There she learned how to solder and how to read build plans, prints, and color codes. Like thousands of others in the Valley, she would one day put this experience gained at a large, established employer into service at a small startup.

After roughly a year, Vineta Alvarez left Lockheed for the production line at Fairchild Semiconductor, the microchip company launched by eight former employees of Nobel Prize–winner William Shockley. She assembled printed circuit boards, but no one told her what they were for— and she did not care enough to ask.[1] Six months after joining Fairchild, she moved on to Sylvania, a military contractor and electronics manufacturer.

With every move, she received a small pay increase, but it was never enough to cover her $130 monthly rent plus food and clothes for Fawn and her three sisters. Vineta Alvarez worked second jobs on the side, often assembling at home, where she could spread out her soldering gun and all the parts she needed on the kitchen table. She would have preferred overtime at time-and-a-half pay, but California law forbade women to work more than eight hours each day.[2] "They said it was so people wouldn't take advantage of women," Vineta Alvarez says today. "Bullshit. They wanted overtime for the men."[3]

In 1966, when Fawn was in third grade, her family moved to Rancho Rinconada, a neighborhood in Cupertino, where the schools were supposed to be the best in the state. Rancho Rinconada was a new development, close enough to a public elementary school, middle school, and high school that the girls walked by themselves every morning.

While her mother worked, Fawn and her sisters did the family shopping. They did the laundry. They did the cleaning. They mowed the lawn, sometimes two sisters together pushing the heavy mower down the slope of grass and then dragging it back up behind them so they could mow down again. There was no "girls don't do this kind of work" talk in Vineta Alvarez's house. The only thing Fawn's mom ever said the girls could not do was sit with their knees apart. Fawn was the only kid she knew with a mom who was gone all day. Most of her friends had dads who lived with them and worked good enough jobs in factories or the service sector that moms worked only part-time or not at all.[4]

Stone-fruit orchards surrounded Fawn's Rancho Rinconada neighborhood. The kids used them as landmarks ("Meet me at the tall post in front of the prune orchard where the guy shoots you with a salt gun if he catches you"), as playgrounds (dilapidating sheds and storage rooms made wonderful hideouts), and as always-open markets with merchandise that was free, sweet, and hanging low.

The kids all knew, too, that some days, especially in the winter, the orchards were scary. An orchard was its own silent, filmy, cloud-cooled world, with yellow mustard grass as high as a child's waist and fog so dense that it was hard to see from one knotty tree to the next.

The orchards had been Fawn's primary source of pocket money before her job stuffing envelopes at ROLM. Rancho Rinconada kids knew that when the fruits grew heavy, it was time to watch for bright orange triangular flags strung in front of an orchard. That meant the harvest was under way and the orchardist would pay for help bringing in the fruit, usually plums. The going rate was fifty cents per crate. On a good day, Fawn could come home with eight quarters in her pocket.

But this new weekend job at ROLM was better—more money, and she got to work inside.

■ ■ ■ ■

Fawn's mother, Vineta Alvarez, had started at ROLM in November 1969. She had first learned of the company several weeks earlier. She had been working on the manufacturing line at Sylvania when her boss came over. "You have a call."

Alvarez stood up. The only phone for the manufacturing area was in the boss's office, and in the five years Alvarez had worked at Sylvania, women on the production line had received calls only for bad news. A husband in a car accident. A kid with a broken arm.

In this case, the caller was Ron Diehl, a technician Alvarez had worked with and who had recently left Sylvania. He wanted Alvarez to join him at his new employer, ROLM. He told her that four Sylvania engineers (Eugene Richeson, Ken Oshman, Walter Loewenstern, and Bob Maxfield) had started the company the month before, naming it by combining the first initials of their last names. They had asked Diehl to find someone with excellent manufacturing skills.

By this point, Alvarez had been working in electronics manufacturing for seven years. She had become expert in many skills, from soldering leads so tiny that she had to work through a microscope, to building circuit boards, to breadboarding, which required a person to build larger, hand-wired prototypes of the printed circuit boards that would later be mass-produced as the guts of electronic devices. She had supervised a group of fifty assemblers, and she even had a security clearance, since Diehl, before he left, had selected her to be part of a team building a sensitive product for Sylvania.

Vineta Alvarez liked Diehl, and she was happy to hear that he enjoyed his new job. Alvarez, however, had four girls at home and no husband. She could not risk going to a company that had nine employees and no products. She needed the stability of a big company like Sylvania. She thanked Diehl but declined his offer. She returned to work, assuring her coworkers that everything was fine.

A few days later, Diehl called again. And after that, again—and then again. Alvarez estimates that he called twice each week for a month. Now when the boss said she had a call, Alvarez was nonplussed. She admired Diehl's guts, using her boss's phone to try to recruit her away. With every call, Diehl offered another reason to join ROLM. They would pay her more money than she was currently making. She would be able to buy stock at

50 cents per share. ROLM's strip-mall office was only a few blocks from her house. The founders were brilliant. The product was revolutionary.

Still she said no. She had been fighting to support her girls ever since the years in the stroller factory, when at the end of every month she would threaten to quit if she didn't receive a nickel-per-hour raise. She could not risk her daughters' future by joining this little startup operation.

Diehl then told her that she would be in charge of all production at ROLM. That was something new for Alvarez. In the five years that she had worked at Sylvania, she had never seen a woman in charge of production, even though some women had been in the department for fifteen years. Women could be supervisors, but when a manager left, he was always replaced by another man—usually from outside the company. It had been the same at Lockheed and Fairchild. "All of the assemblers were women, and nobody was higher than 'just an assembler' except men," she says.

In November 1969, Alvarez told her fiancé that she was beginning to think she might want to take a chance on ROLM, but she worried about what would happen if the upstart company failed. How would she take care of the girls? He assured her that if something happened, he would help with money until she found a new job.

The next time Diehl called, she said yes.

■ ■ ■ ■

Twelve-year-old Fawn did not know it when she sat near the ROLM secretary's desk sealing envelopes, but everyone at the company was learning on the fly. With seed money of $75,000 ($15,000 from each founder and the same from investor Jack Melchor), ROLM was planning to turn off-the-shelf Data General minicomputers into battlefield-ready machines that could withstand extreme heat and cold, as well as hours of strenuous jostling and vibration. The plan was to sell these "ruggedized" computers to the military for data gathering and missile and radar systems. ROLM estimated the market to be $100 million.

But only one ROLM employee, Bob Maxfield, had any real experience with computers. Maxfield, who had worked at IBM, felt that the entire ROLM operation depended on him. ("And I knew how little I knew," he would admit years later.)[5] After Vineta Alvarez joined, she began building the electronics the company needed. Every evening, Diehl, Maxfield, or an engineer would test the board, array, or device Alvarez had built. By the next morning, Alvarez would have a new design to build in time for another round of testing that evening.

Meanwhile, she volunteered Fawn or Fawn's older sister Bobby any time there were menial jobs to be done at ROLM. Over the days making copies or organizing papers, Fawn met almost everyone who worked at the little company. She was especially fond of Ken Oshman, who she knew was in charge but who always stopped to ask how she was doing. Fawn found it hard to believe that this twenty-nine-year-old man she had once seen in Bermuda shorts had made it possible for her mother to quit her second job.

Fawn was almost a teenager. Soon there would be times when she wished her mother still had to be gone nights and weekends. In not too many years, the kids who had played in the orchards would be teens smoking pot in abandoned farmhouses. But she would always be grateful to ROLM for lightening her mother's load.

The Fairchildren

████████████ MIKE MARKKULA

On a warm spring day in 1969, not too far from where Fawn Alvarez picked plums, twenty-seven-year-old Mike Markkula was helping his former boss Jack Gifford plot a new startup. Three years earlier, Gifford had hired Markkula at Fairchild Semiconductor, the company where Fawn's mother had assembled circuit boards whose purpose she did not know. Now Gifford was leaving Fairchild with plans to launch a microchip company that would compete with the pioneering firm. He had seen Fairchild's success—in ten years, the company had grown to 11,000 employees and $12 million in profits—and thought he might be able to duplicate it at a company that would soon be called Advanced Micro Devices. He wanted Markkula to come with him.

Gifford had first noticed Markkula three years earlier, when the younger man was working three hundred miles south of the San Francisco Bay Area, in Culver City, California, near Los Angeles. Markkula was an engineer with the giant defense contractor Hughes Aircraft. Gifford, then in sales at Fairchild, had met him because Markkula bought Fairchild circuits, amplifiers, and components for Hughes's internal electronics stores. Markkula was only twenty-four, but he was already a seasoned employee, hired by Hughes well before he graduated from the University of Southern California.

Markkula was small and soft-spoken, a smoker who kept his desk spotless. He had an orderly mind, a talent for informed guesswork, and a matter-of-fact analytical approach. When he had been an engineering student, he had wanted to start a company to build transistor-based calculators. Rather than rushing off to do it, as most twenty-year-olds might, he planned out every step. He had never taken a business course and had no formal training, but he trusted that the same approach that had helped him through difficult engineering problems would help him plan the business. He broke each problem into pieces and worked them through. Where would he get parts, and how much would they cost? Who would want the calculator, and what would that customer be willing to pay? Would Markkula sell the calculator through the mail? In a store? Planning the business had made it clear that his idea would never work—at $300, the parts alone cost more than customers would want to pay—but even at the age of twenty, he had understood that it was better to be disappointed earlier than later.

Markkula's intense ambition could be hard to discern behind the mild demeanor that people who did not know him well could mistake for blandness. But Gifford saw it. As soon as he had the headcount to hire someone late in the summer of 1966, he had called Markkula to recruit him to Fairchild.

"Meet me at the airport!" Gifford had commanded Markkula over the telephone. "I've got a job that would be perfect for you!"

Back then, Fairchild seemed unstoppable. The company was not yet a decade old, but only the established industry giants Texas Instruments and Motorola manufactured more semiconductor devices per year. Fairchild was brimming with cash, and its engineering teams were responsible for nearly 20 percent of microchip innovations in the technology's first years.

But Markkula had almost refused to consider a job at Fairchild. The very morning Gifford called, Markkula had accepted a job as an engineering manager with Space Technology Laboratories, another Southern California defense contractor like Hughes. Markkula told Gifford that he was excited to work with navy satellite technology and to supervise four hundred engineers and physicists—and to receive four weeks of vacation as well as what seemed to Markkula to be "a salary beyond belief."

Gifford was nonplussed. "Meet me at the airport," he repeated.

Markkula laughed and then warned him, "You better have a really, really fat wallet."

Gifford offered Markkula a monthly salary of $1,800 to join Fairchild. It was not as high as the Space Technology Laboratories offer, but two aspects of the job appealed to Markkula enough to compensate for the shortfall. First, working at Fairchild, which did no direct government contracting, would enable him to move out of the defense industry. He loved his job at Hughes, but the contracting paperwork and bureaucracy frustrated him, as did the limits imposed by his top secret security clearance that meant he could not talk, even to his fiancée, about his work. Moreover, every time he completed a government document, he felt uncomfortable. His legal name was Armas Clifford, but he called himself Mike (as did his father, for whom he had been named). Government reporting regulations classified "Mike" as an "alias," as if he were a criminal.

Markkula also liked that Gifford was offering him a chance to move into marketing at Fairchild, away from straight engineering. Markkula had received his master's degree in engineering while working at Hughes, and he considered himself a good engineer, but he enjoyed trying new types of jobs. The son of a production-line foreman at Lockheed, Markkula had worked since his early teens, saving to put himself first through Glendale Junior College and then the University of Southern California. To raise money, he had built stereos for friends, pumped gas, bagged groceries at Safeway, painted cars, and catered at movie sets. In his last job before Hughes, with the record manufacturer Research Craft, he had engineered a machine that made it possible to press records in half the normal time.[1]

Markkula thought he would enjoy marketing. He was also confident that if he did not like the work or was not good at it, he could go back to engineering. Deep down, he considered himself a "Tommy Techie."[2]

In September 1966, Markkula, his new bride, Linda, beside him, left Southern California for the San Francisco Bay Area. They drove up the freeway in his new silver Corvette, the suburban sprawl giving way to orange groves and apricot orchards before houses began emerging again near San Jose. The drive marked the end of a busy month. Within a few weeks, Markkula had quit a job, gotten married, moved far from home, and embarked on a new career. He was too old to be drafted for the war in Vietnam but young enough to face a new start in a new place with nothing but optimism, primed for adventure.

When Markkula arrived at Fairchild's Whisman Road headquarters in Mountain View, the company was growing faster than it could secure office space. Markkula worked for his first nine months in a tiny eight-by-twelve-foot cubicle shared with the two other junior members of the linear integrated circuits marketing group, Mike Scott and Gene Carter. Their boss called the three young men "individuals who wanted to be in charge of things . . . that Type-A personality," and Carter agrees that they were all "guys looking for a future."[3] But aside from the technical skills they also shared, Markkula's two officemates were quite different from him.

Gene Carter, who preceded the other two at Fairchild by six months, had been working at a meatpacking plant when he resolved to move into electronics as a way into white-collar work. After obtaining an associate's degree in applied science and communication electronics in only fifteen months, he joined Sandia National Laboratories in Albuquerque. His work was so deeply classified that his job interview had been conducted outside the fence surrounding the facility.[4]

Mike Scott, whom everyone called Scotty, was a Caltech grad with powers of concentration so intense that he would sit, elbows propped on his desk and hands at his temples, reading page after page of technical information, oblivious to any level of noise or chaos around him—and then could recall, down to the page number and paragraph, every word that he had read. "Extremely bright and full of energy" in Markkula's fond recollection, "an eccentric" in Carter's, Scotty was also rumored to have once slugged a coworker who teased him about being overweight.[5]

Despite their differences, Markkula enjoyed his officemates—a good thing, since their closet-sized space was so jammed that if one man needed to leave, the others had to stand up from their desks to let him pass.[6] When Markkula went on, a decade later, to help launch Apple, he would hire both Scott and Carter.

Beyond the tiny office, Fairchild was the tech industry's original work hard/play hard operation. The sales team that stayed at work far into the night, practicing mock sales calls, had also been barred from several resorts for bad behavior. (During Markkula's first week of work, nearly every member of the sales team called in sick with dysentery after a raucous week in Mexico during which they had somehow rendered an entire fleet of jeeps undriveable.) Bob Widlar, the mercurial inventor of many of Fairchild's operational amplifiers, was almost supernaturally gifted and productive at some times, but he also enjoyed what he called "having a drink without using my hands." In this trick, he would clutch the rim of a glass between his teeth and toss back its contents, usually straight gin, with a tilt of his chin. As the evening went on, he had been known to bite through the glass.[7] A different engineer once proposed that the company host a party featuring special guests: naked female models, whose bodies the men would paint as a form of stress relief. Though the party never happened, there was much discussion about whether brushes or fingers would be the tool of choice.[8]

Markkula had little use for such antics. Resolved to get ahead through hard work—"Keep your nose to the grindstone!" he admonished himself[9]—he usually skipped the evening drinks at industry watering holes like Rudy's, Chez Yvonne, and the Wagon Wheel. As a teenager, he had chosen not to attend Caltech because he thought that the students there did not take their studies seriously enough. (They were pelting one another with water balloons the day he visited.)[10] When, before coming to Fairchild, he had decided that he wanted to leave Hughes, he hired a professional to help him craft a résumé. Then he had mailed it to every one of the ninety companies with job listings in the newspaper the weekend after the résumé was complete. Mike Markkula was not a man of half efforts.

■ ■ ■ ■

This intensity of focus was one reason why Markkula was telling Gifford that he was not interested in helping to launch AMD in the spring of 1969. Markkula did not want to be distracted by the power struggles he anticipated (correctly) would arise among Gifford and his six AMD cofounders. Moreover, the entire industry was experiencing a near frenzy of startups and spinoffs that felt overheated to Markkula. During the three years he had worked at Fairchild, so many people had left to start or join competitors—he could count at least eight Fairchild spinoffs in only the past two years—that the new firms had garnered a collective nickname: the Fairchildren. Seven of Fairchild's eight cofounders had left. Just the previous year, in July 1968, cofounders Robert Noyce and Gordon Moore had launched a startup called Intel that had already lured away some of Fairchild's top researchers. A year before that, Fairchild's manufacturing lead, Charlie Sporck, had become CEO of National Semiconductor. Within six months, thirty-six Fairchild salesmen, marketers, and engineers—including Markkula's former office mate Gene Carter—had decamped to join National. Markkula had been offered a job there, too, but says he "didn't get the right vibes."

With so many people spinning away from Fairchild, Markkula believed there were advantages to be gained from staying put. So far, his career had been a continuous ascent—he had been promoted three times in as many years. Now the head of all integrated circuit product marketing, he had long before left the tiny office he once shared with Scott and Carter. Markkula thought that one day he might want to launch a company or join a startup, but not yet.

■ ■ ■ ■

When Markkula had started at Fairchild, his engineering focus had been so acute and his sense of customer relations so underdeveloped that he had once critiqued a customer's circuit design in front of the designer, pointing out how it could be better.[11] The customer, an engineer at Zenith, had disagreed. His face blazing purple, he had marched Markkula and the salesman out of the building, growling "Don't you *ever* tell me how to design my circuits!"*

In the years since his Zenith mistake, Markkula had learned to read customers and anticipate how best to position products to interest them. He had also taken a marketing class through the American Management Association, in which he had learned more general principles of marketing: how to understand and target groups of customers (the biggest beer companies, he learned, aimed all their advertising at males who drank a case or more of beer every day); how to spin problems into assets (Listerine mouthwash had built a campaign around its unpleasant taste); how to define markets (get there first, so you can do the equivalent of shooting a hole in a blank wall and drawing a target around it with the hole at the bull's-eye). "It was worth every second I spent in that course," Markkula says. "I used those lessons every day, even at Apple."

In 1968, Markkula, following one of his typically careful plans, developed a marketing program for a line of Fairchild's circuits (called linear integrated circuits) that doubled Fairchild's market share of this important product. He began as he always did, with a series of questions.

---

* Decades later, Markkula laughs at the story of his youthful arrogance. "I thought I was being helpful," he says. After a moment he reconsiders. "I was a damn good circuit designer. My design was definitely better."

What, he asked himself, motivated the engineers who bought circuits? Answer: They loved to tinker. Question: How could he help them tinker? Answer: Give them samples and application notes that described how the circuits could be used. Question: How could he get the samples and notes to the engineers? The answer to this required some innovation. All the circuits were complex, and some were novel. He could not simply drop them in the mail to engineers without some training.

Markkula decided that Fairchild should sponsor a free seminar series. While competitors relied on mailers, advertisements, and one-on-one visits, Markkula and other Fairchild marketers traveled around the country in 1968, hauling cartons of slides, binders, and chips from one company to another. At the end of each two-hour seminar, the Fairchild team would give the engineers in the audience five sample circuits and large three-ring binders of application notes on how to design the circuits into a system. "All the competitors were left in the dust," Markkula recalls. Fairchild's market share jumped from 17 percent to 35 percent in a single year.[12]

He got another promotion at the end of 1969, around the same time Al Alcorn began his job at Ampex. Although several layers of management sat between Markkula and Fairchild's executive suite, he felt his career trajectory was steep and aimed in the right direction.

■ ■ ■ ■

By the middle of 1970, however, Markkula had begun to tire of being one of the few bright spots in a company in which so many other things were going wrong.[13] Fairchild's research and development, engineering, manufacturing, sales, and marketing had been hollowed out as people left for competitor companies. An industrywide downturn in the second quarter left Fairchild's prospects even bleaker.

Markkula had said no to AMD and National Semiconductor when he was offered jobs there, but by the time Bob Graham, formerly of Fairchild and now the head of marketing at Intel, called in the fall of 1970, Markkula was ready to talk. While Fairchild had been disintegrating, Intel had grown stronger and stronger. With nearly three hundred employees and $4 million in revenues, the two-year-old company was already outgrowing the 24,000-square-foot building that served as both its headquarters and manufacturing facility. Moreover, Intel was already profitable, thanks to its superfast version of a popular memory chip—a component that was not particularly sophisticated but provided crucial income while the company pursued more advanced technology.

Only weeks before Graham began recruiting Markkula, Intel had introduced the 1103 MOS memory, which would become the industry standard and account for roughly 90 percent of Intel's profits in the coming years. Intel was also less than a year away from debuting the microprocessor whose descendants still sit at the core of the company's success. With so many important products just launched or about to be, Intel needed to grow its marketing department. As Markkula recalls, the department at the end of 1970 comprised Graham, a product marketing engineer, and a shipping clerk. A sales executive also reported to Graham.

Intel could not have been more appealing to Markkula. It had already begun to rebound from the second-quarter downturn faster than its competitors. Its researchers were among the best in the world, and its leaders—particularly cofounder Robert Noyce, who would become a friend—were people Markkula admired. "I knew they were honest and had no hidden agendas," he says. "Intel was my kind of place."[14] Noyce had staffed the company with care, saying again and again that "a small bunch of people who know what they are doing can accomplish more than a big group of people who don't know what they are doing." Markkula wanted to be part of Intel's small team and agreed to join in the newly created position of North American marketing manager.

Negotiating details of the hire, Markkula surprised Graham. "I don't care what the salary is," he said. More than a sizable paycheck, Markkula wanted stock options.[15] He had not owned stock options at Fairchild or Hughes, but he knew that Intel granted them to every engineer in the company; this practice had become standard among the Fairchildren. The typical grant was 1,000 options. Markkula wanted 20,000—so many that Graham had to lobby for special permission to grant them, particularly since Intel was running out of options to grant.[16]

The approval came at the end of the year. In January 1971, Markkula drove almost directly past the well-appointed Fairchild campus to begin his new job a few blocks down the road—in the bare-bones concrete building, furnished with the previous occupant's cast-off furniture, that served as Intel world headquarters.

What Do We Do with These?

NIELS REIMERS

Some three hundred people, most of them Stanford students, had shown up by 7:30 on that May morning in 1969, determined to shut down a satellite office of the Stanford Research Institute (SRI) a day after the People's Park protests had rocked Berkeley. SRI's much larger headquarters facility, where Doug Engelbart had done the work unveiled six months earlier in the Mother of All Demos, was four miles north, in Menlo Park. This satellite office, located only a few blocks from the southern edge of the Stanford campus, housed a computer that protesters claimed was analyzing activities of Communist insurgents in Southeast Asia. The protesters, many from a radical Stanford student organization called the April Third Movement, wanted that analysis—and any other work associated with the war in Vietnam—stopped.[1]

The group dragged signs, sawhorses, and a steel crane boom from a nearby construction site onto Oregon Expressway, a major east–west thoroughfare. They turned a school bus across two lanes of traffic and deflated every tire so it could not be driven away. A few students stood near the blockade, shouting and waving red flags. Others made their way down the line of cars, handing out flyers explaining the protest. "Turn off your motor and take it easy," one handout suggested.[2]

The group planned to block access to Hanover Street, the most direct entrance to the industrial park that Stanford had built a dozen years earlier in a bid to attract "smokeless industry" to the undeveloped lands near the campus. In addition to SRI, which received nearly half its funding from the Department of Defense, Lockheed and Itek Corporation—two other companies that did substantial amounts of defense-related work— also rented facilities in the industrial park. The companies' low, unassuming buildings were set so far back in their manicured lawns that it was impossible to see inside from the street.

By peak commute time, the blockade had stranded about a hundred cars. One trucker whipped around the barricade, screeched to a stop, and pulled a knife on the protesters who had begun thumping the vehicle's hood. But most of the drivers confined their frustration to shouting that the protesters were "stupid radicals" and "ignorant bloody creeps." One driver told the student who handed her a flyer that she felt "sorry for your offspring."[3]

As cars began detouring around the crane-and-school-bus barricade, cutting across the open fields on either side of the street, the demonstrators abandoned their roadside barricade. Nearly a hundred protesters now linked arms to block the entrance to the SRI building. By that time, police, a handful of whom had been present since the beginning of the protests, had amassed in sizable numbers. Even when fifty officers stood facing the protesters, however, there was none of the rancor that had marked the violent People's Park clashes across the bay in Berkeley. Here uniformed officers, demonstrators, and a small group of counterprotesters even chatted together. Everyone knew the police were awaiting reinforcements to end the protest.

By 11:00, those reinforcements had arrived. Police, 150 strong and armed with tear gas canisters, advanced on the protesters, ordering them to disperse. The students responded by smashing windows at SRI with rocks, a sewer pipe, and a metal sign.

Earlier that morning, the protest organizers had distributed a notice: "What to Do If You Are Arrested." The advice (write the phone number of the legal defense committee on your skin for your call from the station; do not resist; yell your name so the group can track who has been hauled off) came in handy.[4] The police arrested fifteen people—"six girls and nine young men," according to one reporter.[5]

Within an hour, it was over. Police cleared the barricades and reopened the streets. The protesting students could have walked back to the Stanford campus via a few residential blocks, but a large group chose instead to wend their way along the main thoroughfare, El Camino Real. They sang, pumped fists, and accepted beers tossed to them by sympathetic drivers who believed, as one radical paper put it, that the protest had been "a total victory [that] had successfully shut down counterinsurgency work at SRI."[6] That assessment, however, was more wishful thinking than reality.*

Back on campus, a group broke two windows in the president's office before returning to the April Third Movement's headquarters, a geodesic dome on White Plaza in front of the student union.[7]

Protests of this sort—rowdy and defiant, though not as violent as their counterparts at Berkeley—had flared at Stanford over the past few months. Someone ran a Viet Cong flag up the campus post office flagpole. For nine days, students shut down the Applied Electronics Lab, which received Department of Defense funding. A few days before the SRI blockade, the April Third Movement had called for a boycott of classes, suggesting that students instead attend a carnival on White Plaza where visitors could have their finger painting analyzed, throw tennis balls at targets representing the board of trustees, or "Smash the State" by taking a baseball bat to a car that a sign claimed was a former police vehicle. More studious types could attend a series of presentations on subjects ranging from "Southeast Asia, Counterinsurgency and the Empire" to "Imperialism and the University."[8]

At times, the protests seemed silly or counterproductive.[9] Dousing President Kenneth Pitzer with red paint, for example, won few new adherents to the cause. (Pitzer's tenure as president lasted only nineteen months before he was worn down by the protests.) But the protests had real effects, increasing minority outreach in the admissions office, helping to launch

---

* Stanford severed all ties with SRI in 1970, though the move did little to reduce defense-related research at the institute. The May 16 protests came *after* Stanford's announcement that SRI would be sold; the protesters had urged the establishment of a "peace research center," jointly run by Stanford and SRI, that would "cease all classified and secret research" as well as any contract that required a security clearance. The sale meant Stanford lost all control over research at SRI.

interdisciplinary centers at Stanford, and hastening the departure of SRI and ROTC from campus.[10]

■ ■ ■ ■

In the year since thirty-six-year-old Niels Reimers had joined the Stanford staff as the associate director of contract administration, he had come to regard the protests with wary cynicism. He did not like the war, but he also disliked the protests. ROTC had made it possible for him, the son of a Norwegian electrician in the sleepy town of Carmel, California, to attend Oregon State, where he had majored in mechanical engineering. Reimers pitied the students who might not be able to come to Stanford without ROTC support. And Reimers was proud of his navy career, which he had begun as an eighteen-year-old swabby and ended as a division officer on an aircraft carrier. His jobs after the service—first at Ampex, where he worked for two years before the Atari founders got there, and then at the electronics company Philco-Ford—also involved close consultation with defense-related government agencies. He respected many of the people he knew from that work.

In his $13,000-a-year job at Stanford, Reimers worked with what some students had taken to ironically calling "the evil forces": government agencies such as the National Institutes of Health, the National Science Foundation, and the Office of Naval Research, which backed research at the university. This work, along with his office's location in the Encina Hall building that also housed key Stanford administrative and financial groups, made his office a target for activists. A few days before the SRI protesters paraded past, others affiliated with the April Third Movement had broken into his building and rifled through files, carting some away in a Volkswagen Beetle. That time, Reimers, normally a mild man—his expletive of choice is "sigh," as in "Traffic may be a problem. Sigh."—was furious. He was also shocked to hear girls swearing. He had drawn his wiry six-foot, two-inch frame erect, his blue eyes blazing, "ready to smash a guy." He could only watch in disbelief as Stanford's insurance manager dealt with the chaos by collecting the stones that had been hurled into the building and arranging them on a neat pile on his desk.[11]

■ ■ ■ ■

Over the past months, Reimers, finding no inspiration in his official job, had been moonlighting in a position he had created for himself: head of a program to license Stanford inventions for commercial markets. He wanted to turn Stanford ideas into profitable products for the university. He was good at it.

The idea had come to him a year earlier, in his first weeks on campus in May 1968. Opening his mail one morning, he had found an invention disclosure from a chemistry professor who had discovered a new way to synthesize a hormone.[12] An invention disclosure is a confidential descriptive document written by an inventor to help determine whether an invention should be patented prior to publication. Reimers had worked with invention disclosures at Philco-Ford, but he had not seen one at Stanford. Most researchers published their ideas in academic journals without ever filing formal disclosure paperwork with the university.

Reimers took the disclosure to a colleague. "What do we do with these?"

She told him to send it on to Research Corporation, a sixty-six-year-old nonprofit organization based in the Empire State Building in New York City, three thousand miles away. If Research Corporation, which worked with universities from across the country, thought an idea was worth patenting or licensing, it would handle the paperwork. A small portion of the money that came in would be passed on to the inventor's university. An even smaller portion would go to the inventor. The rest of the money would fund scholarships administered by Research Corporation.[13]

With a little searching, Reimers discovered that Stanford had received less than $3,000 from Research Corporation between 1954 and 1967. He found that figure—the precise amount was $2,944.91,[14] all but $44.40 of which had come from Professor William Hansen's patent for an "automatic power bridge"—incredible. How could inventions coming from throughout Stanford together yield only $226 per year for the university?

Reimers began compiling a list of ideas and another list of anyone on campus who had anything to do with patents. An attorney in the business affairs office did occasional patent work. The new SLAC Linear Accelerator Laboratory on Sand Hill Road near the southwest edge of campus employed a full-time patent counsel. An outside attorney worked one day per week for the Hansen Experimental Physics Laboratory.[15] That was it.

Reimers was not an attorney, and his knowledge of patent law was limited. But he thought Stanford could do more with its intellectual property. He knew that Crest toothpaste, for example, had been formulated in a lab at the University of Indiana, which had earned large sums after licensing the product exclusively to Procter & Gamble, the company that had sponsored the university's research.[16] He was also aware of the cautionary tale of the University of Florida's having declined one of its physician's offers to give the school all rights to a hydrating drink he had formulated for the football team: Gatorade.*[17]

Reimers thought that Stanford should create a group focused entirely on licensing inventions and moving them into commercial markets. The notion, though not unprecedented, was daring: in the entire country, there were only nine university offices doing such work. He believed that Stanford should create incentives to encourage researchers to bring their ideas to the university, which should also employ people who knew how to sell them to industry. He was convinced that an idea had the best chance of reaching the public if a company pursued its development.

Almost as soon as Reimers refined these ideas, he knew who he wanted to help him pursue them: Frank Newman. Newman, tall and handsome, worked upstairs in Encina Hall. He had nothing to do with patents or licensing, but he adored new ideas. At forty-one, he had studied economics at Brown and Oxford; earned an MBA from Columbia; and made a spectacularly unsuccessful run for Congress as an antiwar Republican. He had also left a distinguished career at Beckman Instruments, the company that in 1956 had funded William Shockley to start a transistor operation in an old Quonset hut—the same operation that would begin to decline a year later, when eight key scientists and engineers left to found Fairchild Semiconductor.

At Stanford, Newman directed university relations and served on the president's staff. Even this big job did not deplete his prodigious energy. In his spare time, he was earning a PhD in the history of technology. He

---

* In 1971, the University of Florida filed suit against the inventors of Gatorade after they sold the formulation to a private company. In 1972, the inventors and the university settled. As of 2015, the university had received $281 million from Gatorade.

would go on to head a national task force on higher education and eventually serve as the president of the University of Rhode Island.[18]

Even as a relatively young man at Stanford, Newman was one of the busiest, most efficient operators Reimers had ever met. Newman's solution to the student protests outside Encina Hall had been to move his office to his car; his secretary had worked beside him in the passenger seat on those days, and they had not suffered any interruption at all.[19]

■ ■ ■ ■

At their first meeting, Newman embraced Reimers's ideas for a group to focus on patents and licensing. One aspect of the proposal particularly intrigued him. Reimers thought that Stanford should ask companies to assess an idea *before* the university invested money in filing for a patent. This risky approach meant that an unscrupulous company might steal the idea before it could be patented. But Reimers thought it was better to run the risk than to waste time and money protecting inventions that no one wanted.

Newman agreed that the approval of a company that might use a Stanford invention in a product would be a fine proxy for the likelihood of the idea finding a wider market. "I would have loved to have a university do that when I was at Beckman," he admitted. The two men began tossing around ideas and questions. How would they find companies to ask for assessments? Who should they talk to inside of those companies? How would they explain what they were trying to do?

And what about costs? Newman wondered aloud. Stanford did not have money to spare on experimental programs. Reimers volunteered to run the program on his own time, on top of his regular job.

They agreed that they needed to think of this new group as a marketing shop. ("The most important three factors of success," Reimers would later write about the Office of Technology Licensing, "are (1) marketing, (2) marketing, and (3) marketing.")[20] The group would have to market to the faculty to encourage them to submit invention disclosures to the new office, rather than patenting ideas on their own, just noting them in a lab book, or publishing them. The office would also have to find and convince companies to buy a license to use the inventors' ideas, most of which were so cutting edge that turning them into products could take years—and some inventions might never become products. There was a third type of marketing to be done, as well. For ideas produced under federal government contract, Reimers would need to convince the appropriate government office that the rights to the invention belonged to the inventor or Stanford, rather than to the government.

"The office needs to be entrepreneurial," Reimers told Newman. Reimers saw evidence of an entrepreneurial impulse in the companies near Stanford every time he opened his dog-eared membership directory for the Western Electronics Manufacturers Association (precursor of the American Electronics Association). The book's pages were covered with cross-outs, arrows, scribbled new names, and parenthetical notes indicating that a key executive or engineer had changed companies or started a new one.

Reimers wanted the same entrepreneurial spirit for his little in-house Stanford startup. Where the old-line, New York–based Research Corporation took months to evaluate an idea, the office he imagined would move quickly. Where Research Corporation filed for a patent and then stepped away, the Stanford program would work to help a company imagine a market for the idea.

Reimers wanted to offer inventors one-third of the royalties that would come to Stanford after the new office deducted 15 percent for expenses. (The other two-thirds would be split between the inventor's department and the university.) This arrangement was generous. At MIT, which had left Research Corporation in 1963, inventors received 12 percent of royalties; at Caltech and Princeton, they received about 15 percent of gross royalties.[21]

He believed that paying a large share of royalties to inventors and their departments would motivate people to inform the new office of their ideas.[*] At the same time, however, he worried that the program might

---

[*] Always marketing, and well aware that Research Corporation continued to be an option for Stanford inventors, Reimers and Newman asked the provost to issue a memo explaining that any money coming into an academic department from the licensing program would be considered "incremental" funding and would not threaten general funding from the university.

somehow undermine the fundamental mission of the university: to increase the world's store of knowledge. Researchers might become so concerned about protecting an idea or marketing it through the new office that they would hold off on publication until patents had been filed. In Reimers's words, "The thrust of university research is the search for new knowledge and not the search for patentable ideas."[22] He did not want to "make a lot of money and lose a lot of science."[23]

Newman had concerns, too, most of them stemming from his familiarity with university bureaucracy. Stanford embraced a fuzzy line between research and commerce, he knew. Alongside its high-flying goals "to promote the public welfare by exercising an influence in behalf of humanity and civilization," the university's 1891 founding grant spoke of a singular "object, to qualify its students for personal success, and direct usefulness in life."[24] The industrial park targeted by protesters was part of the university's wider effort to tighten connections with what today would be called high-tech companies. Engineers at companies that paid a fee could take classes at Stanford via closed-circuit television. Sponsoring companies would get early looks at Stanford research and access to graduates.

But even Stanford could be slow to welcome change. Newman knew that Reimers could not propose that Stanford leave Research Corporation for an untested in-house operation. "We'll call it a pilot program," Newman advised. "Say it's just for a year, and then we will assess." He must also have suggested that he, not Reimers, make the formal proposal to the Stanford administration.[25]

The meeting ended. Reimers collected his things. By the time he rose to leave, Newman was flying on to the next item on his calendar.

Come with Me, or I'll Go by Myself

SANDRA KURTZIG

While protests blazed at Stanford in 1969, alumna Sandy Kurtzig was three thousand miles away. She and her new husband were working at the Bell Labs campus in Murray Hill, New Jersey. Bell Labs was renowned as the "idea factory" where the laser, the transistor, and radio astronomy had been invented. But up close, Kurtzig thought, the place looked like a prison. Designed to encourage scientists and engineers from various disciplines to cross paths, the campus comprised a series of featureless, low, rectangular buildings veined with long hallways—the one in the physics wing stretched nearly the length of two blocks—and punctuated with tall narrow windows. The buildings lined up. The windows lined up. The grass was clipped as short as the senior researchers' hair. The roads that ringed the isolated complex were nearly empty.[1]

Kurtzig's husband, Arie, who was working on computer memory chips for Bell Labs, flashed his badge at the guard.* Kurtzig waved the pass that proclaimed her a "resident visitor." She had been coming to Bell Labs nearly every day for four months. Not too long ago, she had noticed an empty cubicle and claimed it. Now everyone considered it hers.

Kurtzig did not work for Bell Labs. She sold computers made by

---

* Arie Kurtzig's particular focus was bubble memory, which uses small magnetized areas to represent data bits.

General Electric to the scientists and researchers at Bell Labs. More precisely, she sold time on a time-sharing system run by General Electric.

Time-sharing computers had given Bob Taylor his insight, back in 1966, that computers could build communities. A single time-sharing machine switched among users, usually connected to the computer via dedicated phone lines, so quickly that each user was unaware of sharing the computer with the others.

In 1966, most time-sharing systems were at universities or defense-related companies. But now, three years later, time-sharing computers had become a $70 million business, doubling in size every year.[2] More than fifty companies competed in the market, including giants such as Kurtzig's employer, General Electric, which had a 40 percent market share.[3] Time-sharing companies developed programming languages and applications and sold remote access to powerful computers to businesses that did not want to buy or lease a computer of their own. Customers rented terminals and paid to use software, store data, and connect to the machine.

Competition was fierce. A month after General Electric ran a print ad asking "Would you like to access a million-dollar computer?," IBM (with a 20 percent market share through its Service Bureau Corporation subsidiary) countered with its own question: "Would you like to access a five-million-dollar computer?"[4]

Bell Labs had subscribed to GE's time-sharing service before Kurtzig began her sales position, but the terminals scattered around the campus often sat unused. Researchers either sent their problems to be run on a mainframe, using the same punch-card batch processing that had frustrated Taylor years earlier, or did the calculations themselves, at their desks. Kurtzig's job was to convince the Bell Labs researchers that the unimpressive-looking time-sharing terminals—they resembled overgrown beige typewriters—were useful tools. Every minute that she could convince the scientists to spend on the time-sharing system meant more money for General Electric.

Kurtzig, bright and strikingly attractive, was great at the job. Only twenty-one, she had been working with computers since she was a seventeen-year-old UCLA sophomore with a part-time job in the computer center. By her junior year, she was paying for her classes as a numerical analyst in the fluid physics department of the aerospace company TRW. There, in effect, she had *been* the computer. A math major, she had run calculations for the group of fifteen aerospace engineers. For complex problems, she would use TRW's giant mainframe computer, but she, like Taylor, hated waiting weeks for answers.

When TRW bought a GE time-sharing machine, Kurtzig began writing programs for it in BASIC, thrilled by the instant feedback.* It had taken only a few weeks of working with the new computer for her to decide, as she later put it, that "computers would be a part of my future."[5]

But not immediately. After graduating from UCLA at twenty, and with the encouragement of the engineers at TRW, she had begun studies in a graduate program in aerospace engineering. Her parents, with whom she had lived in their Beverly Hills home while she attended UCLA, forbade her to even consider Berkeley, with its radicals and sit-ins—so she went to Stanford. She was one of seven women among the nearly eight hundred graduate engineering students spread across eleven departments.[6]

Handicapped by not having studied engineering as an undergraduate, Kurtzig spent most of her time at Stanford on her classwork. ("I'm not a genius," she says today, "but I know how to work very, very hard.") The campus protests that disrupted Niels Reimers's efforts to launch a licensing office were background noise for Kurtzig. When she was not studying, she spent time with her boyfriend, Arie, then finishing his PhD under Nobel Laureate William Shockley, the same Shockley who had launched a company around the transistor he had coinvented and then hired the eight young men who would leave to found Fairchild Semiconductor. Shockley had not yet fully descended into the paranoia and eugenic fervor that would later mar his reputation. (He would declare blacks intellectually inferior to whites and donate his sperm with the stipulation that it be used only by a female member of Mensa.) Nonetheless, he struck Sandy Kurtzig, even then, as a "weird duck."[7]

■ ■ ■ ■

When Kurtzig moved to New Jersey to be with Arie after they married, the young couple made a deal: they would live in New Jersey for three years.

---

* Kurtzig does not recall if she taught herself BASIC from a book or had taken a course at UCLA. She already knew FORTRAN.

After that, if she was unhappy, they would return to California, where her family lived and which she loved.

Kurtzig had gotten the GE sales position after she arrived unannounced at the company's Teaneck, New Jersey, office and waited for hours, résumé in hand, until the district manager agreed to talk to her. The manager undoubtedly knew that GE's time-sharing unit had a program in place to recruit women into sales. By the time Kurtzig went into his office, half the national sales team were women.[8] Kurtzig's master's degree in aerospace engineering and her enthusiasm for the GE time-sharing computer that she had used years before at TRW impressed the manager. Before she even left his office, he hired her to sell to Bell Labs.

Bell Labs was a strange place. Kurtzig recalls watching one scientist, lost in his own head, bump into a water fountain, beg its pardon, and keep going without realizing he had apologized to an inanimate object. But for Kurtzig, who had spent her short adulthood in academic math and engineering departments, Bell Labs had a familiar feel. Although most of the other women she saw at the Murray Hill facility worked in administrative roles or the cafeteria, Kurtzig never doubted that she belonged. She had even turned down a job offer from Bell Labs when she graduated from Stanford, not because she felt that she would be out of place but because she could have fit in so well, just another master's-level engineering drone serving the kingdom of science PhDs. She had wanted to stand out more.

Her sales strategy at Bell Labs was simple, as she recalls: "Walk around the halls and ask, 'Can I help you with your problems?' "[9]

Many people thought not. Kurtzig was selling computers, and computers meant mainframes, and mainframes meant setting up a meeting with one of Bell Labs' full-time programmers, talking through the steps and calculations necessary to solve a problem, and then waiting weeks for the programmer to write a program and the mainframe to run it.

"This is different," she explained. She asked about current projects and offered to demonstrate how a program, accessible through the terminal, could help. There were time-sharing programs for statistical analysis, graph plotting, and complex data processing.

Why wait for Bell Labs's staff of programmers when you can do it yourself? she would ask. Why wait for the mainframe computer housed in a different building to chug through its cards, when this terminal next to your desk spits out results almost as soon as you input the data? If a researcher needed a program that did not exist, Kurtzig helped him write a new one.

She was good. "Soon dozens of Murray Hill R&Ders were burning time on their machines while I fed the flames," she recalled.[10] She loved that she was doing her job, and earning her commission, so well, and it also excited her to know that she was bringing Bell Labs into the era of modern computing.

When she got a call telling her to report to her boss's office in Teaneck, she was so energized and nervous—wasn't it too early for a bonus?—that for good luck she paid the toll of the car behind her on the New Jersey Turnpike.[11]

■ ■ ■ ■

Her turnpike altruism had no effect. Her boss greeted her and with little preamble announced that he was taking her off the Bell Labs account. When she asked why, he explained that the mainframe computer at Bell Labs that Kurtzig had been weaning scientists away from was manufactured and maintained by *General Electric*—and the behemoth made much more money for the company than the time-sharing minutes she had been selling with such zeal. If Bell Labs used the mainframe less, how could GE make the case for upgrading to a newer, more expensive machine? Moreover, the programmers at Bell Labs had not been pleased that Kurtzig had been helping researchers write their own programs. ("Solve your problems yourself" was a key phrase in Kurtzig's sales pitch.)[12] Programming was a job for professionals.

Kurtzig was stunned. How could she be punished for doing her job too well?

Angry (GE should have told her about the mainframe!), embarrassed (she should have asked!), and suspicious (had they put her on the account because they thought the pretty young woman couldn't sell anything anyway?), she felt her throat thickening. But she willed herself not to cry in front of the district manager. She could cry later—and she did—but for now she remembered her mother's oft-repeated mantra: "When in doubt, act confident."[13]

What happens next? she asked, hoping that she was not fired.

She was assigned to a new sales territory, with prospective customers that could not be more different from the PhDs at the citadel of science that was Bell Labs. She would cover north central New Jersey, a region of factories, machine shops, smokestacks, catwalks, and motors. The last place anyone would expect to find a computer.

■ ■ ■ ■

Kurtzig proved herself in her new role. She convinced a company called General Foam that a GE time-sharing computer could help maximize the number of sponges pressed from the giant foam sheets the company processed. She sold time-sharing services to the pharmaceutical manufacturer Merck after demonstrating how GE's statistical analysis and graphing programs could help correlate data from clinical trials and lab tests.

Kurtzig was an anomaly twice over: she was a woman selling to factory owners, and she was a sales representative with programming knowledge. But most of her customers had never before had anything to do with computers, so they had few expectations. The sales rep was a woman? Okay. Maybe that was just the way it was done.

By the end of her third year in New Jersey, Kurtzig was a success at work but terribly homesick. Aside from her job, she liked nothing about New Jersey: not the muggy summers, not the frigid winters, not the apartment so close to the railroad tracks that the trains seemed to drive through her bedroom, not the three thousand miles between her and her parents and younger brother.

She had done what she had promised Arie: she had tried it for three years. Now she was finished. She wanted to ask GE for a transfer to California, and Arie, with his Stanford doctorate and his résumé with Bell Labs on the top line, could find a job, too. With the rent they were paying for the noisy nutshell apartment in New Jersey, they could get a nice two-bedroom in a quiet neighborhood in then cheaper California.

But Arie did not want to move. She pushed. He was adamant. At one point, she was so frustrated after one of their fights that she fled the tiny apartment, turning around just long enough to tell her husband, "I'm going back to California. You can come with me, or I'll go by myself."[14]

The next day, he gave in, and the couple went to a New York City newsstand that sold the Sunday editions of the major newspapers from San Francisco, San Jose, San Diego, and Los Angeles. The Sunday papers always had the biggest help wanted sections.

BUILDING

1972–1975

In 1971, a hard-drinking reporter named Don Hoefler published a three-part history of the "men, money, and litigation" behind the semiconductor business. He titled the series, which appeared in an industry newsletter called Electronic News, *"Silicon Valley USA." It was the first formal use of a name that for years had been bandied about inside the valley.*[1]

No one beyond the readers of Electronic News *took notice that a new technology region had been born on the San Francisco Peninsula. Individual companies had begun to attract outside interest—Intel would have a successful IPO in 1971, for example—but three years would pass before any major national publication wrote about the Peninsula as a distinct, technology-based regional economy.*[2] *Years would pass before the name "Silicon Valley" gained widespread use.**

In the early 1970s, Silicon Valley startup companies were largely deemed irrelevant to the real business of the United States, which was concentrated in East Coast financial centers and cities such as Detroit, Pittsburgh, and

---

* "Silicon Valley" was just one of many terms coined to describe the cluster of electronics firms on the San Francisco Peninsula. "Silicon Gulch," "Semiconductor Country," "California's Route 128," and "California's great breeding ground for industry" were some of the others.

Chicago that had manufacturing at their core. The nation's new minicomputer companies, Digital Equipment Corporation (DEC) most prominent among them, were clustered near MIT and Harvard, along Boston's Route 128. A Forbes *editor with responsibility for the West Coast stated unequivocally that the magazine was interested only in publicly held companies with sales of $50 million or more.*[3]

Stanford's reputation was on the rise—it was often called "the Harvard of the West"—but the university attracted mostly students from California. San Francisco was best known for its hippies. Palo Alto was fighting its reputation as "the Peninsula's largest sex-shop center," thanks to an eight-block stretch of El Camino Real near the city's southern border that was home to multiple adult bookstores, an X-rated movie theater, and massage parlors with names such as "The Streaker" and "The Foxy Lady."[4] *None of this pointed to a promising future as an economic powerhouse.****

The rest of the business world might not have noticed, but inside the Valley, things were brewing in 1972. The previous year, Intel had invented the microprocessor, the chip that would soon nestle at the heart of consumer electronics such as the personal computer and video games. Meanwhile, Hewlett-Packard became the first Silicon Valley company to discover that an advanced electronic device could have broad consumer appeal. In 1972, the company introduced the HP-35, the world's first handheld calculator that could do more than add, subtract, multiply, and divide.† *HP sold more than 100,000 in the first year, despite the $395 price. A cheeky chapter called "Shirt Pocket Power" in the owner's manual noted, "We thought you'd like to have something only fictional heroes like James Bond, Walter Mitty or Dick Tracy are supposed to own."*[5]

For the most part, those developments within Silicon Valley happened independently of one another. But like different streams that feed a swift-moving river, the work would soon begin to blend and swirl together. The resulting network of personal and institutional connections would form the foundation of today's Silicon Valley and the high-technology revolution.

---

* In 1976, the city finally shut down its red-light district, closing seventeen massage parlors and banning halter tops as well as massages after 11p.m.

† The HP-35 was the first scientific pocket calculator. It had thirty-five buttons and could perform trigonometric, logarithmic, and exponential functions.

Have You Seen This Woman?

SANDRA KURTZIG

In the summer of 1972, several months after she left New Jersey, Sandy Kurtzig sat hunched low over a table at her new home, a sprawling seven-year-old house with five bedrooms in the bucolic suburb of Los Altos. Kurtzig and Arie had bought the house for $100,000 through a bankruptcy sale. The home was in disrepair and set high on a hill so steep that a friend's three-cylinder Saab could not make the climb.

She was rubbing letters onto a Wanted flyer that her mother, a former Chicago police reporter, had helped design. Kurtzig had typed most of the flyer on an IBM Selectric, but the only way to add the bold letters she wanted was to rub them on, one by one, from a kit she had bought at a stationery store.

She was pleased with the results. Her photo appeared at the top under the bold headline that she had just rubbed on: "HAVE YOU SEEN THIS WOMAN?" Below, the flyer offered a physical description ("female, 5'5" tall, 115 pounds, long chestnut hair, green eyes, fair complexion") and warned "DO NOT ATTEMPT TO RESIST THIS WOMAN!! SHE IS ARMED with facts and figures on how she can save you money and improve your company's efficiency, and SHE IS DANGEROUS to any less-than-perfect computer plan you now are using." The bottom of the page could be torn off and returned to Kurtzig, with boxes indicating either "I could sure use her help" or "Don't bother me! There are all kinds of psychos loose! She's just another one."[1]

The flyer was Kurtzig's first effort at advertising the business that she had launched several months earlier, at the beginning of 1972: ASK Computer Systems. She had chosen the name, a combination of her and Arie's initials, because it would appear early in the yellow pages and she thought the three capital letters looked solid and serious, like IBM.

An entrepreneur to whom Kurtzig had tried to sell GE time-sharing services—GE had allowed her to transfer to a sales region on the San Francisco Peninsula—had inspired Kurtzig's decision to start her own business. The entrepreneur, who had only recently launched a company to manufacture telecommunications equipment, had big plans. He did not like the then current method of tracking his inventory: scrawling information about every part and every step of the manufacturing process on cards stored in drawers like a library card catalog. He wanted a computer to track his inventory. But he did not want to use any of the standard programs offered via the GE time-sharing service. Instead, he asked Kurtzig to write a custom program. "I want to know what I have in stock, on order, and in process," he told her. "I want a list of every part that goes into every assembly. And if we substitute a part, I want it reflected immediately on my bill of materials."[2]

When Kurtzig told him that she couldn't write such a complex program while working full-time for GE, the entrepreneur suggested she quit and start her own business writing custom applications that today would be called software. He pointed to himself as an example of bold venturing. His warehouse was empty now, but he was confident that he was going to grow a $50 million company.

Kurtzig liked the idea of working for herself, running a custom applications company. She talked to Arie, who thought she should do it. Her parents, by contrast, were alarmed—less about her prospective entrepreneurship than about what it might mean for future grandchildren. How, they asked, could she be a mother and run a company? She told them not to worry.

For several nights when she should have been sleeping, she lay awake thinking. One night, she climbed out of bed at 2 a.m. to make a list for and against starting her own business. Reasons to do it included independence; a chance to be her own boss; and a change from GE, where she had spent her entire career. On the West Coast, the company now faced tough competition from one of the roughly one hundred time-sharing businesses around the country: a thriving operation headquartered in Cupertino called Tymshare.[3]

The reasons not to launch her own business numbered only two, but both were formidable. She would not have a guaranteed income, and women don't start their own businesses. The first negative was easy to counter, she decided, if she limited her downside risk. Her commission check from GE, due at the end of the year, would be about $2,000. She could stake her new business with that, and if she ran through the money without making any more, she would fold the company and find a new sales job.

The second concern was tougher. In the early 1970s, women made up roughly 40 percent of the workforce but only 1 percent of engineers[4] and 17 percent of managers[5] at any level. Helen Reddy's song with its smash opening line "I am woman, hear me roar" had been released several months earlier, but in fields like the one Kurtzig was proposing to enter, there were few women to roar.

She thought she could do it, nonetheless. Although women owned only 4.6 percent of the businesses in America, Kurtzig's parents and grandparents had started small businesses (businesses that, in the case of her father, had become successful enough to gain him a house in Beverly Hills and a Rolls-Royce). In her engineering program at Stanford, Kurtzig had succeeded even in the nearly all-male environment. At GE, she had done well selling computer services to manufacturers, precisely what she would continue to do if she went out on her own.

She climbed back into bed just before sunrise. A few hours later, she told the equipment entrepreneur that she would write a custom program for his business. It would cost him $1,200, about half of what he would pay a clerk to maintain the traditional card system each year.[6] She requested $300 up front, with the balance due upon delivery. She had launched her business.

Kurtzig was now a contract software programmer, even if it would be years before she used the word "software" to describe her work or her company. Today worldwide spending on business software alone is $332 billion.[7] More than half of venture capital investments in 2015 went to fund software companies.[8] Apple, Facebook, Google, Twitter, LinkedIn,

Oracle, Adobe, eBay, and Yahoo!—not to mention key companies out-side Silicon Valley such as Amazon, Microsoft, Alibaba, and Baidu—are, at their core, software. The current surge of tech startups is traceable to software, namely the Internet-based services, open-source code, and pro-gramming tools that an entrepreneur can buy or rent at very low (or even zero) cost.

But in 1972, the independent software industry was still in its infancy, born only three years before ASK. In 1969, IBM, under antitrust pressure, unbundled its software from its hardware.[9] This move opened the way for independent software vendors to write for IBM machines, particularly the popular IBM System/360 computer.[*] Like nearly everyone else program-ming computers outside of a large company at this time, Kurtzig thought of herself as a service provider, a writer of applications for computers. She did not see herself as a high-tech entrepreneur.

To launch her business, Kurtzig bought a desk, filing cabinet, chair, and second phone line—"the minimum to get a tax deduction"[10]—and installed them in a spare bedroom. For a $65 monthly fee, she rented one of the overgrown-typewriter-looking time-sharing terminals from General Electric. She composed her program using the terminal, which sent it via a modem over the phone lines to an off-site GE computer. The data moved at fifteen characters per second. Her program consisted of instructions that enabled the computer to display inventory and bills of material and also to retrieve and link information from the databases she was populating with information about the customer's manufacturing business.

Kurtzig asked herself how people would use the information that the computer spat out. What did users need to know? What would they want their invoice and bill of materials to look like? She imagined how someone working in a plant, someone who had most likely never used a computer, might want to interact with one.[11] Since there was no screen or monitor, the user would communicate with the machine via keyboard

---

[*] Before IBM's unbundling, software was usually custom-written by mainframe computer manufacturers for their customers at a cost of about $1 million per pro-gram.

and printout. "The success of the program depended on its being easy to use," she recalled.[12] She structured the program so that for any item in the inventory or bill of materials, the customer could generate a report that listed the part number, quantities on hand, costs, minimum reorder points, and order lead times.[13]

Writing the program was an undertaking for Kurtzig. She reminded herself to approach this more complex task with "concentration and common sense . . . like writing a recipe, with one instruction logically following the next."[14]

After five weeks of round-the-clock work, she delivered the program in February 1972 and collected the $900 balance due.[15] Before depositing the check, she made a photocopy as a keepsake.

Now it was time to find more customers. Kurtzig made copies of her Wanted flyer, sealed them in envelopes whose flaps were emblazoned with fire-engine red print ("WARNING: DANGEROUS WOMAN LOOSE!"), and headed to the post office. Knowing that big companies which could afford their own mainframes paid tens of thousands of dollars for manufacturing software from IBM or Burroughs, Kurtzig sent her flyer to thirty-five small manufacturers whose names and addresses she had found in a directory at the library. Surely, she thought, one of the small local firms would be willing to pay a fraction of a mainframe's cost for a program that could run on a time-sharing computer.

■ ■ ■ ■

The flyers did not yield a single reply. Whether the unconventional design backfired or the recipients did not think they needed computer services, Kurtzig did not know. Nor did she have the time to figure it out. Supplies and rental fees for the terminal were quickly eroding her $2,000 startup stake.

She began calling every company to which she had sent a flyer, figuring out that if she waited until around 6 p.m., the receptionists who refused to forward her calls would have gone home and the people running the business would pick up the phone themselves.

She secured another customer, a company that printed small local newspapers and paid a staff of 1,200 teenagers to deliver them. She wrote a program that sorted and labeled the papers by carrier, tracked subscriber lists, and calculated paychecks. When the owner asked her to maintain his computer records on a regular basis, she agreed, which meant she had entered a new business: updating subscriber lists and printing more than a thousand labels every week on her rented Teletype, which had not been designed to handle such a large job.

Kurtzig soon farmed out the maintenance work to an operation called Optimum Systems Incorporated (OSI), one of several data processing businesses that had popped up in the Bay Area to give small companies access to mainframe computers and printers. She rewrote the newspaper program to run on OSI's computer, and every week, she would deliver updates for one of the "keypunch ladies" at the front of OSI's cavernous office.

OSI's facility was across from the Stanford campus in the Town and Country Village shopping center. Other tenants included a drugstore, an aquarium shop, a record store, and a barbecue joint that specialized in slow-cooked ribs.[16] Today the center features a children's haircut salon and stores selling multithousand-dollar bicycles, designer clothing in toddler sizes, and $6 ice-cream cones.

■ ■ ■ ■

By the summer of 1972, Kurtzig had doubled her customer base to four; two microwave communications companies had hired her to write custom software and run the processing at the data center in Palo Alto. She began to think that ASK might make it as a self-sustaining company.

She bought a new hard-sided briefcase—pink—and printed business cards on lavender stock. She saw no reason to downplay her femininity. "Being a woman often gets you in an office easier than a man," she explained in 1973. "Men, especially executives, are always curious if a woman really knows her field."[17] In the 1970s, she recalls, "Men didn't know how to deal with women who looked and acted like men."[18] She saw her embrace of her femininity as a way to put potential customers at ease, something any good salesperson would want.

On her own, she was discovering tactics that a UCLA professor would three years later describe as essential for a woman to "make it in management." The successful businesswoman was "well-groomed, soft-spoken, slim and youthful," according to the professor's research, which was based on interviews with a hundred women (and a number of men) working in Los Angeles. In addition to several fundamental characteristics that would be essential for any manager (competence, education, realism, self-confidence, career-mindedness, and strategy), a woman also needed to garner the "support of an influential male" and to display "aggressiveness" and "femininity" in equal measure.[19] "Dress like a woman. Conduct yourself like a woman. Be gracious and intelligent. Don't try to imitate a man," recommended one female manager, who added that the "male establishment has been with us too long. We can't change the environment; just adapt to it."

Kurtzig also may have understood that her gender helped make computers, which could seem intimidating, a little less so. IBM hired women to work at the computer on display in the windows of its headquarters building at 590 Madison Avenue in New York City because, as a company representative explained to one of the women, "It's important for IBM sales that it looks easy to use a computer [and] the men will think it's easy if they see women working there."[20]

Kurtzig's faith that being female generally worked to her professional advantage was shaken in the summer of 1972, when she discovered that she was pregnant. In the early 1970s, fewer than one in five mothers of young children worked outside the home.[21] Kurtzig's own mother had quit her job to devote herself to Kurtzig and her brother when they were young.* Kurtzig expected that having a child would "change [her] into a full-time homemaker." That would mean the end of ASK.[22]

∎ ∎ ∎ ∎

The handwritten birth announcement, sent in February 1973, was a clever blend of Kurtzig's personal and professional life. The baby was described as a manufactured product ("our new model: the 'Andrew Paul'"). Height and weight were listed as "floor loading: 7 lbs 13 oz per 20.5 ins." This "first prototype," the announcement noted, was "manufactured by Arie and Sandy Kurtzig."[23]

---

* In 1970, 18 percent of mothers with infants under a year old worked. In 2015, the figure was 58.1 percent.

Kurtzig quickly discovered that she was "not the kind that can stay at home and talk baby talk." Her desire to work outside the home slammed her with what she calls a "double dose of guilt: first, that I wasn't spending all my time with Andy or on the house and second, that I didn't want to."[24]

ASK was making little money, but Kurtzig and Arie agreed to hire a live-in housekeeper who watched the baby and slept in one of the three spare bedrooms while Kurtzig worked in another downstairs. As her son grew, Kurtzig learned to live with the constant threat of his tiny feet ripping through the printouts of her work or the paper tape that stored her programs and data. After he went to bed, she would spend hours at the data center in Palo Alto, often not getting home until after midnight.

In August 1973, a reporter from a local trade newspaper, part of the conglomerate whose computer operations Kurtzig ran, came to the house for an article eventually headlined "Women in Electronics." Kurtzig, and her ability to "mix marriage, motherhood with career," was the subject. After explaining how she had launched her business, Kurtzig credited Arie for supporting her work; he bought the groceries, and until the housekeeper had arrived, the young couple had often eaten on paper plates to cut down on the need to wash dishes.

Alongside the article in a staged shot, a smiling Kurtzig holds a telephone in one hand and her son in the other. She told the reporter that she believed that women should be paid the same as men for equal work but added that "being a woman often gets you in an office easier than a man." The reporter promised readers, "Sandy is no women's libber."[25]

Though many later articles about ASK in trade journals would identify Kurtzig as CEO and say nothing about her personal life—the same treatment accorded men—nearly every article written about her or ASK for general readers took a woman-in-a-man's-world angle. It would be years before the national media noticed Silicon Valley, but when it happened, Kurtzig explains, reporters assigned each successful entrepreneur a unique role. "Steve [Jobs] was the youngest. I was the woman. That's just the way it was."[26]

■ ■ ■ ■

Kurtzig shifted ASK's business model—and planted the roots of the company's success—early in 1974, after a representative from Tymshare, the

Silicon Valley–based time-sharing company that competed with GE, suggested that she write a generic program that could be used by multiple manufacturers. Until that point, Kurtzig had written four custom programs, each tailored to the specific needs of an individual customer. The representative from Tymshare pointed out something that Kurtzig had noticed herself: although their products differed, the manufacturers' processes were similar. Manufacturing operations, whether they made newspapers, electronic equipment, soda pop, teddy bears, or bicycles, had to have parts (or ingredients) in stock and know when to order more. They had to track incoming orders. They had to schedule the different steps of their assembly process. And everything needed to coordinate with everything else, so that if, for example, a big order came in, the manufacturer would know how many parts to order and how to schedule operations so as not to conflict with other orders.

The Tymshare representative suggested that Kurtzig write a universal program. Tymshare would sell it alongside dozens of other software packages to manage financial portfolios, generate reports, analyze survey results, and run payroll. Most Tymshare programs were written and developed by small third-party operations like ASK. The more software Tymshare could offer its customers, the more likely they were to stay online at roughly $20 per hour, plus fees for CPU seconds. In exchange for the rights to Kurtzig's program, Tymshare would pay ASK 20 percent of what Tymshare billed customers to use it—a standard split at the time.[27]

Kurtzig recognized the opportunity. By pairing with Tymshare, which had $24 million in revenues and $2 million in profit in 1973, ASK could reach small manufacturers that already used time-sharing services, but not for inventory control and scheduling.[28] Until now, she had found customers in a piecemeal and labor-intensive fashion. For most software firms in the 1970s, marketing costs accounted for half of all expenses. Working with Tymshare, she could spend less.

Working with Tymshare would also enable ASK to make the leap that, years before, she had urged the Bell Labs scientists to make: from a batch-processed mainframe computer to a responsive time-sharing system. Time sharing gave customers remote access to powerful computers, which meant that they could have terminals on-site, enter data on-site, and even write programs on-site. This arrangement looked like the future of computing. Kurtzig wanted a piece of it for ASK. She paid a contract programmer for three months while he adapted a version of her manufacturing program for the Tymshare platform.

Kurtzig was operating at the fringes of an entrepreneurial support network that was slowly being woven among institutions and individuals in Silicon Valley. The contract programmer she hired was local. The first permanent employee she would hire was a Stanford graduate. Tymshare, the company whose manager was encouraging her to write a generic program, had been founded in Palo Alto and funded in part by George Quist, who went on to cofound, with Bill Hambrecht, the investment bank Hambrecht & Quist. Hambrecht and Quist took Tymshare public in 1970 and would do the same for Apple and Genentech in 1980, Adobe in 1986, and Google in 2004 (among many other companies).[29] Larry Sonsini, the attorney who helped take Tymshare public, was only twenty-nine at the time of the offering, but already he was working to transform the law firm where he worked, McCloskey Wilson Mosher & Martin. An expert in corporate securities, Sonsini imagined a new type of law firm, one that did not specialize in a single area but instead aimed to address "80 percent of the legal problems that a growing business would need."[30] Today Sonsini is arguably the most important lawyer in Silicon Valley. The firm, now called Wilson Sonsini Goodrich & Rosati, has represented more venture-backed companies than any other American law firm and helped launch more than a thousand companies, including AMD, Apple, Google, NVIDIA, Palm, and Seagate.[31]

It would be years before Kurtzig would take advantage of this entrepreneurial support network—or even realize that it existed—but it was coalescing around her. In less than a decade, Hambrecht and Sonsini would be offering to help with ASK's own IPO.

To work with Tymshare, Kurtzig needed to name the manufacturing program and write a manual. She believed that a manual played a vital role in selling something as esoteric as software, which is invisible and hard for people unfamiliar with computers to understand. She pored over manuals from Burroughs, IBM, and Hewlett-Packard to see how established companies explained their programs. She considered calling her package CAMP, for Computer Assisted Manufacturing Program, but settled instead on MAMA, for Manufacturing Management.

Kurtzig took an early copy of the manual to the manager at Farinon Electric, a customer with whom she had worked closely. The manager looked down at the manual, drew a big circle on its cover, and handed it back to her without even turning a page.

He had circled the name: MAMA.

It would never fly, he told Kurtzig. "Can you imagine a hard-nosed executive getting up in front of his board of directors and asking for approval to run the company's manufacturing operations with the MAMA system? No way!"

She was insulted. Who was this fellow to criticize mothers? But she reconsidered. She could not let personal pique rule her business. She thought again about the name's origins—Manufacturing Management—and handed the papers back to the manager, explaining that she would change the name.

She would call the program MANMAN.* After all, she reasoned, it often took two men to do the work of one mom.[32]

■ ■ ■ ■

MANMAN was a success on the Tymshare platform. Fifty companies, ranging from Coca-Cola to Borden Chemical to General Cable, used the program, some paying as much as $30,000 per month. ASK received only a fraction of those payments, however, so when, in the fall of 1974, Kurtzig received a call from a Hewlett-Packard salesman, asking if she would write a version of MANMAN for HP's new minicomputer, she was eager to hear more.

"Minicomputers were so new that I didn't even know how to spell" the word, she later recalled.[33] The machines were rudimentary by today's standards. They had tiny screens, basic keyboards, and no mouse. Text appeared line by line in block letters (green or white) that scrolled up and out of view as the user typed.

But the machines represented a breakthrough in the business

---

* The name MAMA never made it beyond the first draft of the manual. The program was released as MANMAN, and it was under that name that it would be sold, in various iterations, for decades.

world. Before minicomputers, computers were multimillion-dollar mainframes that required their own special air-conditioned room and a dedicated staff of operators and programmers. Minicomputers, at one-tenth the cost and with multiple terminals that could be placed around a building, could bring computers out of the world of specialized, highly technical back rooms into the payroll department, onto the shop floor, and into front offices. DEC, based near Boston, was the largest minicomputer manufacturer, trailed by its nearby competitor, Data General, which made the computer that ROLM was ruggedizing for military use.[34] Hewlett-Packard was third in the market but gaining.[*]

Kurtzig was amazed and thrilled that Hewlett-Packard would consider ASK software for its machines. Hewlett-Packard was the best-known, most established, and most respected company in Silicon Valley in 1974. Founded thirty-five years earlier by two Stanford alumni, Bill Hewlett and Dave Packard, HP, which had branched out from building instruments to calculators and now to computers, had nearly $1 billion in annual sales and thirty thousand employees around the world.[35] Kurtzig likened an offer to write for an HP machine to "being asked to the prom by the golden boy who was both the valedictorian and the varsity's star quarterback."[36]

The Hewlett-Packard salesman told Kurtzig that a small Los Angeles–based manufacturer of power supplies called Powertec Systems was interested in buying an HP computer—but only if the machine could run a manufacturing program. The salesman asked Kurtzig to write a version of MANMAN for the $80,000 HP 2100 computer.[37] Once she confirmed that the machine would have sufficient computing power and memory to run MANMAN, she agreed.[†]

Kurtzig hired a twenty-four-year-old Stanford math graduate to lead the project. Marty Browne had a ponytail and a love of rock and roll that

---

[*] In 1974, DEC, the market leader, sold 30,500 minicomputer systems; Data General sold 11,800; and HP, 8,900.

[†] The maximum memory available on the HP machine was 32k; the cheapest iPhone 7 has 64,000 times as much memory.

he indulged by attending nearly every soon-to-be-legendary concert in San Francisco in the late 1960s and 1970s.[38]

Browne and Kurtzig had budgeted a few months to rewrite MAN-MAN for the HP machine. They had not expected, however, that the computer HP provided would shut down randomly with no apparent cause. Nor had they imagined that the leading scientific minicomputer from Silicon Valley's premier technology company was failing because it shared a power circuit with a copy machine in another room, and every time someone made a photocopy, the computer lost power. They had also not anticipated that the computer's terminal control system and IMAGE database manager—both of which ASK needed to use for the MANMAN implementation—were incompatible. Browne and Kurtzig did not make much progress.

As the deadline for delivering the software neared, and with little to show for the weeks that the ASK team had spent at HP, Kurtzig needed more time. When she phoned the HP sales manager for Powertec's region, Bill Richion, he not only refused to extend the deadline, he kicked ASK off the project. An internal HP team would write the manufacturing program.

But Kurtzig would not accept that she was fired. She switched into the sales mode that one person admiringly described as "Sandy's charming relentlessness."[39] She convinced Richion to let her fly down to Los Angeles to meet with him in person.

When they met face-to-face, he was fresh off the golf course. "So you're the broad who's trying to sell this manufacturing system?" he asked.

Kurtzig said that she was "the very broad." Hauling out her spool of paper tape to load into Richion's computer, she launched into the demo that illustrated how MANMAN would look on the HP machine.

Kurtzig announced that MANMAN would be tracking the manufacturing of an imaginary product called "the Bill Richion." For a product description she entered "one handsome man." The bill of materials included two arms, two legs, and "zero heart."

He laughed. When the demo ended, he called it "clever."

Then he told her that ASK was still out of time. He still wanted to replace the company with an HP team.

"Let me buy you dinner," she countered.

"I used the woman thing where it worked," Kurtzig later explained.[40] Even at HP, which had a program to train and promote women, only 7.5 percent of managers and supervisors were female—and most of those likely held nontechnical jobs.[41] Since demonstrating ASK's technical competence had not done the job, it was time to use "the woman thing."

It worked. At the end of the meal, Richion not only picked up the tab, he gave ASK another month.[42]

ASK had just begun to gain some traction on the project when a new president took over the customer company, Powertec. He did not want to wait for HP and ASK to refine their system. He wanted to bring in IBM. Never mind that the IBM product, based on a mainframe, might cost ten times the HP-ASK solution. As the saying went in the 1950s, "No one was ever fired for choosing IBM."

Powertec canceled the purchase order. The reversal was ASK's first significant failure, and a painful one for Kurtzig, who later learned that many people at Powertec had said from the beginning that "there was no way that a woman could know anything about manufacturing and make it in this business."[43]

Turn Your Backs on the Origins of Computing!

BOB TAYLOR

Bob Taylor's job at the University of Utah, where he had gone after ARPA, had lasted only a year. The same disregard for rules and bureaucracy that had led him to barge into superiors' offices at the Pentagon left him battering heads with university administrators.[1] They did not appreciate his attempts to sell time on the graphics group's computer to students elsewhere in the university.*

So at about the same time that Mike Markkula began his job at Intel, Taylor had again packed up the family station wagon and continued west, going nearly as far as he could this time, almost all the way to the Pacific. Half an hour shy of the ocean, he stopped in Palo Alto, California, the site of what would be his professional home for the next thirteen years: Xerox's Palo Alto Research Center. Nearly every breakthrough for which the research center, known as Xerox PARC, is famous—from the personal computer to networking to printing—can be traced in some way back to the Computer Science Laboratory that Taylor launched in 1970.

---

* Taylor was trying both to earn a bit more revenue for the graphics group (officially the Information Research Lab in the College of Engineering) and to find a way to make the university's machines and computer expertise available to students throughout the university.

By the time PARC was two years old, the computer science lab was as unconventional as Taylor himself. The conference room, for example, was furnished with beanbag chairs. Taylor knew that most conference rooms at Xerox, which was headquartered in Rochester, New York, featured standard-issue rolling chairs and large wooden tables polished to a shine that reflected the clean-cut faces, neat collars, and conservative neckties of the company's managers, salesmen, engineers, and executives. Taylor also knew that other Xerox offices did not have bicycles leaned up against the interior walls or the work-when-you-want policies that kept his lab busy through the night.

Taylor knew it, but he didn't care. His suspicion of bureaucracy had grown to encompass many aspects of American business, which he believed was too concerned about appearing to do something rather than actually doing it.

So when someone in the lab suggested the beanbag chairs, other people liked the idea, and the chairs proved cheap, Taylor was happy to have his unconventional meeting room in PARC's low-slung, glass-faced building in the Stanford Industrial Park, an easy walk from the spot where students had overturned a school bus during the anti-SRI protests three years earlier.[2] Now he had some twenty-odd corduroy-covered chairs in the colors of the 1970s rainbow: burnt orange, goldenrod yellow, avocado green, midnight blue.[*]

In the spring of 1972, a young reporter named Stewart Brand came out to write about the lab for a magazine article on the "youthful fervor and firm dis-Establishmentarianism of the freaks who design computer science."[3] The article would be published in *Rolling Stone*, established five years before with support from Arthur Rock, the venture capitalist behind Fairchild, Intel, Scientific Data Systems, and soon Apple. The magazine sent a young photographer named Annie Leibovitz to accompany Brand to PARC. She took many shots of the beanbags, along with candids of the researchers in their plush beards and sideburns measured in inches. Taylor loved Leibovitz's portrait of him: clean-shaven, his hair carefully parted at one side and just grazing the tops of his ears, the bottom of his

---

[*] By the end of 1972, the computer science lab had a staff of thirty-one, most of whom were conducting research, with a support staff of roughly a half dozen.

face almost obscured by the pluming smoke from the deep-bowled pipe in his left hand.[4]

Liebovitz also shot the mocked-up model of the computer that one researcher, Alan Kay, had been urging the group to build: a notebook-sized machine, mostly screen and keyboard, that would be easy enough for a child to use. Kay, handsome and luxuriously mustachioed, was a former professional musician who built a pipe organ in his home and would go on to win a Turing Award, the highest honor for a computer scientist. Kay worked in the systems science lab, but Taylor had recruited him to PARC, and he collaborated closely with Taylor's group.

Kay's idealized computer—he called it the Dynabook—was as different from the roughly 150,000 computers humming away in the world's back offices, banks, and universities as the beanbag room was from its executive counterpart.[5] In 1972, computers no longer needed to be room-size mainframe behemoths that cost millions of dollars and ran batches of punch-card programs. But the new minicomputers, the type of computer for which Sandy Kurtzig's ASK was writing software for HP, could still fill multiple cabinets. Minicomputers ran the scoreboard at Red Sox games and the lights for Broadway's *A Chorus Line*, but at $50,000 or more, the machines remained a specialized—and costly—business expense.[6]

Meanwhile, the notion of a computer small and easy enough for a child to use was preposterous. In 1969, the Neiman-Marcus Christmas catalog had offered a $10,600 Honeywell "kitchen computer" aimed at the homemaker whose "soufflés are supreme, her meal planning a challenge." The catalog also featured a baby elephant, and perhaps the animal seemed no more foreign, intimidating, or useless than the computer. Not a single machine was sold, despite the apron and cookbook included in the purchase price.[7]

As things would turn out, even Taylor's lab—which was stocked with so many bright computer scientists that the president of MIT blamed PARC for causing faculty shortages at the top universities—could not build anything like the notebook-sized Dynabook in 1972.[8] What PARC built instead was a machine that would be recognizable today as a personal computer. It had a large monitor, a mouse, menus, a word processing program, and multiple windows. It could compose and edit documents and send them to a printer (also developed at PARC)—and the printout would look like the document laid out on the screen. The PARC personal computer could store files, documents, and images. It could connect through an Ethernet network, also developed at PARC, not only to printers but also to other computers to send emails and files.

Taylor named the revolutionary computer "Alto." To him it was "a continuation of the work I was doing at ARPA," the logical progression in the vision that he and Licklider had laid out years before, the vision that they had separately funded at ARPA: the notion that the computer could be a communication device connecting people and ideas.[9]

At ARPA, Taylor had funded the vision in two ways. He had supported efforts to develop time-sharing computers that were more responsive to their users than ever before, and he had launched the Arpanet, which connected computers to one another (and, by extension, connected the people using those computers to one another).

At Xerox PARC, Taylor's team, working with the systems science lab, would weave those two strands together into a single Alto system that combined the world's then easiest-to-use computers with a network that connected them to one another. It would take a decade for anyone to catch up to them.

■ ■ ■ ■

Taylor first learned of Xerox PARC in the summer of 1970, shortly after Xerox's chief scientist, Jack Goldman, convinced the company to launch a research facility supported by a $6 million annual operating budget and staffed with "the best in computer scientists; electrical engineers; systems analysts; operations researchers, mathematicians and statisticians; and biophysicists and biochemists." The goal was nothing less than the invention of "the Xerox of the future."[10] The original name of the research center, possibly rejected for its unfortunate acronym, was the "Xerox Advanced Scientific and Systems Laboratory."

Goldman proposed that Xerox locate the research center in a university town with an "intellectual night life" that would attract and benefit researchers. Within the center, he envisioned three laboratories to target basic sciences, systems science, and computer science. He also had a few specific ideas for future products that might emerge from the labs—"a machine which is half xerographic printer and half computing machine," for example. But the essence of his plan could be distilled to seven words: hire great researchers and leave them alone. He told Xerox's top management to expect nothing useful from PARC for at least five years.[11]

Taylor arrived at PARC in typical damn-the-torpedoes fashion. His first visit came within months of the lab's founding, when the physicist George Pake, the new director of PARC, invited him to Palo Alto. Pake was a former Stanford physics professor and provost at Washington University in St. Louis. A member of the National Academy of Sciences, Pake, an expert on magnetic resonance, would go on to receive the National Medal of Science. Pake's goal was not to recruit Taylor to PARC but to see who Taylor thought *should be* recruited for the computer science lab. "That would depend on what the lab is going to do," Taylor said. When Pake told him that the lab would support Scientific Data Systems, a computer company that Xerox had just bought for a breathtaking $918 million, Taylor informed him that Xerox had "bought the wrong computer company, an incompetent organization." He went on to call the latest offering from Scientific Data Systems, a Sigma computer, "an abomination." He told Pake that no one worthwhile would want to join a lab in support of such a second-rate company.[*]

Taylor left his meeting with Pake expecting never to hear from him or PARC. Soon, however, Pake invited him for a second visit, this time to talk about a job. As the two men sat together in Pake's PARC office, Pake told Taylor that he admired his ARPA connections and would like to have him at PARC's Computer Science Laboratory—but not as its director. Taylor did not have a PhD, Pake reminded him, and a lab aspiring to world-class status needed a man with a doctorate at its head. Moreover, Pake believed

---

* One possible source of Taylor's animosity: years earlier, Scientific Data System's founder, Max Palevsky, had told Taylor that the time-sharing efforts Taylor was supporting with millions of ARPA dollars were bound to fail—a comment that had inspired the irate Taylor to throw Palevsky out of his office. "It was as though an idiot comes into your house and starts telling you what's wrong with the way you're living. You know, after a point you can't tolerate it any longer," Taylor says. His reservations about SDS, which were shared by Chief Scientist Goldman, proved reasonable. In 1975, Xerox would write off the Scientific Data Systems purchase for $484 million.

that only someone who had conducted high-level research could manage an advanced lab.[12]

Taylor never got over the insult that Pake saw him as unqualified. Even after Pake hired him and let him hand-select his own PhD-bearing boss (Jerry Elkind, a former student of Licklider's who had been a senior manager at Bolt Beranek and Newman, the company that built the routers for the Arpanet); even after Taylor had decided that he was happy for Elkind to handle the administrative side of the lab while Taylor led the research team; even after Pake supported Taylor's bid to pay PARC computer scientists 15 to 20 percent more than scientists with comparable degrees at Xerox's research lab in Webster, New York;[13] even after Pake left the job as head of PARC only to be replaced by someone Taylor thought was worse—even *forty years* after Pake told Taylor he was not qualified to run PARC's computer science lab, the sting of the initial insult burned.[*]

Despite his reservations about Pake, Taylor accepted the job of principal scientist at PARC with the understanding that he would be named assistant manager of the Computer Science Lab as soon as a proper manager was found.[14] Taylor liked the climate and "aliveness" of the San Francisco Bay Area, which he had been visiting since his earliest days of supporting Doug Engelbart at SRI, and he liked Stanford, which he thought produced good "computerists." The heart of the sprawling campus was only a mile from the PARC lab, the spire of the university's Hoover Tower jutting high above any other structure in the area. Taylor moved his family into a large Craftsman-style house in Old Palo Alto, where the streets are named for Romantic poets. The neighborhood teemed with young families. Taylor had three sons.

The general mandate from Xerox corporate headquarters was to focus on "the architecture of information."[15] For Taylor, it was a matter of simple semantics to repackage the vision of a network of interactive computers to fit this mandate. Whatever the effort was called—the computer science lab's focus on building "the office of the future" came later—Taylor believed that "it was time to get rid of centralized machines . . . and give everyone their own machine."

---

* Taylor mentioned his dislike of Pake in four separate interviews.

"And I knew who to hire to do it," Taylor adds. "I knew, personally, who the strongest young computer scientists in the country were."[16]

■ ■ ■ ■

Taylor, who lacked formal technical training, knew where he wanted to end up but not how to get there. Severo Ornstein, a member of the inner circle of PARC researchers that Taylor called his "graybeards" and a man who respects Taylor, calls him "a concert pianist without fingers."[17] Taylor could hear a faint melody in the distance, but he could not play it himself. He knew whether to move up or down the scale to approximate the sound, he could recognize when a note was wrong, but he needed someone else to make the music.

There have been many great technical visionaries whose ideas never reached full expression under the visionary's guidance. In Taylor's own day, indeed within a few miles of the Xerox PARC office, there were two. In the 1970s, Ted Nelson, who coined the word "hypertext," wrote about a complex information architecture called Project Xanadu that never came to fruition, despite anticipating and in some ways exceeding the World Wide Web. Likewise, many of Douglas Engelbart's ideas were not realized until they were refined at PARC, in the computer science and systems science labs.

Taylor was different. He could recruit to PARC an outstanding group of researchers—selected based on his belief that "a very good researcher was worth two dozen good researchers"—and keep them working together for years.[18] He also had the support of his boss, Jerry Elkind, who handled much of the lab's administrative work. As one researcher, who respected both men, put it, "Bob was my leader. Jerry was my manager."[19]

Taylor's work at ARPA had pointed him to the researchers he desired, and many of the young graduate students he knew from those days fondly remembered him—and the generous ARPA funding he offered. As a result, he could handpick his team.* "All it took was a few phone calls from

---

* One early hire that Taylor wanted but did not get—Larry Tesler—fell through because Taylor was not directly consulted in negotiating his offer. Tesler later joined PARC's Systems Science Laboratory.

Taylor and the chosen researchers signed up," recalled one Xerox executive.[20] Taylor began recruiting at the University of Utah. He swept in several employees from Bolt Beranek and Newman. At Harvard, he pursued the networking expert Bob Metcalfe, even after Harvard balked at giving Metcalfe a PhD.*

Closer to home, Taylor bypassed the young microchip and electronics operations to pursue clusters of talent with academic, not business, leanings. From Berkeley, he plucked a group that had worked together on an ARPA project at the University of California and then launched a commercial operation near the campus. The Berkeley Computer Corporation was about to go bankrupt, and many of its young computer scientists were excited to leave business concerns behind to join a research lab. As Butler Lampson, who would go on to gain recognition as one of the world's greatest computer scientists, put it, "We started BCC because it was the only way we could think of to carry out this particular research program."[21] Joining Lampson were Chuck Thacker, who would, like Lampson and Kay, win a Turing Award. The team also included the expert system designer Peter Deutsch; the top design engineer Richard Shoup, who would go on to win an Emmy for his work in color graphics; the elite programmer Jim Mitchell; and Charles Simonyi, who is best known for writing Microsoft Word after he left PARC. Taylor would call the Berkeley contingent "the cadre of my computer science lab" at PARC.[22]

Back near the PARC offices, Taylor paid a call to Engelbart's lab at SRI and tapped Bill English, who had built the prototype of the first mouse and organized the technical production of the 1968 Mother of All Demos. English, in turn, recruited other members of Engelbart's lab; fifteen would join PARC, most in the systems science lab. Taylor never considered wooing Engelbart because Taylor wanted what he called "hands-on engineers," and Engelbart was anything but. "He was a visionary, and if there was anything this group didn't need, that was another visionary," Taylor says. He thought that Engelbart "could not explain what he wants."[23] Engelbart, meanwhile, was not comfortable even visiting PARC. "I went over there a couple of times to visit but people were always showing me what they

---

* Metcalfe would resubmit his dissertation and be granted the degree while at PARC.

were doing and it was in a sense almost like 'See this is the real way to go,' "
he said.[24]

Every one of his recruits, Taylor knew, could play a different, important role in his campaign to transform the computer from a glorified calculating device into a communication tool. He had hired hardware and software experts, engineers and computer scientists, specialists in programming languages, and authorities in human-machine interactions—all of them blessed with an agility of mind and network of contacts. "Be willing, where necessary, to turn your back on the origins of computing!" Taylor exhorted his twenty-person team.[25]

The recruits had one other trait in common: Taylor's vision was theirs, as well. Reminiscing with Butler Lampson, Alan Kay recalled, "We already came in as born-again interactive computerists. Taylor was pretty vigilant about not inviting people who weren't really into that thing." ("Didn't sign up for that, right," Lampson agreed.)[26]

Taylor was a draw, but PARC also offered high salaries and excellent resources. PARC researchers could easily interact with others at SRI and Stanford, particularly the Stanford Artificial Intelligence Laboratory (SAIL). They could attend or give talks on campus, or meet with people visiting one of the other institutions. Several of the researchers from Taylor's lab taught classes at Stanford. A job at PARC offered the intellectual stimulation of an academic career without teaching or publication demands.[*]

There was only one inviolable administrative requirement in Taylor's lab. Once each week, everyone had to be in the beanbag room for a meeting. Taylor did not care when his researchers got to work or what they wore while there. It did not matter when they went to lunch or whether they shaved. But he wanted them in the beanbag room every Tuesday—and he expected them to stay for hours. The meetings served as the intellectual

---

[*] The first publication out of the lab came three years after its founding—and it was a doctoral thesis. Another option for researchers, to pursue ARPA-funded research, required them to explain how their work supported the country's defense and to accept money from a Department of Defense still engaged in the unpopular Vietnam War.

pulse of the lab and a way for him to keep the team moving in the same direction. The gatherings also reveal the inner workings of what both Thacker and Lampson have described as Taylor's "magical" leadership.

Taylor opened the sessions with administrative issues such as available billets and announcements of upcoming visitors. He then moved on to solicit project reports and ask for social announcements. Many members of the group, young and single, biked together in the hills or grabbed lunch at the Alpine Inn (fondly called "Zotts") in the rural town of Portola Valley, a few miles from PARC. Researchers shared long weekend hikes among the coastal trees in the misty coolness above the valley. They barbecued at one another's houses and worked on one another's basement workshop projects. Taylor, a competitive tennis player, enjoyed challenging the best players at PARC to hard-core, sweat-streaming matches on the court behind his neighbor's home. Afterward, the players would adjourn to Taylor's house. "The Dr Pepper was cold, the doors were open, a breeze was blowing through, and everyone was always welcome to come in," recalls Bob Metcalfe.[27]

During the weekly lab meetings, Taylor also spent time fielding complaints. The cafeteria food was bad. Equipment was inadequate. Secretaries were overworked. Surprisingly often, grumblings about balky Xerox copiers led the list of problems.

Taylor listened and took careful notes. He knew that truly creative computer scientists tend to be opinionated individualists. He catered to that tendency. When the group was preparing to move to a new building, he polled every researcher to determine the ideal phone system. All twenty researchers wanted Touch-Tone phones rather than the old-fashioned rotary dial, but beyond that, there was little consistency. One did not want any secretarial backup, but nine wanted the so-called Alan Kay switch that transferred incoming calls immediately to an attendant without interrupting the intended recipient. Five shunned the message waiting light, while seven wanted an intercom line.[28] Customized phones may seem trivial, but they are an example of a feeling that Taylor created at PARC—the sense that he was working for the researchers, rather than the other way around.

Every week, once the social and administrative housekeeping were out of the way, Taylor brought the group to the heart of the meeting. "A time for half-baked ideas," he called the sessions. A researcher would present work in front of a blackboard and set all the rules of engagement. Could he be interrupted with questions?* Were people allowed to interject at all? Would he talk for an hour and invite discussion for two more? Or would he toss out only a few ideas before opening the floor? Taylor called the featured presenter the "dealer" because the speaker set the terms much as a dealer sets out the rules of a poker game. Soon enough, the meetings themselves came to be called Dealers.

"Dealer" also had a more antagonistic connotation. Taylor had read the popular 1962 how-to guide on card counting, Edward O. Thorp's *Beat the Dealer*. He hoped that meeting participants would use their expertise to poke holes in a presenter's work in the same way that card counters used continuous computations to get the edge on a blackjack dealer. Nearly every presentation was interrupted by mutters or even shouts of disagreement from people sprawled in the soft chairs on the floor. Taylor believed that the confrontations, which he tried (not always successfully) to keep focused on the work, not the presenter, yielded better results.[29] In a phrase that could serve as a personal motto, Taylor once said that "controversy is healthy and should not be inhibited."[30]

The dealer sessions owed much to the principal investigator meetings that Taylor had directed when he was at ARPA. Then, as now, he wanted the researchers to ask one another questions he might not know how to ask himself. Then, as now, he believed that tough questions would help clarify thinking and illuminate each participant's viewpoint and particular expertise.

Throughout his time at PARC, as at the Dealer meetings, Taylor framed the conversation in the lab and then let the technical experts take over, interceding only to correct course. "The *theme* that assembled [the team] was Bob's notion of interactive computing in the service of people," wrote Bob Sproull in 1977, when he left the lab for an academic job at Carnegie Mellon. "Whenever the theme fades among the clamors of daily activities and crises, Bob works to raise all our sights out of the details."[31] That, too, had been Taylor's approach when launching the Arpanet: set the goal, hire the right technical people, and then step in only when necessary.

---

* Later, women would be hired into the computer science lab, but early on, the researchers were all men.

Allowing the research team to define its own path to the goal did not mean that Taylor abdicated control. He cut off discussions if he felt they had reached the end of their usefulness.[32] When he thought the quality of Dealer presentations was slipping, he threatened to move the weekly meetings to a monthly schedule, drawing the objections of his staff.[33] He mediated disputes between researchers, insisting again and again that each person needed to be able to explain his antagonist's point of view to that person's satisfaction, even if he did not agree with it.* He was also willing to make the least popular decisions in the lab: shutting down research efforts (advanced work on color graphics, for example) that he considered tangential.

At the same time, he deferred to his staff's technical opinions, particularly those of Butler Lampson. Twenty-nine years old when he came to PARC, Lampson was wire thin and pulsed with energy. Brilliant and kinetic, he had once been scheduled to speak at a computer science conference, and when the previous speaker had gone on too long, he had delivered his entire hourlong talk in a whirlwind but cogent fifteen minutes before rushing through the door to catch his flight.[34]

Lampson was also as hardheaded as Taylor. He refused to write abstracts for his long memos at PARC; if there were a way to summarize what he needed people to know, he said, he would have written a shorter memo.[35] He once declared he would rather sit on the floor of his office than have ugly furniture in a new lab Xerox was building.[36]

"Taylor listened to him as though God was speaking," says Severo Ornstein. "It would have been terrible if Butler had had any crazy ideas because they would have been followed. But Butler was pretty much right about things."[37]

■ ■ ■ ■

Several of Taylor's managerial innovations were radical. The lab had no formal organizational hierarchy, aside from everyone reporting to Taylor.

---

* Taylor distinguished between what he called "Class 1 and Class 2 disputes." In the first, the two sides are so estranged that they cannot even hear, much less understand, what the other is saying. In Class 2 disputes, the two sides disagree but understand each other. Taylor's goal was to move all Class 1 disagreements to Class 2, even if resolution was not possible.

For any new hire, nearly every employee had to raise a hand indicating that he—or she; a few female researchers joined over time—would be "deeply disappointed" if the person did not come to PARC. The group internalized Taylor's own standards. One unstated criterion for hiring, explains Chuck Thacker, was for "every person we hired to increase the average IQ of the group."[38] The need for near unanimity could at times encourage what Butler Lampson called "groupthink" and lead the lab to miss important hires. (Alvy Ray Smith, for example, who went on to cofound Pixar, worked briefly at PARC but was not offered a permanent job.) But the people who joined the staff at PARC did so with the support of the entire lab.

If Taylor's team collectively felt something was a good idea, he saw it as his job to push that idea as far as he could, to "give it a lot of air cover," according to Chuck Geschke, who would go on from PARC to cofound the software giant Adobe. Taylor cared less about his own political position within the wider Xerox hierarchy. In the first months of Taylor's tenure, his research team's most urgent need was for a computer to design and test their ideas about the machine they wanted to build. The lab computer would serve as the digital equivalent of a chemistry lab's collection of beakers, chemicals, and spectrometers: the basic equipment that made research possible. Xerox corporate wanted the lab to use a machine built by the recently acquired Scientific Data Systems. Taylor's researchers thought the computer—the one Taylor had called "an abomination" in his first interview at PARC—was not capable of supporting their work. They wanted a PDP-10 machine made by Xerox's competitor DEC.

Taylor backed their request, taking it to Pake, who passed it on to corporate. Eventually a Xerox vice president flew out from the company's headquarters to see if, after listening to the cases presented by the researchers, he could justify the purchase of a million-dollar PDP-10 system that competed with Xerox's own product.[39]

Unpersuaded, he denied the lab's request for a DEC machine. When the news reached Taylor's lab, several researchers threatened to leave, rather than use the Scientific Data Systems machine. Taylor stood behind them. Pake brokered a compromise: Taylor's lab could not buy a DEC PDP-10, but they could build a clone.

Designing and assembling the cloned computer, which Taylor's researchers named MAXC in mocking homage to Scientific Data Systems founder Max Palevsky, proved an extraordinary team-building exercise.

Constructing a computer in 1970 was not a simple matter of welding or screwing together some kit parts and loading in an operating system and a few software programs. Components had to be designed and built. Code had to be written and programmed. People in the lab had to agree on the type of memory to use (they ended up going with a new semiconductor memory, Intel's 1103, whose marketing the newly hired Mike Markkula was helping with). PARC employees had to review proposals from suppliers. In wrestling with those decisions, people who had never worked together learned how to do so, and a disparate group of what Taylor calls "soloists" became an orchestra. The group also developed a network of trusted local suppliers and fabricators.

But from the corporation's vantage point, the decision to allow the researchers to build their own machine was a terrible capitulation; Xerox employees were permitted to spend months and hundreds of thousands of dollars copying a competitor's computer instead of buying a Xerox product they could use immediately. Xerox also lost an opportunity to have an exceptionally qualified group of computer scientists develop software for the Scientific Data Systems machines.

Taylor and his research team say that they could not have done the work they wanted to do on the Scientific Data Systems (soon to be re-named Xerox Data Systems) computer. True.[*] But many Xerox corporate executives thought that point was irrelevant.[40] What Taylor's researchers wanted to do to further computing progress did not matter. What mattered was what they could do to make money for Xerox. They should use a Xerox computer and build products for it.

The corporate executive who had visited PARC as part of the computer-purchasing debacle later noted, "This small incident set the tone for everything that would be coming in the future. PARC was allowed to operate in complete isolation and to build technologies that were designed to demonstrate the brilliance of PARC researchers without any consider-

---

[*] The group that came from Berkeley Computer Corporation had intimate knowledge of the SDS computer line, since before coming to PARC they had modified an SDS computer to serve as a time-sharing machine. The first customer for the modified SDS940 computer was Tymshare, the company for which Sandy Kurtzig wrote MANMAN.

ation of how this could possibly relate to the future of Xerox." Taylor's team, he says, were "people filled with messiah [*sic*] zeal. Their loyalty was never to Xerox but to the intellectual challenge of taking information technology research out of the stranglehold of mainframe computing."[41]

Taylor did indeed want to break "the stranglehold of mainframe computing," but he, and nearly everyone in his lab, also felt loyal to Xerox. Again and again over the course of the next decade, Taylor would try to interest the company's leadership in the technology coming out of the computing labs at PARC. His efforts would meet with little success.

■ ■ ■ ■

In early 1972, the lab was at a critical point. The MAXC computer was almost complete. Soon Bob Metcalfe would connect it to the Arpanet, the computer network that Taylor had kick-started in 1966 when he was at ARPA. Researchers were beginning to complain that the final-stage refinements were "depressing" and too far removed from basic research.[42]

Around this time, Jerry Elkind, Taylor's PhD-bearing boss, asked whether Xerox should "buy the ARPANET." ARPA was looking for an outside institution to run the network as a public service akin to the national phone network.[43]

After some discussion with Taylor, Elkind, along with Xerox's head of research, Jack Goldman, put together a group "to analyze the opportunity for buying ARPANET and to recommend the action Xerox should take."[44] In the end, Xerox, like AT&T (which had also considered a purchase), declined to bid for ownership of the Arpanet, and in 1975, the Defense Communications Agency took over operational responsibility.

What would have happened if Xerox or AT&T had bought the Arpanet in the early 1970s? The network almost certainly would have been much less independent and freewheeling than it—and the Internet that followed in its wake—later proved to be. A corporate owner would likely have exerted more control over who could join and might also have adopted a more rigid definition of acceptable behavior.[45]

While Xerox contemplated buying the Arpanet, Taylor's researchers were clamoring for a new project. He needed to figure out what would come next. A talk by Lampson at a February 1972 Dealer meeting provided a spark. "We should turn our attention to building a simple machine. They are not expensive, and they are not hard to build, given what we have now," Lampson said. "Simple machines will give us most of what we want," and if a number of them were networked together, "almost all jobs that people will be wanting to do in the next ten years could be taken care of."[46]

At the next week's Dealer meeting, Lampson and Alan Kay, who saw Lampson's proposed computer as an "interim Dynabook"—a first step toward the portable kid-friendly machine he imagined—gave a more detailed talk about "Alan Kay and Butler's $500 Machine." The two men laid out a few technical thoughts and invited interested people to a meeting to be held several days later.[47]

Taylor was thrilled. He had wanted to build a small, easy-to-use machine when the lab had launched eighteen months earlier. At the time, he could not describe his ideas about what would come to be called a personal computer in a way that Lampson or Chuck Thacker, whom he also approached, thought reasonable.[48] "He was waving his arms, talking about interactive mumbo-jumbo," recalls Lampson. "We interpreted this to mean he was describing something completely impractical." Researchers in other places, many of them supported by ARPA funds, had built hugely expensive one-off prototypes of personalized interactive computers. That was not the agenda for PARC. From the beginning the lab aimed to build machines that could be used by large groups of real people in real environments. Lampson was interested only in building systems for one hundred or more users.

By the end of 1972, a team headed by Lampson, Kay, and Thacker was planning an entire *system* built around "ten to thirty" Alto computers. In December, Lampson circulated a seminal memo titled "Why Alto?" That memo moved beyond Taylor and Licklider's vision to a blueprint for reality, laying out hardware and software specifications for a machine that would fit under a desk and use a graphical display, a keyboard, and a "mouse or other pointing device." The memo also placed the machine within a network that included other computers as well as a printer.[49] It was not enough for an individual to possess significant computing power in a dedicated, easy-to-use, affordable machine, Lampson's memo made clear.*

---

* This systems insight, that the network was as important as the computers in it, recalls Taylor's original vision of the Arpanet as a means of connecting remote computers. As Taylor told Chuck Geschke, "The Arpanet had opened the door

"If our theories about the utility of cheap, powerful personal computers are correct, we should be able to demonstrate them convincingly on Alto," Lampson wrote, using the word to refer not only to the computer but also to what he was now calling "an Alto system." He continued, "If [the ideas] are wrong, we can find out why."[50]

Over the next few months, the computer science lab and Alan Kay's group in the systems science lab worked together at a frenzied pace to build the computer described in Lampson's memo and in a companion hardware note from Chuck Thacker. Larry Tesler, who worked in Kay's group, recalls tag-teaming with Tim Mott on the graphical user interface. They shared a computer, one coding all night, the other coding all day, with an hour of overlap scheduled at the shift change so they could update each other.[51] "It had the intensity of a startup," says Chuck Thacker. "I feel very sad that I didn't get to know my first daughter except at the two o'clock [a.m.] feeding."[52]

The group was grappling toward their shared vision of what it meant, in silicon and wire, in operating systems and microcode and routers and zeroes and ones, to have a truly interactive, networked machine. The memos they produced reveal the shared sense of mission and intensity: "These problems seem worth solving." "Here's my take." "Comments welcome." "This is just a draft." The subject lines of two memos are noteworthy: "Vented Frustrations" and "What Am I Doing Here."[53]

Discipline was tight. Every memo was stamped "Read and understood," signed, and dated, in accordance with standard protocol in scientific labs around the world.

Taylor remained a warm and welcoming presence in the midst of the excitement. "You could go into Bob's office with a problem, any problem," recalls Larry Tesler, "and he would shut the door behind you and help you."[54] Bob Sproull wrote, "It was Bob who always had more faith in me than I had, and who nourished every shoot of confidence I demonstrated."[55]

Tesler recalls that Taylor would occasionally hand him one of the

---

to communications, but there were a lot of things that were undone if you really wanted a computer system that was interactive and focused on communication. So we set about doing that in PARC."

cartoons that illustrated "The Computer as a Communication Device" article Taylor had coauthored with Licklider in 1968: "He'd say, 'This hasn't happened yet.'" But Taylor emphasizes that he did not lay out a comprehensive vision for the Alto in advance. "As we finished one piece, it would be clear, I thought, what would be next pieces," he says. Since interactive computing mandated that a computer be optimized for communication above calculation, the Alto was built around a display rather than an arithmetic unit. The focus on the display required a new type of operating system. A new OS required new applications, and since one goal was communication, a word processor was an obvious program. And so it went. "As you posited one element, the next would become apparent," Taylor says. "You knew you wanted to send characters over a network (by the way, you need a network), to a printer (so now you need a printer). And these characters you were sending, you might want to save them, so you needed a file system. Another thing you could do with these characters: you could compose them into an email, so you needed an email system. These were all separate components in a way, but in another way, they were part of a system. One of the things I did was guide this progression."[56]*

Step by step and piece by piece, Taylor and his team were laying the groundwork for the modern personal computer industry.

---

* Both Thacker and Lampson agreed with Taylor's characterization of his work, though Thacker amended it to "*help* guide this progression."

Hit in the Ass by Lightning

■■■■■■ AL ALCORN

Al Alcorn knew he was being wooed. It was June 1972, around the same time that Xerox was considering buying the Arpanet. Several miles south of the PARC building, Nolan Bushnell, the tall, brash, young engineer from Alcorn's work-study days at Ampex, had shown up at Alcorn's Sunnyvale office. Bushnell was driving a new blue station wagon. "It's a company car," he said with feigned nonchalance. He offered to drive Alcorn, recently hired as an associate engineer at Ampex, to see the "game on a TV screen" that Bushnell and Ted Dabney had developed at their new startup company.[1]

The two men drove to an office in Mountain View, near the highway. The space was large, about 10,000 square feet, and looked like a cross between an electronics lab and an assembly warehouse. Oscilloscopes and lab benches filled one area. Half-built cabinets and screens with wires protruding from them sat in another.[2]

Bushnell walked with Alcorn to a sinuous, six-foot-tall fiberglass cabinet with a screen at eye level. Bushnell was proud of what he called its "spacey-looking" shape. He had designed it in modeling clay, and Dabney had found a swimming pool manufacturer willing to cast the design in brightly colored fiberglass. The cabinet housed a shoot-em-up-in-outer-space fantasy game called *Computer Space*.

Computer Space was striking enough that the makers of the film *Soylent Green* would commission a special white one for the movie.[3] But

Alcorn paid the lovely cabinet no attention, aside from noting the vague stink of the fiberglass. He thought the most interesting feature of this, the first video game he had ever seen, was Bushnell and Dabney's decision to use an off-the-shelf television set as a screen. Had they asked him, he would have said that the thirteen-inch black-and-white General Electric model with balky wiring would be most useful for starting fires.

Watching Bushnell demonstrate the game, Alcorn grew excited. *Computer Space* was based on an iconic game called *Spacewar!*, written in 1963 by an informal group at MIT led by Steve Russell. Across the country, programmers played and constantly modified *Spacewar!* on time-sharing machines in fledgling computer science departments that ARPA had seeded. (Bushnell had played *Spacewar!* when he was an undergraduate at the University of Utah.) Most technical people who saw *Spacewar!* were entranced by its computing implications: it demonstrated that a computer could draw on a screen, calculate trajectories, and detect when a ship was hit.[4]

But Alcorn knew that there was no computer inside Bushnell and Dabney's *Computer Space*, even if the promotional literature bragged of a "Computer (Brain Box)." Computers were far too expensive to use in a scenario like this one.* Something else must be controlling the patterns and movement on the screen. Alcorn wanted to know what.

He opened the cabinet, glanced at the wiring, and fell in love. Bushnell and Dabney had tweaked the dedicated logic circuits within the wiring of the television so that they could produce the same effects as the time-sharing computer in the original *Spacewar!* game. "A very, very clever trick," Alcorn called it.[5] Without a computer, without software, without a frame buffer, a microprocessor, or even memory chips beyond a few

---

* Bushnell and Dabney had originally planned for *Computer Space* to run off a computer with an attached monitor—they had even invited a computer programmer from Ampex, Larry Bryan, to join them in building it—but the computer-based design almost immediately ran into problems. Bushnell says that no computer was ever purchased.

   Spacewar! purists will note that *Computer Space* was a diluted version of the original game; gravity was not taken into account in *Computer Space*, for example. A much truer homage, developed at the same time as *Computer Space*—and containing a real computer—was Bill Pitts's *Galaxy Game*, which was installed at Stanford's Tresidder Student Union.

flip-flops, Bushnell and Dabney had made a dot appear and move on the screen. Even to Alcorn, who had repaired televisions since he was a teenager and was now working on high-resolution displays at Ampex, the trick seemed "almost impossible."[6]

Alcorn had a gush of questions. Bushnell waited for him to calm down. Then he offered Alcorn a job at $1,000 per month and 10 percent of the startup company that he and Dabney had each kicked in $350 to launch.[7] Bushnell and Dabney called their company Syzygy (a word that refers to the alignment of three celestial bodies) but soon renamed it Atari, after discovering that another company had incorporated under the name Syzygy. In Bushnell and Dabney's favorite game, Go, "Atari" means roughly the same thing as "Check" in chess. Or, as Bushnell later chose to define it, "Atari means you are about to be engulfed."

Syzygy, the soon-to-be Atari, designed games for manufacturers such as the pinball giant Bally to manufacture and sell. Syzygy had designed *Computer Space*, Bushnell explained, but a small operation called Nutting Associates, which owned the office in which they were standing, was manufacturing it.

Bushnell and Dabney's chutzpah impressed Alcorn almost as much as the electronic trick. He had never known anyone who had left a job at a big company to start a new business, as Bushnell and Dabney had left Ampex.[*] The move felt right, though, he thought—another way in which young, bright people were writing new rules for themselves in the wake of the 1960s.

Then again, the salary Bushnell was offering was a 17 percent cut from Alcorn's Ampex paycheck. The 10 percent ownership stake, he figured, was worthless since Atari would probably fail. When he mentioned the job offer to his parents, they were shocked that he would even consider a move.[8] Did he not understand that taking this job might be a form of slow suicide? The fly-by-night operation would disappear, and then who knew where more work would come from? During the Depression of Alcorn's parents' youth, men without jobs had starved.[†] Big, established Ampex was the safer choice.

---

[*] Memorex had spun out of Ampex in 1961, before Alcorn's time there.

[†] Bushnell says his mother's reaction was similar to Mrs. Alcorn's: "But Nolan, you had such a good job!"

Alcorn, a child of the Cold War and Vietnam War, did not share what he called his parents' "naïve paternalistic belief" that large institutions or companies offered his best hope for the future. Ampex, he knew, was facing financial problems; its future success was not a foregone conclusion. The US government, the largest and most stable institution he knew, had sent thousands of young men to die in the Vietnam War. Police had not protected the unarmed protesters at People's Park—they had shot at them. Alcorn saw no reason to trust established organizations.

Alcorn's then girlfriend (and future wife), Katie, encouraged him to "take a chance on a flyer." After all, they had no kids and no mortgage. And if, as Alcorn predicted, Syzygy/Atari failed, he would find another job at one of the many businesses in and around Mountain View that were hiring electrical engineers.

In the end, Alcorn, the careful adventurer, decided that he "had nothing to lose" by joining Bushnell and Dabney. "Life is short," he thought.[9] It was time to create his own chances.

■ ■ ■ ■

When Alcorn reported to work at Atari's newly rented offices on Scott Boulevard in Santa Clara, he learned that Bushnell's entrepreneurial risk taking that had so impressed him was a sham. Though it was true that Bushnell had launched the startup company with Dabney, he had done so with a safety net that Alcorn did not have: he was a full-time salaried employee at Nutting Associates, the company that licensed and built *Computer Space*. Bushnell's salary was higher than what he had earned at Ampex—and on top of it, he had negotiated licensing fees from Nutting as an independent contractor.

Bushnell had told his wife that he would be running his own company within two years of coming to California. He decided to consider the Nutting job "kind of a rounding error" that he could "edit out of conversations" when he talked about his new video game business. "Entrepreneur" sounded "more glamorous," he later explained when asked why he had not told Alcorn about his job with Nutting.[10] Appearances mattered to Bushnell; his first hire at Atari was a receptionist, his children's seventeen-year-old babysitter, whom he told to place all callers on hold with a promise to "see if Mr. Bushnell or Mr. Dabney was available," even if the men were right in front of her. Years later, he would call his early success in business "a matter of being enthusiastic and glib."[11]

Alcorn soon learned about a second misdirection. Bushnell and Dabney had built *Computer Space* using spare Ampex parts. Before Alcorn had joined Atari, he had asked if the cofounders had offered the game to Ampex, which likely had rights to it. Bushnell had assured him that Ampex had turned down the offer. Now Alcorn learned that Bushnell had never offered *Computer Space* to Ampex.[*]

Soon Bushnell misled Alcorn a third time, though Alcorn would not know it for weeks. Bushnell told his new engineer to build a Ping-Pong game for a contract with General Electric. He described how he wanted the game to look, specifying details down to the line dividing the screen and the rectangular paddles on either side. The game needed to be cheap, he said, and ideally, it would contain no more than twenty chips. It needed to use the clever video-positioning technique that Alcorn so admired.

Alcorn, determined to impress General Electric, drove to a department store on El Camino Real and bought its best black-and-white television. Back at the office, he designed segmented paddles, with each segment sending the ball careening back at a different angle. The sync generator inside the television, he discovered, already contained certain tones, and with a bit of manipulation, he came up with a satisfying *pong* sound when the ball hit the paddle.[12] He configured the game so that play would speed up after a few rallies. He decided not to try to fix a bug that kept the paddles from reaching the top of the screen, since it meant that a ball could slip above or below even the most skilled player's reach, making for a more challenging game. When Alcorn went to the founders for additional ideas, Bushnell pushed for sounds of crowds cheering for good shots. Dabney suggested boos and jeers for misses. It was a perfect encapsulation of the differences in the two men: Bushnell all enthusiasm, Dabney more guarded.

---

* "I may have told Al that I did [approach Ampex]," Bushnell told me. Bushnell's boss, Kurt Wallace, who would have been the one to receive the licensing offer at Ampex, told me that no such offer was made.

After only three months, Alcorn had a working prototype of the game, which either he or Bushnell named *Pong*. (When asked in 2016 who had come up with the name, Alcorn and Bushnell each pointed at the other.)[13] Alcorn thought the game played well, but he worried that he had failed in his assignment. With more than seventy chips, rather than the twenty Bushnell had requested, there was no way the game would meet General Electric's specifications.

Telling Bushnell that *Pong* was finished but too complex, Alcorn offered to redesign it. Bushnell suggested that they play. He had played the game while Alcorn was developing it, but this time, he grew increasingly excited with each rally. *Pong* was a "great game," he declared. The phrase had a specific meaning for Bushnell: easy to learn but hard to master.[14] When Alcorn again worried aloud that General Electric might reject the game due to its high chip count, Bushnell seemed to smile to himself.

Then he let Alcorn in on a secret: there was no General Electric contract. Bushnell had lied. *Pong* was an in-house exercise that Bushnell had thought would help Alcorn master the video-positioning trick.

Alcorn was surprised but not angry. He would feel the same way three years later when he learned that Bushnell had been able to describe the Ping-Pong game he wanted in such fine detail because he was describing a table tennis game sold by Magnavox for its Odyssey system.* In essence, he had assigned Alcorn to reproduce the Magnavox game. "It's like the movie *The Producers*, you know?" Alcorn reminisced years later. "We're going to steal this idea from Magnavox, but it's a turkey so what's the problem? [But] all of a sudden it's a success."[15] (Magnavox later sued Atari for patent infringement, eventually settling out of court.)

Bushnell's misdirections and exaggerations freed Alcorn to achieve technical feats he otherwise would have talked himself out of attempting. "'It can't be done! You don't want to do that!': I used to say that a lot in my life," Alcorn later explained. "I fortunately had Nolan to goad me into doing it anyway."[16] Alcorn, who had the technical skills to build just about

---

* Bushnell had seen the Magnavox game in May 1972 at the Magnavox Profit Caravan at an airport hotel in Burlingame.

anything but was not a dreamer as a young man, needed someone like Bushnell to spark and channel his talent. Bushnell recorded so many new ideas every day that little sheets of paper covered in his scrawled handwriting regularly dropped from his pockets.[17]

And Bushnell, with his nearly limitless imagination and more limited technical ability, needed Alcorn to help realize his visions. "Nolan is a dreamer," Alcorn says. "I get the dirty end of the stick and have to make these things happen."[18]

■ ■ ■ ■

Bushnell told Alcorn that *Pong* was so good that they should try to license the game to Bally, with which Atari had a contract (a real one) to build a video game and a pinball machine. Bushnell booked a flight to Bally's Chicago offices, taking a red-eye to save the cost of a hotel room. After showering at the airport, he donned his favorite suit, and, using the prototype board he had brought with him, he tried to convince the Bally executives that *Pong* would be a good game, even if, unlike the pinball games Bally usually made, two players were required to play. "We could probably put a one-player version in," he offered.[19]

Bally agreed only to consider buying the game. A firm answer would have to wait.

Back in California, Ted Dabney built a cabinet for *Pong*. Far from the elegant sculpted fiberglass that encased *Computer Space*, *Pong*'s cabinet was a simple wooden box painted orange, with two silver knobs to control the on-screen paddles. A metal panel with P-O-N-G on the front offered the only nod to aesthetics. Onto the side of the box, Dabney welded a coin slot of the sort used in Laundromats and kiddie rides.

As soon as Bushnell brought the prototype board back to California, Alcorn connected it to the black-and-white television, shoved the entire contraption into Dabney's cabinet, and drove with the founders to a nearby bar. Andy Capp's Tavern was dim, smoky, and, like many bars in Sunnyvale in the summer of 1972, notable only for cheap beer and pinball machines. Bushnell and Dabney knew the owner. Atari ran a small side business servicing pinball machines for a percentage of the take, and Andy Capp's was a customer.

The three Atari employees plunked *Pong* down on a decorative barrel.

It was not much to look at, particularly next to the slickly packaged, bling-ing and flashing pinball machines and the beautiful *Computer Space* Bush-nell had convinced the bar owner to put on the floor.

Nonetheless, two guys soon separated themselves from the crowd of muttonchopped men at the pinball machines and began inspecting *Pong*. After a minute, one man dropped a quarter into the coin box.

The prototype *Pong* had no directions, but the players figured it out. They seemed to enjoy their few minutes of playing, their heads pushed together in front of the screen.

When the game ended, they did not put in another quarter. They walked away.

Bushnell stood up. He had to go talk to those guys, he said. He wanted to know how they'd liked the game. Alcorn followed him across the bar.

Bushnell said hello to *Pong*'s first-ever paying customers and then, nodding toward the game and keeping his voice neutral, asked, "What do you think of that?"

"Oh, yeah. I've played these things before," one player replied. "I know the guys who built these things."

No one corrected him. There was some satisfaction in having built a game so cool that people were pretending to have a connection to it.[20] "Watching people play your game," Bushnell later explained, "is like get-ting a standing ovation."[21]

About a week later, the bar's manager called Alcorn. There was something wrong with the *Pong* machine. Alcorn drove over in his sec-ondhand '63 Cadillac Fleetwood and was greeted inside the bar by a small group of *Pong* fans. Explaining that he would need to play a few games to diagnose the problem, Alcorn bent to unlock the coin box so he could throw the inside switch that would grant unlimited free games.

As soon as he pulled the door open, he saw the quarters. Coins had filled the coffee can that served as a coin box and overflowed onto the wooden floor of the cabinet. There had to be $100 in quarters. *Pong* had not been starting because the coin box was too full to trip the start mech-anism.

Alcorn swept up Atari's half of the take and handed the manager the balance and a business card. "Next time this happens, you call me at

home right away. I can always fix this one," Alcorn promised.[22] His immediate solution was to replace the coffee can with a larger receptacle: a
milk jug.

Alcorn told Bushnell, who was getting ready to fly back to Chicago
to meet again with Bally, about the cascade of quarters. He assumed that
Bushnell would share the story with Bally, to convince it to buy the game.[*]

Wrong.

■ ■ ■ ■

When he returned to California, Bushnell asked Alcorn and Dabney to
join him at Andy Capp's. The three men sat at their favorite table, watching
as person after person lay a quarter on the *Pong* cabinet top to secure a
spot in line. After a few minutes, Bushnell said something that Alcorn had
never expected to hear.

We don't have to license *Pong* to Bally, he said. We could offer Bally a
different game.

But what would happen to *Pong*? Alcorn asked.

Bushnell had been waiting for that question. Atari could build *Pong*,
he said. Atari could have its own factory assembling *Pong*s as fast as possible, maybe even a hundred every day. *Pong* could become a big game, and
Atari could become a big company. Bushnell could oversee business operations, Alcorn could handle engineering, and Dabney could take charge
of manufacturing.

Bushnell seemed unconcerned that none of them had run a company
or supervised a factory, much less built one from scratch. He believed that

---

* The chronology is a bit murky. Bushnell, as well as video game chroniclers Steve
Kent, Marty Goldberg, and Curt Vendel, all say that Bushnell was in Chicago, meeting with Bally, when Alcorn dealt with the overflowing coin box at Andy Capp's.
The chroniclers further say that Bushnell had taken the guts of the *Pong* prototype
machine with him to Chicago. Alcorn reasonably points out that there was at this
point only one *Pong* prototype in existence; it could not have been simultaneously
in Chicago and in Andy Capp's bar. Alcorn thinks Bushnell went to Bally after hearing about the money spilling out of the coin box—and went with the specific goal of
dissuading Bally from buying the game. Bushnell recalls at least two meetings with
Bally, as the text recounts.

"a bright person should be able to fundamentally master any discipline in three years."[23] He was also perversely encouraged by the example of Nutting Associates, which had built *Computer Space*. The guys at Nutting could get games out the door, and they "couldn't find their butts with both hands," in his colorful phrasing.[24]

Alcorn listened to Bushnell and watched him gulp down his beer.

Had Bally refused to license *Pong*? Alcorn asked.

Bushnell dodged. He did not admit that without checking with either Dabney or Alcorn, he had decided he would try to undo his previous sales efforts and convince Bally *not* to buy *Pong*. The overflowing coin box that Alcorn had thought would excite Bally had excited Bushnell instead. He knew that if Atari licensed *Pong* to Bally, Atari would receive only 3 percent of the sales revenues of the game.[25] Bushnell wanted 100 percent for Atari.

Bushnell said none of this to Alcorn and Dabney, though. Instead, he repeated that Atari might want to build *Pong*, rather than licensing the game to a known manufacturer.

No, Alcorn said. He did not want to build *Pong*. He wanted to design games, not to be part of some big company or some factory operation. The former Berkeley hippie "had no aspirations of being a capitalist pig."[26] He was an engineer. They were *all* engineers. They should stick with the original plan, focus on design, and leave the building to somebody else. Dabney agreed.*

Bushnell appeared not to hear Alcorn and Dabney. "We *are* in the manufacturing business," he stated after a minute. He had already made his decision. Atari, not Bally, would build *Pong*.† He turned back to his beer.

_____

* In some accounts, it is Dabney who suggested that Atari become a manufacturing house. Alcorn and Bushnell, on the other hand, both recall the decision as Bushnell's alone, though Alcorn does say, "Nolan made that decision on his own and then convinced us it was our decision." In the absence of documentary evidence, it is hard to know the origin of the idea. It is true that Dabney headed manufacturing for Atari, but Bushnell was running the company, as the stock holdings make clear. At that time, Bushnell owned 60 percent of Atari, Dabney 30 percent, and Alcorn 10 percent.

† Although Bushnell delivered prototype designs of both a pinball game and a different video game, per the terms of their contract, Bally declined to produce either.

Alcorn raced home and burst through the front door to yell to his girlfriend, Katie, "Nolan's nuts! He wants to build a hundred machines a day!'" Even at Ampex, Alcorn had never seen products being churned out at the rate Bushnell imagined.

After a few hours and a bit of perspective from Katie, he decided to play along with Bushnell's crazy idea. He would go to the office and do what Bushnell said. And then, he thought, "when [the manufacturing idea] flops, I'll say to Nolan, 'I told you so.'"[27]

Bushnell said the first step was to build a few more prototype machines and send them to other bars for testing before deciding on the exact features to include in the final version of the game. Dabney found a local shop, P. S. Hurlbut in Santa Clara, to build a freestanding tall cabinet to house the game's screen and components. Alcorn drove over to Andy Capp's to get a clearer sense of the demands *Pong* faced. He counted the coins that had been deposited through the coin slot. If each quarter represented twenty or thirty turns of a knob, *Pong* needed a potentiometer that could rotate a million times in three months without failing. He set about to find one.

Bushnell, meanwhile, worried about game play. He told Alcorn that the game needed instructions. Alcorn thought that was absurd. The players at Andy Capp's had figured it out, hadn't they? But again he decided to play along. He wrote three commands to appear on the game's faceplate:

- Deposit quarter
- Ball will serve automatically
- Avoid missing ball for high score

■ ■ ■ ■

Within a few weeks, ten bars had *Pong* games. Alcorn, Bushnell, and Dabney were confident that they had built machines that could survive semi-intoxicated players with sloshing beer cups, but they had underestimated the abuse that the games would face. Players chucked pool balls at the cabinets, figuring that if a certain spot were hit just right, the reward would be a free game. The machines shorted out when they were shaken, or even just played often, because quarters would fall on the printed circuit board under the coin mechanism. Even well-intentioned bar owners broke the *Pong*s. The owners were accustomed to pinball machines with mechanical relays, flippers, and lights that could be fixed with a screw-driver or a file. If a *Pong* machine was not loud enough or the screen not bright enough, the bar owners would open the back and start looking for something to adjust. More often than not, they settled on an appealingly accessible dial—and began turning, not realizing it was the game's external power supply. Every prototype came back to Atari with the power blown.

Despite the problems, the *Pong* machines brought in some $150 per week, roughly three to five times as much as the typical pinball machine.[28] The game was simultaneously intuitive (turn knob, move paddle) and as-tonishing in 1972, when most Americans had only seen screens display images sent from a broadcast network or projected from slides or a reel of film. *Pong* was different. It was interactive, viewer-commanded television. Bushnell would grow accustomed to people asking how the television net-works sensed that *Pong*'s knobs had been rotated.[29]

Alcorn began hearing stories of lines outside the bars at nine in the morning—not to drink but to play what Alcorn sometimes called "this stupid *Pong* game." In Berkeley, Steve Bristow, an engineering student who had done the same Ampex rotation program as Alcorn (and who had helped build *Computer Space* using Ampex parts) worked part-time for Atari, maintaining pinball machines and collecting Atari's weekly take.[*] He began to fear for his safety after *Pong* was installed at a bar on his route and his canvas bags earmarked for Atari swelled to hold some $1,000 in quarters. When the police refused to issue him a gun permit, he scared up a novel mode of protection: the hatchet he had used in a previous job roof-ing houses. He gave his wife the hatchet to carry while he walked behind her with the heavy money bags. "Even in Berkeley, people would part for a crazy woman with hatchet," he said with satisfaction.[30]

The success of the prototype *Pong*s lit a fire under Bushnell and Dab-ney. "We got hit in the ass by lightning with *Pong*. Holey moley!" Alcorn says. Atari rushed into large-scale manufacturing.

---

[*] The East Bay Atari route included pinball machines at Boalt Hall on the Berkeley campus and games at Larry Blake's and the Pelican on San Pablo Avenue, also in Berkeley. True Believer's #3 Soul Food near the Greyhound bus terminal in Oak-land also had a game.

Bushnell had never run a company, but he possessed a number of gifts that would serve him well as an executive. He carried himself like a leader. "I expect one day to be working for him," his Ampex boss had written on Bushnell's evaluation.[31] Bushnell's monumental enthusiasm, which led one early video game journalist to call him "about the most excited person I've ever seen over the age of six when it came to describing a new game," would also inspire customers and Atari employees.[32] Bushnell loved games of all sorts; he even created them out of everyday circumstances. One former Atari employee says, "If there were two flies on the wall, Nolan would be betting on which fly would take off before the other one."[33]

But Bushnell was alone. He had no mentors, no venture capitalist backing him, no business school professors or consultants watching over his shoulder. There were no video game industry leaders to ask for help or analysts to measure Atari's performance against its competitors'.[34] Atari had an attorney who had helped with the incorporation but did not seem useful for much more. Dabney knew no more about business than Bushnell, and Alcorn knew even less.

So Bushnell read books. He consumed business tomes in the same way he had once devoured guides to chess and Go, looking for classic strategies and unexpected moves. He read about how to name companies and how to find customers. In essence, he decided that if video games could be a big business, then business could be played like a big game. He needed to unravel the motivations of all the players: the employees, the customers, the suppliers, the banks that granted Atari $2.5 million in lines of credit at usurious rates.[35]

Bushnell put this gamesmanship into practice almost immediately. He talked suppliers into giving Atari between thirty and sixty days to pay for the televisions, chips, harnesses, and cabinetry inside *Pong*. At the same time, he insisted that Atari's customers, game distributors who were buying the machines as fast as Atari could build them, pay on delivery. With *Pong*s selling for roughly $1,100 and costing roughly $600 to build, it was a classic bootstrapping operation: the high gross margin allowed Atari to self-finance its growth.[36] To save the time and expense of posting job openings, Bushnell and Dabney hired manufacturing employees out of the lobby of a nearby unemployment office and from a training center whose president brought in a few students and suggested, "You wanna hire these guys."

Some seven thousand *Pong* games were sold in six months—not Bushnell's hundred per day, but impressive nonetheless. Most *Pong* games ended up in bars or arcades. But airports, hotels, and high-end department stores that never would have considered a pool table or pinball machine also hosted the game, prized for its relative quiet, novelty, and cutting-edge feel.[37]

■ ■ ■ ■

Atari moved its manufacturing operations to an abandoned roller skating rink in Santa Clara. Alcorn, the head of engineering, was working on new games along with his small team of engineers, some of whom were drawn to the company because it was one of the few places doing computer graphics work with no connection to the military.* The war in Vietnam was still raging, and there were also mounting concerns about President Nixon's involvement in the break-in at Democratic National Committee headquarters at the Watergate office complex.

Steve Bristow, the former ax-wielding Berkeley student and an excellent engineer, had come to work full-time at Atari as soon as he received his degree. Alcorn and Bristow wore their hair and beards long and their pants bell-bottomed, but they saw themselves as professionals in a sophisticated engineering operation. When their workdays ended, they went home to their families.

That kind of professionalism could seem unusual at Atari. *Pong* was born into an industry already censured, or even banned, by polite society. The game was classed with pinball machines, and in 1972, pinball was still illegal in New York and had only recently been legalized in Chicago. Pinball machines were considered games of chance with a "payout" in the form of a free game: a small step removed from gambling. Even as the occasional *Pong* sneaked into a nice lounge, the seedy feel of sticky bar stools and dingy pool halls hung over the video game industry. The first

---

* Among the early games Alcorn helped design and develop: *Space Race*, *Gotcha*, and *Pong Doubles*.

major profile of Bushnell, written by Alcorn's fraternity brother Bob Wieder, appeared in the skin magazine *Oui*. Alcorn had a gun pulled on him by a distributor who claimed that *Pong* was encroaching on his territory. At one trade show where Atari displayed its games, another product on offer was "The Duke," "the First, Original, And All New Adult Movie Machine." For a quarter, a patron could step into a phone-booth-like cabinet (designed for "ease of cleaning"), shut the door behind him, and watch a short eight-millimeter film in privacy.[38]

Some aspects of Atari culture had a similar unsavory tone. The company newsletter, *The Gospel According to St. Pong*, ran a short piece of fiction about a "Beauti Bust" machine, developed by the "famous European Breastologist Wolfgang Tittleboob," that grew women's breasts to "melon size" via a technique of "light sucking" so enjoyable that a woman undergoing treatment could only moan "oohhhh myyyy" in a "warm glow of pleasure"—until she was sucked into the machine and died.[*]

One man was quietly unhappy within the frenzy that was Atari: cofounder Ted Dabney. He felt overlooked and underappreciated. Bushnell referred only to himself as Atari's founder, as if Dabney had not been there since the beginning. Bushnell patented the video-positioning technique that so impressed Alcorn without informing Dabney or including his name on the application. Dabney felt that the ideas behind the technique were at least as much his as Bushnell's. Bushnell asked him to take a low-level job without direct reports and kept him out of important meetings.[39]

In March 1973, Bushnell called Alcorn into his office. Dabney was there. With Alcorn looking on, Bushnell began hurling basic questions at Dabney, who ran manufacturing. What are our run rates? What's the total manufacturing capacity? How does this week's performance compare to last week's? Last month's?

Alcorn was surprised and heartbroken to see that Dabney had no answers. "That was a very sad moment. I really loved Ted," Alcorn recalled

---

[*] *The Gospel According to St. Pong* also printed the lyrics to "Stairway to Heaven" as a poem and called for the company to "always be a cool, far-out, special and peculiar family of coworkers who love each other and their product."

years later. ("I engineered that epiphany on Al's part," Bushnell says with satisfaction.)[40]

Dabney left Atari that month.[41] He says he quit. Bushnell says he was fired. Either way, once the severance paperwork was signed, Dabney disappeared from Atari history. Until recently, almost every interview and article about the company identified Bushnell as Atari's sole founder.

■ ■ ■ ■

As Atari continued to grow, the corporate culture became even more freewheeling and experimental. An employee council with representatives from departments throughout the company was given the breathtaking power to review firings and unwanted transfers—and even to overturn management's decisions in these areas.

Out of space again, Atari moved to a building on Winchester Boulevard in Los Gatos, seven miles down the road from the Winchester Mystery House, a sprawling 160-room mansion built by the widow of the heir to the Winchester rifle fortune who believed that the ghosts of the rifle's victims would haunt her if she stopped building.[*] Bushnell had the entrance to the new Atari building paneled in wood, with a curved overhead arch: an homage to the barrel on which the original *Pong* machine had sat. He decorated his office with ferns and hanging plants. In one corner he kept an oak beer tap, filled with Coors, for late-afternoon meetings.[42]

The company gained a reputation as a fun place to work, with great parties. Bushnell was notorious for making business decisions in after-work hot tub "meetings," but he also saw the raucous atmosphere as a strategic move in the business game. The kind of employees he wanted "responded to a good party better than bonuses."[43]

Bushnell wrote of "the Atarian Philosophy," which he described as "the nebulous feeling that we want to make Atari a different, exciting and dynamic corporation that dares to be different." Atari would strive, he

---

[*] The Winchester mansion covered six acres and contained 160 rooms, 2,000 doors, 10,000 windows, 47 stairways, 47 fireplaces, 13 bathrooms, and 6 kitchens.

wrote, to satisfy every employee's need "to learn and progress . . . to make a contribution to an object of worth . . . to be treated with dignity by his fellow man." He promised to pay fair salaries and "share corporate successes," treating "all employees with love, dignity and respect, independent of job classification." He vowed that Atari products would not "kill, maim, or degrade society." He also reminded his employees that the company must "make a profit in order to survive and grow."[44]

In June 1973, Atari followed through on these promises, introducing a stock ownership program in which the company granted small amounts of stock to all full-time employees, including those in manufacturing. Atari also provided other uncommonly generous benefits, such as fully covered medical and dental care—Bushnell recalls looking at the assembly line one day and realizing that nearly everyone on it had Atari-funded braces on their teeth—as well as a half-day paid vacation on birthdays and a bonus check when an employee became a parent.[45]

For all its strangeness, Alcorn liked the feel of Atari, with its bizarre combination of half-baked hippie love and high-stakes business pressure to build more *Pong*s or develop new games. Bushnell's seat-of-the-pants management style seemed to be working. By 1973, Atari, just over a year old, was on track to record sales of $3.2 million, with 18 percent profits. The company established subsidiaries to sell games in Japan, Canada, the United Kingdom, and Hawaii. An official company document declared that the market for "computer video games" was limited "only by supply and not demand."[46] The demand, it seemed, was infinite.

■ ■ ■ ■

Success was short-lived, and so, too, was Alcorn's satisfaction with the company he had helped to build. Within six months of *Pong*'s release, copycat competitors pounced, replicating the *Pong* game through reverse engineering. Some even bought authentic printed circuit boards, smuggled out by unscrupulous Atari employees, on the black market.[47] Trying to make light of the competition, Atari ran an advertisement mocking competitors that were attempting to climb on "the Atari bandwagon." But the consequences were devastating. Within two years of *Pong*'s release, five cheaper, copycat versions of the game were sold for every legitimate Atari *Pong* game.[48]

Several of Atari's subsidiary companies went bankrupt or were liquidated for cash.[49] A push for union representation failed, but as Atari reduced its benefits, noisy old-fashioned labor disputes over everything from wages to company cars roiled the company whose leader and employees had, in better times, pledged each other love and dignity.

Attempting to gain some control, Bushnell promoted himself to board chair in 1973 and hired a new president for Atari—his wife's sister's husband, a psychiatrist and industrial consultant who was as close to a real executive as he knew.[50] Shortly after his hiring, the president told a reporter that he did not feel like a businessman. When the new president brought in an engineering manager from Ampex, Alcorn was pushed out of engineering to head research and development.

It was a more business-oriented role than the title would suggest, and Alcorn was miserable. All he wanted from a job was the opportunity to engineer new products, but he now spent time thinking about shipments, buybacks, and board of directors' liabilities. Meanwhile, the new engineering manager, dressed beautifully and working late, seemed more interested in flowcharts and formal plans than in developing new products. Alcorn was only twenty-eight, but he had recently found a gray hair. When it became clear that his mother was in the final stages of lung cancer, he left Atari. He called it a sabbatical, but he was not sure that he would return.[51]

■ ■ ■ ■

While Alcorn spent time with his mother, Atari's struggle continued. The company sold nine games, but several were variations of the easily copied *Pong*, and the newest, a driving game called *Gran Trak*, had been priced too low.[52] Money was so tight that on the Fridays when paychecks were cut, the parking lot emptied as people raced to the bank to cash their checks while Atari still had funds to cover them.[53]

Bushnell shoved his own uncashed paychecks into a desk drawer to conserve funds. His marriage fell apart. Alcorn, visiting Atari during his sabbatical, came upon Bushnell in the parking lot, sitting behind the wheel of his locked car, crying.[54] In 1974, Atari lost $600,000.

Bushnell asked Alcorn to come back to Atari in the spring of 1974, at the end of the fiscal year in May. Things would be different, he promised. The brother-in-law president was out, replaced by Joe Keenan, who had run Atari's only successful subsidiary, Kee Games.[*] (Kee was folded into Atari, and Bushnell served as chairman of the combined company.) Alcorn knew and liked Keenan, who, although only a year older than Bushnell, nonetheless served as what one early employee called "as close as we had to an adult in management."[55] When Keenan was asked if his primary job was to say no to the ever-exuberant, try-anything Bushnell, he responded, "Sometimes I say yes and just don't do anything about it."[56] Bushnell considered Keenan "a brilliant hard head."[57]

Twenty-four-year-old Steve Bristow, who had headed engineering at Kee Games and had emptied the take from early *Pong* machines in the company of his ax-wielding wife, would run engineering at Atari. He, too, developed ground rules for dealing with Bushnell's perpetual idea generation, telling the engineers, "Be nice, be respectful, listen to Nolan—but don't change anything until we talk."[58] With Bristow, another person Alcorn liked and respected, in charge of engineering, Alcorn's title would still be VP of research and development. But Bushnell promised that it would be a real research job with real engineering.

So Alcorn returned. His work would help turn Atari's fortunes.

■ ■ ■ ■

Before his sabbatical, Alcorn had received a note from Bushnell outlining a "charter for engineering." The wish list contained eight items and an admonishment that "statements concerning our manufacturing capacity are inapplicable to the above design schedule." That note was vintage Bushnell: big dreams with little regard for the reality of what Atari could actually build. Alcorn, who had a budget to think about, had asked if he should take "financial capability" into account. Bushnell's response was a handwritten single word: "NO."[59]

When Alcorn returned to Atari, he turned his attention to item seven

---

[*] Bushnell and Keenan had established Kee Games to supply Atari games to a second distributor network, since Atari had an exclusive agreement with the distributor of Bally pinball games.

on Bushnell's list: "Packaging and P.C. [printed circuit board] for color modulated consumer Pong." In other words, Bushnell wanted a version of *Pong* that could be played in the 69 million American homes with television sets, rather than as a freestanding coin-operated game.[60]

Alcorn had slipped easily back into his work at Atari after his sabbatical, and the mandate to build a version of *Pong* for the home felt like a familiar overreach from the boss. "Absurd" was his initial response to Bushnell's latest vision. Atari had never sold directly to consumers. The customers for the coin-op arcade games were distributors, who then resold or rented the games to arcade owners. Moreover, Atari had no idea how to build in the kind of volume that a consumer business would require. But again, as had been the case when Bushnell had suggested they find a way to manufacture *Pong* themselves, Alcorn saw Bushnell's confidence as an opportunity to prove his boss wrong.

Alcorn asked an engineer named Harold Lee to help him design a version of *Pong* that would be cheap enough to sell to individual consumers and difficult to copy. Lee pointed out that microchip technology was now so advanced that they could reduce the entire printed circuit board inside the coin-operated version of *Pong* to a single, custom chip.

"So we set out to do it," Alcorn says. He anticipated a prolonged and ultimately fruitless development process, with months of designing, testing, and retesting before, most likely, he would have to abandon the project.

Instead, the chip worked almost immediately. Atari now had the functioning guts of a home video game—with no idea how to package it, market it, or sell it. Alcorn says, "It was like a dog chasing a car. What do you do when you catch it? Where do you go from here?"[61]

■ ■ ■ ■

One person knew what to do with the caught car: Don Valentine, who had been consulting with Atari since the company's disastrous 1974 fiscal year.[62]* A former marketing executive and self-described "designated

---

* Alcorn and Lee used an AMI chip for *Home Pong*, but they had originally considered an Intel chip instead, and Alcorn invited Intel cofounders Gordon Moore and Bob Noyce to visit Atari. When Don Valentine, who consulted for Atari, heard

talker" at Fairchild Semiconductor and National Semiconductor, Valentine now worked as a venture capitalist.[63] He managed a fund called Sequoia at the venerable investment house Capital Group. Most Capital partners were conservative attorneys and finance experts who enjoyed working in the staid, predictable, highly regulated world of mutual funds.[64] Valentine, by his own admission, "was totally different and totally inappropriate." His $5 million Sequoia fund also differed from the typical Capital Group fund: he was charged with investing for clients who, in his description, "wanted to be in private companies with a lot of upside and maybe disastrous down."[65] While still at National Semiconductor, Valentine had personally invested in a number of small microchip-based companies (including SDS, the computer company acquired by Xerox) that had done quite well. His boss at the Capital Group, Bob Kirby, knew of those investments and admired Valentine's ability to identify risky small companies with the potential to soar. He nicknamed Valentine "Rocket Man."[66]

Valentine's work at Capital Group had brought him valuable contacts in the East Coast financial community. His work at National and Fairchild had given him a deep understanding of the microelectronics market and a place near the center of the growing network of Silicon Valley electronics executives. The best builder of prototype chips or supplier for quantity orders, the strongest attorneys, the most effective publicity and marketing people, the recruiter with the finest track record, the distributor who could be negotiated into the best deal: Valentine knew most of them through shared ties back to Fairchild or one of the other early chip-making companies.

In 1975, with the encouragement of Capital Group, Valentine spun the Sequoia fund into an independent venture capital firm of the same name and with a similar group of investors—not individuals but deep-pocketed institutions such as pension funds and universities with large

---

about the visit, he, in Alcorn's description, "went ballistic, saying 'You told them what we're doing?'" When Alcorn assured him that Noyce and Moore had signed a nondisclosure agreement, Valentine, who had both worked for and competed against the Intel founders, replied, "That's like going after a tank with a flyswatter."

endowments. ("Go where the money is," Valentine likes to say, quoting his mentor Bob Kirby and the bank robber Willie Sutton.) The independent Sequoia Capital also retained a focus on high-risk, high-reward invest-ments. Valentine rented an office on Sand Hill Road, near the venture firm Kleiner & Perkins, started by Fairchild cofounder Eugene Kleiner and Hewlett-Packard alumnus Tom Perkins. Sequoia Capital and Kleiner & Perkins were pioneers in a trend that would later prove an essential ele-ment of Silicon Valley's longevity: an older generation of successful tech entrepreneurs and executives mentoring and funding later generations. Between them, Sequoia Capital and Kleiner & Perkins would go on to fund, among others, Amazon, Apple, Cisco, Dropbox, Electronic Arts, Facebook, Genentech, Google, Instagram, Intuit, and LinkedIn–and that is just the first half of the alphabet.

One key to Valentine's success as a venture capitalist was his willing-ness to overlook behavior and appearances that others of his generation considered unpleasant or even deviant. A conservative whose favorite author was Ayn Rand, Valentine also had no problem seeing past an en-gineer's long hair, bare feet, unpolished manners, or questionable hy-giene to ask what one of these "renegades from the human race" (his own term) might have to offer. At Fairchild, he had watched the same salesmen who drunkenly drove golf carts into water traps at midnight sell tens of thousands of dollars' worth of products the next afternoon. At National, he had worked with Bob Widlar, the engineer who had bit-ten through glasses when he was at Fairchild. At National, Widlar kept a bottle of Jim Beam in his desk drawer, deigned to speak only to a select few people in the company, and occasionally engaged in office-destroy-ing outbursts so prolonged and explosive that Valentine describes him as "certifiable." But when Widlar was calm, he designed analog circuits that at one point accounted for roughly 75 percent of the worldwide market. "Bob steeled me from trying to homogenize great technical genius," Val-entine says.

Compared to Widlar, Bushnell was, in Valentine's fond term, a "jester," with an amusing propensity for exaggeration, self-aggrandize-ment, and over-the-top "clown suits" that Valentine had to teach him not to wear in important meetings.[67] Every time Bushnell tried to rattle Valentine, he failed. When Bushnell insisted on holding meetings in his backyard hot tub, Valentine stripped down and climbed in, dodging the floating bottles of Ripple.[68] When Bushnell took him on a tour through an Atari manufacturing line so hazy with pot smoke that a contact high was almost guaranteed, Valentine tapped the lung capacity from his days as a competitive swimmer and held his breath.[69] He saw a huge market for video games and other consumer products based on microelectronics. If he had to take off his shirt or hold his breath to get to that market, he would.

When Valentine learned of the *Home Pong* chip, he knew how to sell the game Atari would build around it. Bushnell had thought that toy stores would be logical, but Valentine thought bigger: Sears, Roebuck, which had finished construction on its iconic Chicago tower one year before. Fifty-seven percent of US households held a Sears card, and the company was so large that its annual revenue at times approached one percent of the gross national product.[70] Sears sold products at a higher price point than toy stores and had a much better distribution network. Moreover, endorsement by Sears—a "real retail company," Valentine stressed—would help "position Atari as a real company." Valentine asked one of the investors in his Sequoia Fund, who was also a major shareholder in Sears, to arrange an introduction. He also, in his recollection, "had to dress Nolan up in one of his nonclown suits, script the meeting, and ask him to not attempt to be too humorous."[71] In March 1975, the sporting goods buyer at Sears ordered 75,000 copies of the *Home Pong* game that Alcorn and Lee had reduced to a single chip. Sears gave Atari six months to manufacture the games for the Christmas season and also loaned the company $1.5 million through Sears Bank.[72]

Shortly after the Sears deal was inked, Al Alcorn and the engineering team at Atari learned a lesson that surprised one high-tech company after another in the early 1970s. The "softer" side of a consumer product—packaging, in particular—can prove more difficult to master than the technical innards. ("How hard can it be?" may be the five most foolish words in high technology.) When Intel tried to sell digital watches built around its chips, the technology worked fine, but the business was lost on watchbands and display boxes. That experience, Intel cofounder Bob Noyce would later say, had taught him that "when the other guy's business looks too good, you don't know enough about it."[73]

For Atari, the problem was the plastic case that held the *Home Pong* electronics and attached to a television. Alcorn, who was responsible for the case, could not find anyone who could build it. He flew to Los Angeles to visit a company that sounded promising. No luck. With Sears's holiday-season deadlines approaching, Bushnell grew so panicked that he assigned a group to start designing a wooden case that would never have worked for large-volume production. Finally Alcorn found a small company down the road in Santa Clara that made the plastic tooling that surrounds semiconductor chips. Normally, the company would have refused the Atari job, but the semiconductor industry was in a slump. The little plastics company needed the business, and Atari, like many other Silicon Valley companies of its generation, took advantage of the infrastructure that had arisen to support the microchip entrepreneurs.[74]

It was Atari's good fortune that Sears handled the marketing of the *Home Pong* game, featuring it in its popular "Wish Book" holiday catalog for $98.95 and running advertisements.[75] Moreover, the decision to sell *Home Pong* with sporting goods, rather than toys, meant that the game attracted not only children but teenagers and adults, as well.[76] Sears was a master marketer, *Pong* was a great game, and Atari sold more than $10 million of *Home Pong* games to Sears and another $3 million directly to consumers. Even after established companies familiar to consumers—Magnavox, Coleco, and Midway Games—began selling in-home consumer video games, and Magnavox filed a lawsuit claiming that Atari had infringed its patents, *Home Pong* soared.

Atari's arcade division revived around the same time. The problems with the *Gran Trak* driving game had been resolved, and *Tank*, a game written by Steve Bristow and Lyle Rains, was a runaway hit. *Time* dubbed video games "space-age pinball" and declared that at some colleges, playing the games was "the second-most popular pastime after streaking."[77]

Thus the low that had followed a high climbed into another upswing. The good times returned to Atari. By the end of 1975, the company was designing and marketing more than thirty games, occupying eleven separate buildings, and employing 725 people, 55 of whom were considered important enough to receive stock options.[78]

Home Pong had become such a hit that one day Alcorn's son came home from school and announced that Bushnell's daughter had told the teacher that her father had invented the game. Alcorn usually had little time for who-did-what arguments, but he now made an exception. He told his son, "Go to Nolan's daughter and say, 'If your daddy invented *Pong*, how come he had to ask my daddy to come fix his machine when it broke down?'"

Make It Happen
███████████ NIELS REIMERS

Niels Reimers was at his desk, paging through the mail. It was the summer of 1974, and his pilot experiment to encourage Stanford inventors to license their inventions was now a four-year-old official operation called the Office of Technology Licensing. Reimers's office had moved from Stanford's stately Encina Hall, once invaded by protesting students, to a small trailer behind the building. He and his assistant Sally Hines had hung ferns and covered the walls with memorabilia and a sign in all caps: MAKE IT HAPPEN.[1]

The Office of Technology Licensing had moved from pilot to permanent in June 1970, after bringing in eighteen times as much revenue from licensing fees and royalties in a single year as Research Corporation had in all previous years combined ($55,000 versus less than $3,000).[2] In the ensuing four years, the office had brought in $461,000 and licensed twenty-two inventions,[3] including a laser device licensed to a company interested in using it for eye surgery; a fluorescent activated cell sorter; and a sound-generating and -processing system, licensed to Yamaha, that would, in 1983, help launch a wave of synthesized music with the introduction of the Yamaha DX7.[4]

Reimers, determined to run "an entrepreneurial enterprise where the . . . yardstick of profit can be used to measure success," faced the excitement and challenges familiar to leaders of any startup.[5] His efforts to encourage inventors to license their ideas had been successful, but filing for patents was expensive. His office, which kept only 15 percent of the money that came to the university, was running an $11,000 annual deficit. So far, the office had filed for patents on only 5 percent of the 343 disclosures it had received.[6]

In the day's mail one June afternoon, alongside letters from companies Reimers had approached about licensing and memos from government agencies that supported research at Stanford, was a photocopied news clipping, sent to him by the head of the Stanford News Service. Reimers, once characterized as "less patent officer and more talent scout," always read the News Service clippings carefully.[7] They often summarized or publicized university laboratory work that he otherwise might not know about.

This clipping was a five-column article from the *New York Times*. "Animal Gene Shifted to Bacteria," the headline trumpeted. Two labs, one at Stanford and the other at the University of California, San Francisco (UCSF), had jointly developed a practical way to clone genetic material by transplanting foreign genes from a complex animal (in this case, a frog) into bacteria. Bacteria are cells that are very simple and multiply quickly. As the bacteria multiplied, they created exact copies of the frog genes and expressed them to make proteins. The bacteria, in other words, had served as tiny DNA-reproducing factories.

The implications were staggering. If bacteria could be engineered to replicate insulin-producing DNA and express genes coded for insulin, for example, it might be possible to make a limitless supply of the lifesaving hormone for diabetics. The same technique could be applied to make antibiotics to fight off infection or microorganisms that could reduce the need for fertilizers. The *Times* quoted Joshua Lederberg, the Nobel Prize–winning chair of Stanford's Genetics Department, who called the breakthrough "a major crossroads."[8]

Reimers did not know much about biology or genetics—"I'm just a mechanical engineer," he often said—but Lederberg's endorsement and the potential commercial applications impressed him. Moreover, Reimers knew and liked the Stanford scientist behind the breakthrough. Stanley N. Cohen, the head of the Clinical Pharmacology Division at the Medical School, had worked with Reimers to license a computer-based system to monitor potential drug interactions.

Reimers called Cohen. "Stan," he said, "this looks like important and interesting work." Had Cohen considered patenting the DNA technique?[9]

Cohen, thirty-nine and balding, with horn-rimmed glasses and a clipped beard, approached the world with deliberation and moral rectitude. When something bothered him, he was direct: "Listen, Chief," he'd begin. His powers of concentration were so intense that once, deep in thought, he walked into a sliding glass door and shattered it with his forehead.

Cohen also had a gentler side. He was a careful listener and a musician who sometimes gave impromptu banjo performances during scientific conferences. He had paid his way through college, in part, with royalties from a song that had a brief appearance on the Hit Parade. Most Friday afternoons, he left campus for a run to Baskin-Robbins, where he would buy ice cream for the people in his lab.[10]

Cohen told Reimers that he had not considered patenting the breakthrough and was not sure he wanted to. But Reimers was welcome to come make his case.

■ ■ ■ ■

Reimers grabbed his bike and clipped on his helmet. Five minutes later, he was weaving through a construction site around the medical school and hospital, speeding past heavy machinery, hard-hatted crew members, and the handful of striking workers from the hospital union who stood outside holding signs.[11]

Once in Cohen's office, Reimers reiterated that he thought the scientist should patent the process to make what would come to be known as recombinant DNA. Cohen explained why he thought he should not. First, the process depended, in part, on earlier discoveries by other scientists. What right did he have to patent just the latest breakthrough?[12] It felt unethical. Second, the process was important to scientific progress because it made it possible to "manufacture" large amounts of identical genetic material for experiments at a time when the scarcity of available natural DNA limited research; Cohen worried that a patent would restrict availability. Finally, the National Science Foundation, the Department of Health, Education, and Welfare, and the American Cancer Society had provided grants that had made some of the research possible. Was it right for Stanford to take public money but keep the spoils for itself?

Every one of the concerns that Cohen raised in this first conversation would become a significant issue that the men would have to confront in the years ahead. But Reimers, who had been talking to Stanford inventors about the benefits of patents and licenses for nearly five years, knew how to address Cohen's points. "I had to work almost one-on-one with each scientist to explain the system to them," he later recalled.[13]

"It's not like an academic publication," Reimers said in an effort to counter Cohen's concerns about the breakthrough building on previous work. Academic papers include detailed references to prior work by other scientists, but a patent application refers to work by others only in certain cases. Reimers also explained that because only commercial entities, not nonprofit research institutions and universities, would pay royalties, a patent would not restrict academic use of the recombinant DNA process. "Patents are *intended* to ensure that technological discoveries are not kept secret," he liked to say.

As for the concern about public funding: there was an established process, called an institutional patent agreement, which allowed a university to petition for rights to an invention that had received public funding.[14] Some people objected to those agreements—Ralph Nader's Public Citizen organization filed unsuccessful civil actions against them in 1973 and 1974—but Reimers believed that the agreements were essential. Universities would be less inclined to pursue applied research, he said, "if the government would march in to take back any patent with significant commercial potential."[15] Reimers also explained that a patent application would cost Cohen no money and that Stanford had a good record of receiving the rights for which it petitioned.[16]

"Why don't we proceed and let this all sort itself out?" he suggested.[17] If the recombinant DNA process were not patentable, the application would be rejected.

Cohen, still hesitant, said, "There's a co-inventor on this, Herb Boyer, and he will have to agree as well."[18] He would give Boyer a call, and if his coinventor agreed to patent, Cohen would, too.

■ ■ ■ ■

Herb Boyer, a biochemist at the University of California, San Francisco, was a year younger than Cohen and as ebullient as Cohen was deliberate. "Stan doesn't tell a lot of jokes," Boyer once said of his coinventor. "On the other hand, I tell a lot of jokes, and some of them you don't want to hear about."[19] Boyer had been a regular participant in the antiwar protests in the Haight-Ashbury district just a few blocks from his lab. When he first peered through his microscope and saw evidence that the recombinant DNA process was working, he started to cry.[20]

Two years earlier, in 1972, Boyer's lab had isolated an enzyme that could clip apart a strand of DNA.* Once clipped, DNA from another source could be inserted.[21] At the time, Boyer had not known if a molecule with its new "recombined" DNA was a lab curiosity, or if, inside a cell, the molecule (or gene(s) in the recombined DNA) would be reproduced intact as the cells divided.

Cohen, meanwhile, had developed a method that could help resolve Boyer's uncertainty. He had figured out how to make bacterial cells take up and propagate foreign DNA, specifically a type of DNA called a plasmid. He had also isolated plasmids that could confer resistance to antibiotics. That trait made it possible to identify which bacteria carried the recombinant DNA.

On a break from an academic conference in Hawaii, the scientists had devised a way to help each other. Cohen's lab would isolate the plasmids and get them to Boyer, who would clip them, insert new DNA, and send the now-recombinant plasmids back to Stanford, where Cohen would introduce them into bacterial cells and see if the recombinant structure reproduced as the cells divided.

Forty miles lay between the scientists' labs at opposite ends of the San Francisco Bay. It was simple good fortune that Cohen's research technician Annie Chang lived near Boyer's UCSF lab and was willing to carry DNA between the labs. She would pack the test tubes in ice inside thermoses and then load them into her Volkswagen Beetle for the drive up the freeway and through the city's famous hills; and she would do the reverse, as well, carrying materials from Boyer's lab back down to Stanford. Chang was a key contributor in Cohen's lab—she plated bacteria, analyzed DNA,

---

* The restriction enzyme, called EcoR1, cuts DNA at a specific place in the molecule. Janet Mertz in Paul Berg's lab at Stanford used EcoR1 to develop a highly efficient genetic "cutting and pasting" technique.

offered suggestions, and, in Cohen's words, "played a central role in the experiments"[22]—and she, along with Cohen, watched the bacteria multiply under a microscope, waiting to see if any cells reproduced an exact copy of the hybrid DNA from Boyer's lab.[*]

"We worked literally almost day and night. Things were too exciting for us to get very much sleep," Cohen recalls.[23] Even the hyperrational scientist wished that there were some way to get the bacteria to grow more quickly.[24]

■ ■ ■ ■

At the end of July 1974, Stan Cohen called Reimers. Herb Boyer at UCSF had agreed to apply for a patent.[25] Cohen would file an invention disclosure with the Office of Technology Licensing so that Reimers could move forward.

Cohen made one unusual request: he did not want to receive the 33 percent of royalties that would be due to him under the one-third each to the inventor, the department, and the school arrangement that Reimers had devised.[26] Cohen remained deeply ambivalent about patenting the recombinant DNA process and agreed to do so only after again reiterating to Reimers that the process he and Boyer had devised "has resulted from the efforts of many." He would allow Reimers to move forward with the patent application only if it were "made perfectly clear to all concerned that the matter was pursued at the initiative of the University and that I would receive no personal gain from the patent." Boyer likewise declined any "personal gain."[27]

The scientists' reluctance is understandable. When Reimers asked Cohen and Boyer to pursue a patent, he asked them to break from their discipline's cultural norms. Although physicists and engineers had long traditions of conducting research with an eye to practical ends, for biologists in the academy, basic research was the sine qua non. In the biological sciences, one scientist recalls, "an aspiration to do something applied . . . was a clear admission of intellectual weakness."[28] Cohen worried that his scientific peers would interpret the patent application as an unseemly overreach, an attempt to claim credit for work that built on the

[*] Cohen later promoted Chang to the position of assistant.

breakthroughs of others. He told Reimers that he would donate his share of any royalties to Stanford.[29]

At the same time that he agreed to file for a patent, Cohen shared another bit of news with Reimers: along with several colleagues, he and Boyer had published an academic paper on the recombinant DNA process.[30]

The publication presented a challenge. Inventors have only a year from the first public disclosure of their innovation to file a patent. The paper, published in November 1973, meant that Reimers would have to file an application by November 1974. July was almost gone, and in four days, Reimers was leaving for a long-planned family vacation that would steal another two weeks.

Reimers wasted no time. As soon as he hung up with Cohen, he called the patent office at the University of California. He needed the university to agree to patent the recombinant DNA process, since Boyer, a UCSF professor, would be named as coinventor. Reimers introduced himself to the patent administrator, Josephine Opalka, and asked her to send a letter stating who had sponsored Boyer's research. Two days later, he received a note naming the sponsors. Opalka would contact them and then "be in touch with you sometime next March."

March? March was eight months away! Reimers had to file in four.[31]

He dictated a reply to the UC patent office, suggesting how to approach the sponsors. He also offered to have Stanford administer the patent, with any royalties split fifty-fifty after deducting a percentage to cover Stanford's patent filing and licensing expenses.[32]

He then left for vacation. But he did not leave the patent behind. Still unsure of the science behind the breakthrough, Reimers shoved into his suitcase a book Cohen had loaned to him: *Molecular Biology of the Gene*, a 662-page textbook written by James Watson, one of the discoverers of the structure of DNA.[33]

Much like Bob Taylor pulling together computer scientists to develop the hardware and software that would eventually underpin the personal computer industry, Niels Reimers was beginning to knit together the legal, academic, and commercial connections that would underpin the biotech industry. Throughout Silicon Valley, the ground was being prepared for the birth of major industries that would shape the modern world.

■ ■ ■ ■

When Reimers returned from vacation, he wrote to Stanford's staff counsel, asking for help in drafting an agreement with the University of California. "I am most anxious to have the arrangement with UC consummated soon and will appreciate your early attention to this," he wrote.[34] He believed that recombinant DNA, like the transistor or silicon microchip, could serve as "the foundation of a major new industry in which the U.S. has the potential for being the leader."[35]

Reimers had never before coordinated a patent application with another university. At first glance, the patent office at the University of California shared several features with Reimers's program at Stanford. Both organizations were small (three people at Stanford, four at UC); both worked with outside counsel, rather than keeping attorneys on their office staff; and both offered a generous share of royalties to inventors. The University of California shared net proceeds with the inventor fifty-fifty, a concession made, in part, because disclosure to the university patent office was mandatory.[36]

The similarities ended there. Where Reimers asked companies about an invention's commercial prospects, the University of California's first step upon receiving an invention disclosure was to send a form letter to faculty experts in the field, asking for comments. The focus at UC thus was more on identifying a contribution to knowledge than on determining practical applications. When university technology licensing officers held their first conference in 1974, the head of the UC patent office warned that "you have to police your licensees" while Reimers urged an "entrepreneurial emphasis" and preached the importance of finding "the best way to market technology" because "inventions are very perishable."[37] The UC office reported to an eleven-person Patent Board that sat inside the UC systemwide office, deep in the bureaucracy endemic to any large public university system. Reimers reported to the dean of research and was in regular contact with top Stanford administrators.

Stanford inventors could choose how or if to patent their inventions, but faculty and staff at all nine University of California campuses had to file invention disclosures with the patent office. That requirement meant each year, the UC staff of four received some three hundred disclosures— about five times what Reimers dealt with. The volume left the UC office so "overloaded with technology that they did not move," Reimers recalled.[38]

Administrator Josephine Opalka told Reimers that the UC office had no objection to filing for a patent for the recombinant DNA process, but it could offer only limited assistance. Stanford would have to shoulder the risk. UC would not cover any up-front costs.

Reimers pounced. He reiterated his offer that Stanford would pay all filing and related costs. In exchange, if fees or royalties ever did come from the patent, Stanford would deduct 15 percent off the top to cover the up-front expenses. The remaining 85 percent would be divided among the universities and inventors.[39] After a bit of back-and-forth, Opalka agreed.

Reimers was placing a large bet. If the patent was denied or did not yield licensing fees or royalties, Stanford would never recoup its costs, and the University of California would not have lost a penny chasing a failed idea. If, however, the patent proved valuable, Stanford would receive 15 percent of an unlimited upside on top of the 42.5 percent of royalties that would go to the university and its inventors. UC and its inventors would receive only 42.5 percent in all.

In the end, Reimers would win big. The patent ultimately yielded some $255 million in licensing fees and royalties, earning Stanford roughly $40 million from its 15 percent, even before receiving another $107 million from its share of the 85 percent.[40]

Several years after their initial conversations, Reimers says, the University of California complained that Stanford's 15 percent was "an over-recovery." Stanford, it said, was getting too much money off the top before the balance was split fifty-fifty between the two schools. "You had your chance," Reimers told them.[41]

■ ■ ■ ■

As late as September, and with the November patent deadline looming, Reimers was still searching for a clear sense of the commercial viability of the recombinant DNA process, aside from the theoretical "it may one day be possible" list that he had first read in the *New York Times* article alerting him to the invention. He could not use the practice that had worked so well for the office since its beginning: asking a company in a relevant industry to offer an assessment of the invention. There was no biotech industry with members he could consult. The industry would be born of the patent.[42] Instead, Reimers had a business school student whom he had hired for the summer talk to Cohen and Boyer about commercial possibilities. [43]

He had two months.

■ ■ ■ ■

"Reimers called me and asked me if I knew what a plasmid is," recalls Bertram Rowland, a San Francisco–based attorney whose firm had worked with Reimers on several earlier patent applications.[44] The lawyer had a PhD in organic chemistry and a good knowledge of biochemistry, but his degree was twenty years old, conferred before the discovery of plasmids or the popularization of the term "molecular biology."[45] When Rowland admitted that he did not know what a plasmid was "but would find out," Reimers said he would find another attorney. Rowland countered that because molecular biologists did not tend to take up patent law, his own scientific training was more than Reimers would likely find elsewhere. Reimers gave him the job.

"I had about three weeks to write the case," Rowland recalled. Working from three academic papers and several conversations with Cohen—"he was teaching me about patents, and I was teaching him about biology," Cohen recalls[46]—Rowland drew up the application. He sent it, thirty-five pages long, with four carbon copies, to Cohen and Boyer for approval. [47]

Cohen soon called with a question: Why was Rowland limiting the claims only to recombinant DNA in bacteria? Plasmids occurred in complex organisms, as well. Why not make the patent as broad as possible?[48]

This brilliant suggestion presented a pedestrian problem. Had Rowland been at Xerox PARC, he could have changed the application with a few keystrokes on his Alto. But the personal computer had not made its way down from PARC's empyrean perch, which meant that making Cohen's suggested changes would require retyping the entire application. That was "not an alternative in view of the time pressure," according to Rowland.

His solution was to add a new claim to cover "a cell." He was confident that this generic term, along with earlier language in the application, "could be interpreted to support [claims] other than bacteria." He added the claim near the end of the document, to minimize the need for retyping. With that addition, the patent application covered the use of recombinant DNA in applications as varied and broad as possible.

While Rowland drew up the application, Reimers worked to convince the National Science Foundation and the American Cancer Society to waive their interest in the invention to a third sponsor of the recombinant DNA research: the Department of Health, Education, and Welfare. After a single agency had the rights to the invention, it would be much easier for Reimers to petition for the rights to revert to the universities. Without those rights, even if a patent were granted, Stanford and the University of California would receive no royalties.[49]

On November 11, 1974—one week before the deadline—Stanford filed a patent application for "Process and Composition for Biologically Functional Molecular Chimeras." It was a patent of notable breadth, claiming title to what the Stanford biochemist Paul Berg (who had done pioneering recombinant DNA work himself) later criticized as "techniques for cloning all possible DNAs, in all possible vectors, joined in all possible ways, in all possible organisms."[50]

Cohen, the Stanford inventor, calls the patent application "truly remarkable," given the lack of precedent and the need to prepare the application with little time and assistance from only Cohen, who, in his own words, "had little or no understanding of patent law at the time" and wanted nothing more than to "get the application out of the way, and go back to my research."[51] At the time of the application, Cohen told Reimers that he did not want to be kept informed about Stanford's licensing efforts and left for a sabbatical in England, staying in touch with his lab via a teletype machine and an Arpanet connection.[52]

■ ■ ■ ■

While Reimers, Cohen, and Boyer saw the promise in recombinant DNA, many others saw only danger. The science was new and opened a field that would potentially give humans godlike powers to create never-before-seen hybrid life-forms. Cohen and Boyer's patent uses one common term for recombined DNA cells: "chimeras," a reference to the mythical monster with the head of a lion, the body of a goat, and the tail of a serpent.

Months before Reimers heard about Cohen and Boyer's breakthrough, Boyer had mentioned it at a scientific conference. Within days, concerned conference organizers had asked the National Academy of Sciences to create a committee to assess the recombinant DNA process. "New kinds of viruses with biological activity of unpredictable nature may eventually be created," they cautioned. "Certain of these hybrid molecules are potentially hazardous to both laboratory workers and the public."[53] One year later, the National Academy of Sciences committee, whose members included Cohen and Boyer, recommended a moratorium on certain recombinant DNA experiments until the risks were better understood.[54]

In February 1975, 150 top scientists from thirteen countries, along with a number of invited journalists and attorneys, convened at the Asilomar Conference Grounds near Monterey, California. The conference-goers wrestled with a monumental question: how to proceed safely in the hitherto unimaginable world in which genes could be swapped between species and easily reproduced. The terrifying implications included the possibility of pathogens or drug-resistant genes infecting large segments of the human population. With some scientists urging caution and others eager to press on with research, harsh rhetoric and accusations punctuated many sessions. ("You fucked the plasmid group!" was one comment offered on the floor.)[55] Both Cohen and Boyer attended the Asilomar conference, and both deplored its unprofessional fractiousness.

On the final day, a majority of the assembled scientists, possibly influenced by a panel of attorneys who had presented the previous afternoon, proposed a set of guidelines for minimizing safety risks when conducting recombinant DNA research. *Rolling Stone*, which dubbed the Asilomar meeting the "Pandora's Box Congress," claimed that the safety guidelines marked the first time that scientists had proposed self-regulation since early in the Second World War, when some physicists had agreed to keep nuclear data from German scientists.[56] One biologist was so alarmed by the risks that he wrote in *Science* that the world was now facing "a pre-Hiroshima situation."[57] The Stanford biochemist Paul Berg, one of the conference organizers, recalls, "It was the period just after the Vietnam War. People were concerned about doing things that would come back to haunt us."[58]

■ ■ ■ ■

The safety concerns presented a special challenge for Reimers. Paul Berg, the leading voice of the scientists urging self-regulation for recombinant DNA research, was also an esteemed member of the Stanford faculty. Berg was not concerned that recombinant DNA research posed dangers to the general public, but he feared that if the scientists did not regulate the research themselves, the science-unfriendly Nixon administration would do it for them.[59] To anyone looking at the situation from outside, however, it appeared that Stanford, led by Reimers, was attempting to garner profits from a breakthrough that deeply concerned one of the university's own top research scientists.

When Reimers joined several contentious meetings in Berg's office, the subject was not safety or self-regulation but patent law and claiming credit. Berg, along with two Nobel laureates on the Stanford faculty, Joshua Lederberg and Arthur Kornberg, had shared Cohen's concern from the beginning—that the process, a basic building block of science, never should have been patented in the first place. The mission of a university is to increase public knowledge, but a patent, by definition, is restrictive. The scientists also raised another objection, again familiar to Reimers from his conversations with Cohen: the recombinant DNA process built on research by many more scientists than just Cohen and Boyer. (To this day, Berg calls the patent claims "dubious, presumptuous, and hubristic.")[60] This objection echoed an anonymous reviewer of the patent application at the University of California who wrote, "I am concerned that given the fundamental nature of the work and the number of scientists involved, either directly or indirectly, that this patent will not reflect favorably on the public service ideals of the University."[61]

These issues were of such importance to Stanford that the university provost attended at least one of the meetings in Berg's office. In February 1975, Reimers agreed to discuss with the vice provost for research whether the patent application should be turned over to Research Corporation or even abandoned.[62]

It was a delicate balancing act for Reimers. He wanted to patent the process. He needed to do so without undermining the protocols of open science, skewing incentives for Stanford research, alienating top faculty, or undermining the very premise of the university.

■ ■ ■ ■

Reimers, who had built an innovative, entrepreneurial office that embraced the get-it-done spirit of the nearby startup companies, now juggled two inventors, two universities, three sponsoring agencies, a legal team, the U.S. Patent and Trademark Office, objecting scientists, advocates convinced that Stanford had no right to patent an invention supported by public money, and a global research community with some members concerned that the invention could irrevocably alter life as we know it.

At the same time, he was running an office, sifting through roughly six patent disclosures each month, helping to launch the Association of University Technology Managers, and trying to convince companies around the world to license other Stanford-patented inventions.[63]

Reimers was certain that the DNA patent was worth the extra demands, even as he claimed to be working on another 130 projects.[64] He told one Stanford administrator, "This is an invention of the sort that, if it pays off, will pay off big."[65] He told another that "the recombinant DNA invention has the potential for being Stanford's most important-*ever* invention in terms of income potential [. . . and] *the* most important in terms of potential public impact."[66]

The desire for "public impact" motivated Reimers in everything he did at the Office of Technology Licensing. In general, he believed that corporations were the best vector for transferring academic ideas to the broader public, and he was willing to fight anyone—the university, the patent office, government officials—who disagreed with him. The patent counsel for the Department of Health, Education, and Welfare, who called Reimers "the number one university technology licensing officer in the country," attributed much of Reimers's success to "his aggressive, outgoing personality."[67]

A few years after filing the recombinant DNA patent application, Reimers claimed, with evident frustration, that there were "over 28,000 unused patents that the government has accumulated mainly due to a lack of incentive in developing them."[68] He wanted the Office of Technology Licensing to provide the necessary incentive by connecting interested companies to important ideas that could improve people's lives. Mindful of the sign on the wall of his trailer office, he was determined to MAKE IT HAPPEN.

That's What I Did on Mondays
MIKE MARKKULA

It was December 1972, a few months after Sandy Kurtzig had launched ASK and Bob Taylor's group at PARC had begun their work on the Alto. Mike Markkula, now finishing his second year at Intel, was at home for the holiday break, doing what he did every year between Christmas and New Year's: he was calculating his net worth and planning for the coming year. He had gotten into the habit as a student when, at twenty, he had prepared a detailed financial plan with the goal of reaching financial independence within fifteen years. From college, through his jobs at Hughes and Fairchild, to this ranch house on Sunderland Drive in Cupertino, he had sat with his bank statements and bills every December, adding machine close at hand. The task had been simple until a year ago, at which point he'd had the pleasant additional work of factoring in the generous stock option grant he had negotiated at his hire at Intel. Now, with the calendar turning to 1973, he looked up and thought, "I never need to work for anyone else again." He and his wife, Linda, could stay in this nice house, send their daughter to the excellent schools nearby, and live off his stock and the savings he had set aside with every paycheck. He had reached his goal four years early.

The realization that he could stop working had been pleasant—wonderful, really—but nothing to act on. He was only thirty-one. Only half of his stock options had vested. Every year he stayed at Intel meant another quarter of his grant vested. Already this year the share price had hit $55 for shares Markkula had bought at $6.22 each. He had every reason to expect that their value would continue to go up.[1]

Moreover, he had no desire to leave Intel. "Most of the folks at Intel were completely consumed developing the products," he explains, which meant that his charge, in consultation with his boss Bob Graham, was "to take care of all the rest of it."[2] Because the company had been so small when he joined, Markkula was responsible for not only product marketing but also areas that a more mature operation already would have segmented off into different jobs: forecasting, planning, shipping, and customer service. Those business fundamentals were not sexy, but done poorly, they could kill a company. And Markkula found the meticulous, important work as engaging as any he had done before. He wrote data sheets, the baseline documents that provide specifications and performance parameters. He combed through trade journals and reports to track Intel's competitors. He talked with the engineering team until he understood the products so well that he felt he could answer almost any question about them.

A few months after he joined, Intel introduced its 4004 microprocessor, the so-called computer on a chip. "I was so excited about the microprocessor I could hardly see straight," Markkula says.[3] A programmable, general-purpose logic device, the microprocessor was revolutionary. Before, designers at customer firms had built their systems by choosing and connecting individual microchips, each with a different dedicated function, on a board. Changing the system required changing the physical arrangement of the chips, or hardware. Intel's new microprocessor systems required something very different—changes made not by moving physical objects but by reprogramming the instructions stored in program memory.[4] The microprocessor, in other words, brought software to the semiconductor industry. In doing so, it placed new demands on customers, most of whom were experienced hardware designers but unfamiliar with using computer programs to solve their systems problems.

In 1971, Intel hired Regis McKenna, formerly of National Semiconductor and now the principal in an eponymous marketing and publicity firm he had launched with $500, to develop advertising and publicity campaigns highlighting what a microprocessor could do. McKenna had grown up with six brothers, which meant that he was easy to get along with but also knew how to defend his ideas. He was an expert in explaining technology to the media and the general public.

Markkula, meanwhile, needed to teach circuit designers how a microprocessor worked. He hired an engineer from Fairchild, Hank Smith, to write the technical documentation for the chip. This task was monumental—the manual alone ran well over a hundred pages at a time when other Intel chips were boxed with only a ten-page data sheet. But Intel initially shipped more manuals for the 4004 than actual processors. As the head of microprocessor marketing would later put it, "There were just lots of people who wanted to read about microprocessors, independent of whether they were buying any."[5] Intel first sold the microprocessor, now best known for use as the heart of a computer, as a control device. Early uses included a library bar code system, liquid chlorine control monitoring, a dialysis machine, an elevator control system, sawmills, and farm irrigation systems.[6]

Like Kurtzig, who was witnessing and participating in the birth of the independent software industry, Markkula was gathering the skills and connections he would use to help launch the personal computer revolution. His familiarity with the microprocessor, as well as his relationships with Regis McKenna and Hank Smith, would prove essential for launching Apple.

Looking back, however, Markkula calls something more mundane than personal connections or helping to market one of the twentieth century's greatest inventions "the most important thing I did at Intel." He wrote a computer program that handled all of Intel's order processing. If a company does not have accurate, real-time knowledge of the products that have been ordered, who ordered them, the promised delivery dates, and the ability of the manufacturing operation to meet that schedule, the results can be devastating. The company will miss deliveries. It will not have good data on which to base future forecasts. It will either build products no one wants or fail to build products that are in high demand.

When Markkula arrived at Intel, he went down to the small shipping area and asked the clerk for a report on the backlog (the number of products ordered but not yet shipped). He was expecting her to pull out a computer printout. Instead, she said, "Wait a minute. I have to add it up"—and grabbed a handwritten accounting ledger and a pencil to do the math.

Markkula resolved to computerize Intel's order processing. He had been using computers since college. His first program, written in FOR-TRAN on punch cards, was a betting scheme to maximize the amount of money he could win and minimize the amount he could lose in any craps game.* Computers had helped him design his circuits at Hughes. At Fairchild, he and his office mate Mike Scott had calculated their forecasts using a Teletype Model 33 connected to a PDP-10 computer owned by Tymshare, the company for which Sandy Kurtzig wrote a generic manu-facturing program.

At Intel, Markkula rented a Model 33 Teletype, installed it in the cus-tomer service area, and spent the next several weeks writing a program using BASIC and Tymshare's database, Retrieve. The work could be tricky, the technology balky. He was happy that the Tymshare offices were on his drive home, since he occasionally needed help with solving a program-ming problem or restarting the processing after his application crashed.

Markkula watched and learned as Intel accelerated. In his first year, the number of customers nearly doubled (from fewer than five hundred to nine hundred), overseas sales quadrupled, and the company went pub-lic at $23.50 per share in October 1971. Intel moved to an expansive new campus on what had been a twenty-six-acre pear orchard in Santa Clara. For a few months, employees picked pears off the orchard's remaining trees—until the trees were plowed under for a second building.[7]

Because Markkula oversaw forecasting, he occasionally presented to Intel's senior executives and the board, which included some of the most successful and powerful men in Silicon Valley. Markkula met Robert Noyce, Gordon Moore, and Arthur Rock that way. Rock was the legendary venture capitalist behind Fairchild and Intel, as well as the SDS computer company sold to Xerox. He would later help fund Apple and serve as a director, but at Intel he shared with Markkula his secret for staying awake during the boring parts of board meetings (parts, he had to admit, that occasionally included Markkula's areas of responsibility): swallow a few caffeine pills.

---

* To this day, Markkula uses the lessons from his FORTRAN betting scheme when he plays. He says he has "been way ahead for a lifetime."

In the four years that Markkula would spend there, Intel introduced fifteen memory components, seventeen logic circuits, and eleven memory systems.[8] The number of employees rose from three hundred to more than two thousand. And profits soared—from $2 million in 1972 to $9.2 million in 1973 to $19.8 million a year later. Markkula was learning how a startup becomes a giant.

■ ■ ■ ■

Markkula's satisfaction with his job was all the more impressive given that he was working at a significant disadvantage. Andy Grove, the third-most-powerful person at Intel and the founders' heir apparent, did not like Markkula. In general, Grove did not value marketing in Intel's earliest years. He liked to say that engineering developed products, manufacturing built them, and sales sold them—so what did marketing do? [9] It did not help that Markkula was a protégé of Intel's first marketing vice president, Bob Graham. In July 1971, barely six months after Markkula hired on, Intel fired Graham after a him-or-me ultimatum from Grove.

Markkula was convinced "Andy was not only anti-marketing, he was anti-Markkula."[10] Outside observers, including Arthur Rock, agree with that assessment.[11] Grove would later change his mind about Markkula, telling Rock, "I must have missed something in Mike" and investing in Apple while it was still privately held.[*] In 2015, Grove said, "There would be no Apple without Markkula."[12] But at Intel, Grove felt differently about Markkula, summarizing his opinion with a dismissive "He was very nice to me, but I had no use for him."[13]

This assessment undoubtedly affected the trajectory of Markkula's career. In sharp contrast to his rapid rise at Fairchild, at Intel he advanced only incrementally. Hired as North American marketing manager, he later also managed product marketing in Europe and Japan. Ann Bowers, the head of human resources at Intel, says that he was "a junior marketing guy" throughout his tenure at the company.[14]

But Markkula was content. "I loved Intel," he says, describing his job as a "fun combination of my technical ability, along with some business

---

[*] Grove bought 14,000 shares of Apple in a 1979 private investment round.

planning and marketing sense."[15] Meanwhile, three-for-two stock splits in 1973 and again in 1974 meant that he now held options on more than twice the number of shares in his original grant. He could have comfortably retired, but he loved the work.

■ ■ ■ ■

Then came 1974. Intel began the year as one of the most profitable public companies in the country—*Forbes* praised it as a "debt-free money machine"—and ended it by laying off 30 percent of its 2,500 employees, most of them in production. The entire semiconductor industry suffered as computer and television companies around the world canceled orders in the wake of the OPEC oil embargo and resulting oil price shock.* Nearly 20 percent of the semiconductor industry's production workers in Silicon Valley lost their jobs in the fourth quarter of 1974 as shipments dropped by 25 percent in a few months.[16] Protesters outside the annual Western Electronics Manufacturing Association meeting carried signs reading "Fairchild Workers Unite!" "Indigestion to Noyce from Intel Workers!" and "70,000 Electronics Workers Say No Vacations Without Pay!"[17]

Markkula was initially unaffected by the turmoil. But early in 1975, Intel restructured its marketing group as part of a companywide reorganization away from functional lines (such as marketing and engineering) to divisional ones (such as components and systems). Markkula's boss, Ed Gelbach, who had replaced Graham, was promoted to general manager of the components division, with responsibility for all Intel products aside from the microprocessor and a digital watch subsidiary.

Gelbach's move meant that Intel needed a new vice president of sales and marketing. Markkula wanted the job. He was among the longest-tenured employees in the marketing area. He was also an expert in marketing memory chips, Intel's core business. He was a logical candidate.

But he did not get the promotion. Instead, in the first quarter of 1975, Intel brought in an experienced marketer who had worked with Gelbach

---

* It was the same downturn that had freed up the small company that normally made plastic tooling for microchips to instead manufacture the plastic case for *Home Pong*.

at Texas Instruments, Jack Carsten.[18] After a few weeks working with Carsten, Markkula decided that he did not want to continue.[19]

Carsten's arrival at Intel coincided with the vesting of the last of Markkula's large original option grant—wiping out in a few months both incentives (pleasure and money) that had kept Markkula pulling his silver Corvette into the company parking lot every morning. Thanks to Intel's skyrocketing value and multiple stock splits, Markkula's original stock option grant alone was worth $2.25 million ($10 million in 2016 dollars), sixteen times what he had paid for it. He had also bought every share available to him at a discount under Intel's generous employee stock purchase plan.[20]

He knew what he wanted to do next. Ever since he had realized that he could retire, he had planned how he would spend his free time once he had more of it. Confident that "reaching goals is what life is all about," he did not dream idly.[21] He strategized—on a three-by-five-inch note card he kept tucked in his date book. He had been playing guitar since his teens, but only by ear or using tablature; he resolved to learn to read music. He loved woodworking; he would build real furniture that the family could use. He was a good athlete; he would become an even better tennis player and skier. He wanted to do something to "give back" to the community; he had yet to determine how, but he would do it.

For two years, he had been pulling out the note card to jot down a goal as it occurred to him. Even if he wrote small on the front and back, the card could hold only about fifty goals, which meant prioritizing and erasing old resolutions when he thought of new ones. He appreciated the discipline imposed by the card's physical limits. "If it can't fit on the card, then I'm never going to get there," he explained years later. "If it's that important, cross something else off."[22]

For two years, the card had been a pleasant diversion. But by early 1975—after he had been passed over for the job he wanted, begun working for someone he did not find compatible, and watched the last of his original options vest—the card had become something different: the road map for his future.

Once he and his wife, Linda, agreed that he should leave Intel, Markkula did not let anyone convince him to stay. He did not ask for a transfer to a new job within the company. He did not look for a new employer. He told anyone who asked that he was retiring from paid work—for good. "Well, I guess it's time to start on that list," he told himself. He was thirty-three years old.

■ ■ ■ ■

He began working through the plan on his note card. He went to the music shop near his house and found a teacher. He devised a set of overlapping paper wheels that showed all possible positions for a given chord on a guitar's fret board.[23] He volunteered to teach fourth-grade math at the local elementary school. He joined the Cupertino Planning Commission.[24] Setting up a woodworking studio in his garage, he built deck furniture and planters that his family would use for years. He converted a tiny bedroom into an office, installed a teletype machine that connected to a PDP-10 computer owned by Tymshare, and wrote a program to balance the family checkbook. He crossed one goal after another off his list. He and Linda soon fell into an easy, comfortable rhythm, eating lunch and playing tennis together. For the first time, he was home during the day to spend time with their daughter, and soon, their infant son.

He found it all so satisfying that he did not miss his paying job—with one exception. He missed what he called "bright, fiery-eyed, fire-in-the-belly people wanting to accomplish things."[25] He remedied this in the best way he knew: he added another item to the note card. "I decided that every Monday I would be a consultant," he explains. "You could call me up, and I would do the best I could and critique your ideas. No charge. But don't call me on Tuesday because I'll be playing tennis or skiing or building furniture in my garage."[26]

Soon after telling several friends in the semiconductor and venture capital industries that he would be happy to talk to any aspiring entrepreneurs, he began getting calls and making appointments. Every Monday, smartly dressed men, usually in their thirties or early forties, came to meet Markkula in his converted-bedroom office. He estimates that he heard about two dozen pitches for companies: semiconductor operations, equipment suppliers, a soap business. A few entrepreneurs asked him to join their founding teams; a few more wanted him to write their business plans. He always declined. "It's a lot of work," he explains. Instead, he would "tell them how to do it." He offered advice on how to conduct

market surveys and calculate distribution costs, how to think through the
pros and cons of sales teams versus distributor networks, how to plan for
the economies of scale that could come from rapid growth.[27]

In early fall 1976, some eighteen months into his retirement, Markkula received a call from Don Valentine, the venture capitalist who had
worked with Atari. Markkula and Valentine had known each other as colleagues at Fairchild and later as competitors when Markkula was at Intel
and Valentine at National Semiconductor.

"Don called me up and said, 'There's two guys over in Los Altos that
could really use your help, and you ought to go see 'em,'" Markkula recalls.
The young men, both named Steve, had started a partnership called Apple
Computer.

"And I said, 'Okay.' 'Cause that's what I did on Mondays."[28]

CHALLENGES
1976–1977

By 1976, both the New York Times *and the* Wall Street Journal *had cited Silicon Valley by name. A few products, companies, and ideas with Valley roots were generating occasional excitement. The stock prices of Intel and Hewlett-Packard were soaring. Atari's video arcade games and* Home Pong *were hits.*

More often, however, innovations from Silicon Valley faced indifference or even fear. Steve Jobs and Steve Wozniak struggled to find a backer for their fledgling computer company. The independent software industry was so tenuous that Sandy Kurtzig wondered if her tiny contract programming business could survive. Bob Taylor and his team endeavored, with little success, to excite the executives at Xerox about PARC's personal computer. There were pockets of backlash against violent video games. Protests and research bans followed reports about the recombinant DNA process, some people mistakenly fearing that it threatened the very existence of humanity. Reimers's patent application was criticized for its potential to undermine the Stanford's fundamental mission to increase the world's cumulative knowledge.

Silicon Valley itself faced criticism as it transformed from an agricultural hamlet into a congested metropolitan region. San Jose grew by 24 percent between 1970 and 1976, making it the fastest-growing major city in the country. As freeways filled, commutes lengthened, housing prices soared, and all but a few orchards disappeared, residents began to fight back. In San Jose, three frustrated building contractors used a crane to lower a Chevy onto an unfinished freeway interchange in an effort to persuade the state to release highway funds.[1] *In San Mateo County, voters passed a measure to acquire and preserve in perpetuity a regional greenbelt of open space land.*

The Arpanet that Bob Taylor had kick-started a decade before when he was at the Department of Defense also gained unwelcome attention. In June and September 1975, Senator John Tunney, a Democrat from California, called a series of congressional hearings, citing "our concern that powerful new technologies—reminiscent of those described almost 50 years ago by George Orwell—will destroy the Constitution's delicate balance between the powers of the State and the rights of individuals." NBC News had run a series of reports alleging that the Pentagon, using a "secret computer network . . . made possible by dramatic breakthroughs in the technique of hooking differ- ent makes and models of computers together so they can talk to one another and share information," had built "a secret electronic intelligence network that gives the White House, the CIA, and the Defense Department instant access to computer files on millions of Americans."[2]

This "secret network" was the Arpanet. It was no secret, but its having been backed with military money made it particularly suspect on the heels of the unpopular Vietnam War. Moreover, the country was still staggering less than a year after learning of President Nixon's abuses of power, many of them aided by technologies such as wiretaps, secret tape recordings, and elec- tronic bugging devices. Senator Tunney hinted at the possibility of "strength- ening the Bill of Rights" to cover the new computer networking technology. *

After three days of testimony by nearly a dozen people, the specific NBC alle- gations were proven unsubstantiated. The concerns they raised about privacy in the face of electronic records, databases, and networks remain open to this day.

---

* Congress had been wrestling with the question of how to protect individual privacy in the new technological era for more than a decade, ever since the IRS had begun using computers in 1963 and shortly thereafter began to require taxpayers to include Social Security numbers on their returns. By the late-1960s, many gov- ernment departments and agencies—not just the IRS but also the army and the FBI—maintained databases with information about US citizens. In 1968, Congress, fearing an Orwellian situation in which the government would know everything about every citizen, blocked a proposed National Data Center that would have consolidated information from various databases.

I Needed to Land Behind a Desk

FAWN ALVAREZ

Fawn Alvarez, who had picked plums and stuffed envelopes for pocket money as a child, graduated from Cupertino High School in 1975. A few months after tossing her tasseled cap, she began work on the assembly line at ROLM, the company that also employed her mother and older sister Bobby. Fawn Alvarez had gotten the job on Halloween, when, dressed in Minnie Mouse gear down to the gloves, she had come by to drop off her mother's forgotten lunch. Alvarez's neighbor was the hiring manager, and she asked if Fawn would be interested in full-time work in production.

Alvarez had been stocking shelves and working the floor at Mervyn's or another department store ever since she was old enough to lie convincingly about her age. She had always looked a little askance at assembly jobs, even if they were dressed up in high-tech clothes. She could have asked about a different job at ROLM, but she knew that the other obvious position for a young woman without a college degree would be as a clerk in the document control group. That group tracked the thousands of pages of documentation that ROLM generated on its products as they moved through manufacturing and into the market.

Alvarez did not want to work in document control at ROLM because her mother, who had left production work several years earlier, managed the group for the division that built ROLM's ruggedized computer. Fawn, just past her 18th birthday and living on her own for the first time, loved her mother, but she also wanted to carve out an independent life for herself. Moreover, she was paying $100 monthly rent on her own apartment around the corner from the house her mother had bought for $22,000 shortly after joining ROLM. With that bill and others looming every month, Alvarez had to admit that a line job looked pretty good. She would make only minimum wage—$2.00 per hour—but the job came with sick leave and the promise of steady work.[1] Her hours in retail varied depending on the season.

Manufacturing jobs were plentiful in Silicon Valley in the mid-1970s. Local companies were building chips, calculators, computers, peripherals, video games, and electronic equipment. In the two decades after 1964, Silicon Valley added more than 200,000 manufacturing jobs, 85 percent of which were in high tech.[2] On average, experienced workers earned about $5.00 per hour. Unlike higher-level engineers or managers, even the most senior line workers did not receive stock options. A few progressive companies, ROLM among them, offered assemblers profit-sharing plans or allowed them to buy stock at a reduced price.[3]

Alvarez joined a production line at ROLM that built a fully electronic, computer-controlled telephone system (a private branch exchange) called CBX. The CBX was a new product for ROLM, which had sold more than $30 million worth of ruggedized computers in the six years since its founding. The company had entered this new telecommunications market shortly before Alvarez joined in 1975, seven years after the landmark Carterfone regulatory decision had forced AT&T to allow telephones and other equipment made by outside companies to be connected to the AT&T network.[*] ROLM's CBX system, which had a minicomputer at its core, was cheaper and more flexible than the legacy AT&T systems with electromechanical phone switches. The ROLM CBX could also do things

---

[*] The Carterfone decision opened the way for answering machines, voice mail systems, and other such equipment. Before Carterfone, AT&T, which since 1908 had used the motto "One system, one company," had not only owned the phone lines but also built and rented every phone connected to the system. ROLM cofounder Bob Maxfield likened the old way to being required to rent lamps and lightbulbs from the electric utility company.

that once had been the task of operators: restrict certain extensions to local numbers only; identify the duration and dialed number of any call; forward, conference, and queue calls; and redial numbers automatically.[4]

Developing the CBX system had been a substantial technical challenge for the engineering team. The product had also created a terrifying ordeal for management and engineers alike when nearly every system began failing at customers' offices simultaneously. (The culprit was the early Intel DRAM chip inside the system.) And the CBX had kicked off a thrilling chesslike game of strategy for ROLM's lawyers, who successfully battled a massive AT&T effort, both in Congress and on the state level, to block ROLM and other competitors.[5]

Fawn Alvarez soon discovered, however, that for people on the production line, there was nothing exciting about assembling the breakthrough CBX. The room where she worked was the size of a football field, bright with fluorescent lights and ringed with large cabinets and tall racks like the ones used to cool bread in a bakery. A conveyor belt ran between two rows of desks, and it was her job to pull off it a box of green circuit boards that the woman in front of her had already loaded with electronic components. Following the engineering schematic at the front of her desk (or more likely, doing it by rote), Alvarez would then add more components, some Chiclet-sized, some the shape and heft of a Lego block. When she finished, she would put the boards back into the box and place it on the conveyor belt. A gentle push, and the box would travel on to the woman behind her. Then Alvarez would grab another box of boards.

When the last person in Alvarez's line had put in the final parts, the boards were moved to a different conveyor belt that pulled them over a vat of solder. The solder, which flowed over a metal bar to make what looked like a tiny waterfall of silvery mercury, stuck to the metal and secured the parts in place. The boards were then washed in a chemical bath, after which workers hung them to dry like photographs in a darkroom. ("That would be so illegal now that I can't believe I don't have cancer," Alvarez says.) Later, her coworkers would load the boards, key technical innards of the CBX phone system, onto the tall metal racks. Other people would test the boards off-site and install them into refrigerator-looking bays that ROLM painted a distinctive burnt orange.

When Alvarez hired on, half of ROLM's employees worked in

manufacturing, all of them in Silicon Valley.[6] Alvarez describes her co-workers as "a mini United Nations." ROLM's rapid growth coincided with changes in the ethnic makeup of California. Many of the women who worked with Alvarez were from Mexico, Vietnam, or Cambodia. Though Alvarez heard no outright hostility, there was some sotto voce grumbling among some workers about people speaking to each other in languages other than English and about the "stinky lunches" that were being heated in the new microwave and making it hard to enjoy a brown-bagged bologna sandwich.

■ ■ ■ ■

Alvarez quickly decided that eight consecutive hours of work she described as "plug the right piece in the right hole in the right direction" was mind-numbing. Even dipping pizza slices or pencils into the solder wave machine did little to relieve the monotony. But when one woman began asking if the others had ever considered unionizing, she was "laughed out of the room," according to Alvarez.[7] Many other production jobs in Silicon Valley ran on strict schedules and required special attire or equipment; Intel, for example, implemented its first clean room and "bunny suit" uniform in 1973. Intel workers were forbidden to wear makeup—even a wayward flake of mascara could contaminate the chips—and worked covered from head to toe in mandatory hats, goggles, long sleeves, pants, and gloves. The dress code at ROLM, on the other hand, consisted of little more than "you must wear closed-toe shoes," Alvarez says. She and the other workers also had more control over the pace of their work, since it could take hours to add parts to all the printed circuit boards in a box.[8] "It wasn't Lucy in the chocolate factory," she explains, referring to the classic 1952 episode of *I Love Lucy* in which the title character is overwhelmed by the onslaught of candies that she is supposed to wrap as they speed past her on a conveyor belt.[9]

A quarter century before Google included "Don't be evil" in its code of conduct, ROLM listed "Be a great place to work" as a corporate goal alongside profits and growth. Aside from stock options, which were reserved for those cofounder Bob Maxfield calls "our most creative and top people," every benefit or perk at ROLM was available to every employee in the company.[10] Everyone could participate in generous medical and dental plans, as well as profit sharing and reduced-rate stock purchase plans. ROLM reimbursed tuition for employees attending school while working. Every permanent employee—even those on the assembly line—was entitled to a twelve-week sabbatical with full pay after six years with the company. When the executives learned that many manufacturing employees, who spent nearly all of their paychecks on food, rent, and other necessities in the increasingly expensive Silicon Valley, did not have the spare cash for even a short trip during their sabbaticals, ROLM introduced a new option: six weeks off at double pay, rather than twelve weeks at full pay.

The four ROLM founders shared a dislike of unions, which they believed fostered adversarial relationships among people who should be working together. "The way I always looked at it, any company that got unionized was doing something really, really, really wrong," recalls Maxfield. "If you're not treating your employees well enough so that they feel like they have to join the union and have somebody represent them to management, then management has failed."[11]

There was never even a union organizing drive at ROLM. As Alvarez puts it, "We had good chairs, good lighting. We could go to the bathroom without raising our hands. We had dignity. What could we get from a union that we didn't already have?"[12]

Even at Silicon Valley companies with less generous benefits or less comfortable working conditions, unions were rare. In 1974, the United Electrical Workers created an organizing committee specifically to target Silicon Valley production workers who performed highly repetitive tasks for relatively low wages, but the effort had little effect. Union organizers made significant inroads only at defense contractors, and even with those numbers factored in, fewer than 5 percent of electronics workers in the Valley were represented by a union in the 1970s.[13]

Union membership throughout the country dropped by 36 percent between 1972 and 1982, and organizing in Silicon Valley was further complicated by a number of local factors. As was the case on Alvarez's line, most production workers were women, many were immigrants, and almost half were members of minority ethnic groups. None of these was characteristic of people who traditionally joined unions. It was not unusual, moreover, for workers to change employers. Alvarez's mother, Vineta, had worked as an assembler for three different companies in her first two years in Silicon Valley. Turnover rates among Silicon Valley assembly workers topped 50 percent annually; the near-constant churn made workers difficult to organize.[14] Moreover, electronics industry associations offered legal aid to companies facing unionization efforts and multiday seminars for "companies that are non-union and wish to remain so."[15] The seminars advised that the best way to keep out unions was to offer production workers benefits and compensation comparable to what they would receive with union representation. Many companies, including ROLM, did precisely that.[16]

■ ■ ■

After a few months on the assembly line, Alvarez decided that the work could be done better. "I had worked in retail, and I knew a lot about organizing stockrooms and how to move inventory around. I thought the way that they would do it at ROLM was crazy," she says. Soon she was spending her breaks and lunches with the three production engineers who laid out the work flow and drew the schematics for the women on the manufacturing line. Maybe a single task needed to be broken down into two tasks, she might suggest. Why not install a ramp on rollers to make it easier to move items around the floor? Or maybe someone could scoop one more component than she was being told to grab from her supplies. "I started telling them, basically, how to make our lives easier just out of self-preservation," she says. "I just wanted it for me." She recalls flat-out telling an engineer that his plan for organizing the assembly work was "stupid."*

To Alvarez's surprise, the engineers, all college-educated men in their twenties, listened. Looking back, she guesses that they paid attention, in part, because she was spirited, attractive, and unthreatening: "tiny, cute, 18, and often not wearing a bra" is her description of herself at the time.[17] They also paid attention, it seems, out of surprise. Most line workers, a generation older than Alvarez, had come of age before the women's

---

* When Ron Raffensperger, Alvarez's boss later in her career, heard that, even at eighteen, she had been telling higher-ups that their ideas were stupid, he started laughing. "That's Fawn, all right."

movement. Many did not speak fluent English. It would have been rare for someone in Alvarez's position to question the engineers.

But Alvarez, who had protested the Vietnam War and believed Gloria Steinem was "a goddess," was not the sort to keep quiet. She took after her mother, who had fought for every nickel and had moved from the production side of ROLM to a managerial role in document control, thanks to a deep intelligence and a personality that one admirer described as "take no prisoners and make sure things get done well and on time."[18] Fawn Alvarez had initially been a bit afraid to share her thoughts with the male engineers who were so much better educated than she was. But then she reassured herself that if the men dismissed her, she would just take her ideas directly to Ken Oshman, the cofounder whom she had met when she was a little girl.

There was one more reason why the engineers paid attention when Alvarez offered a suggestion or questioned one of their directives. She was often right, and ROLM was growing so quickly that employees adopted good ideas wherever they came from. "You didn't have this corporate bureaucracy to help you do your job," recalls production engineer Jeff Smith. "We had to just be bullheaded about it and get it done however we could figure out how to do it."[19] As Alvarez puts it, "I was a risk taker surrounded by other risk takers." It was hard for the engineers to pinpoint bottlenecks or other problems simply by looking at reports or troubleshooting the line. Alvarez could help them.

One engineer suggested, for example, that the individual workers' desks be replaced by large central tables, each with several boards on it. He described how every assembler could pick up a handful of parts and walk around the table, plugging the parts into board after board.

For eight hours, no one is going to sit down? Alvarez asked him.

He hadn't thought of that.

One memorable day, Alvarez received a check for $1,000—the equivalent of three months' pay—after one of her suggestions for reconfiguring the line was implemented. Soon thereafter, she was promoted to line leader.

In September 1976, around the same time Alvarez was promoted, ROLM went public. (Larry Sonsini, the young attorney with a vision for a law firm that could serve as a one-stop shop for entrepreneurs, drew up the registration statement.) Although ROLM would become a Fortune 500 company and create many millionaires among its stockholders, its IPO was a disappointment. The initial offering price of $14 per share dropped to $10 a few days later.

Not that it mattered to Alvarez. "Young and dumb," in her own description, she put her paycheck toward rent, food, clothes, and fun. If she had any money left over, she might buy some stock at the discounted employee rate, but that rarely happened.

In 1977, she was promoted again, to manufacturing supervisor. She was twenty years old, supervising fifty women, many of them twice her age. She was responsible for hiring and firing and for helping to train new workers. At the same time, she continued to try to impress upon her own supervisor and the process engineers that she was more than just another assembler with a couple of promotions. She admits that for a while, she drove production too hard. Embracing her interpretation of Henry Ford's mass production model, she repeated to herself, "If a person does one little thing as fast as possible, and does it over and over again, it maximizes production." One day she overheard two assemblers referring to someone as "Little Hitler"—and then she realized that the person they were talking about was her. For the first time, she wondered if her model might just "maximize production right up until 'person 78' turns around and kills 'person 77.'" Alvarez began varying each worker's tasks, having people switch jobs and check one another's work.[20]

The move to supervisor suited her. She was learning. She felt that her opinion was respected and she had influence. She was paid fairly. Every few Fridays, ROLM held a companywide beer bash, and she was at ease chatting with managers and executives. "Everyone mixed with everyone," she later said.

Alvarez was also excited by the news that ROLM was about to move to a new twenty-one-acre campus in Santa Clara. The campus had a six-lane lap pool; volleyball, racquetball, and tennis courts; a fitness room, a sauna and steam room, waterfalls, a jogging trail, and ponds stocked with fish.* Goodwin Steinberg, a student of the Bauhaus architect Ludwig

---

* Alvarez recalls that since the new buildings were surrounded by fields, in the earliest days, mice overran the campus. "You could hear screams all day long coming

Mies van der Rohe, designed the headquarters building. A precursor of the elaborate amenity-filled campuses popular in Silicon Valley today, the ROLM site, now a parking lot for the San Francisco 49ers' Levi's Stadium, also housed a cafeteria with heavily subsidized lunches, which meant the battles over the microwave would disappear.

Alvarez loved ROLM, but she also felt that over the long term, manufacturing, in its repetitiveness and demand for perfect consistency, would always be a "soul-sucking job." She wanted to leave production. But she had heard of only one woman who had left a manufacturing job for professional work: her mother. Alvarez had never asked how she had made the transition, nor did Vineta Alvarez offer any advice. ("I never told the girls how to get ahead," says Vineta Alvarez. "They did that on their own.")[21] Fawn Alvarez was determined to repeat her mother's success. "I wanted an office job," she says. "I didn't care where I landed; I just needed to land behind a desk."[22]

---

out of these beautiful offices," she says. Soon the employees were throwing the poor rodents into the pond and betting on whether the mice could make it out before one of the fish ate them. She explains, "Nobody wanted to kill the mouse, but no one cared much if the bass got it."

This Is a Big Fucking Deal

AL ALCORN

"I sent a memo that said, 'I have seen the future,'" explains Manny Gerard, one of three members of the Office of the President at Warner Communications and the company's self-described "filter for acquisitions."[1] Gerard, who would soon become a vital force shaping Al Alcorn and Atari's future, recalls thinking when he sent the note in 1976, "This is something structural. This is going to be a big business. This is a big fucking deal."[2]

Warner Communications, with offices in Rockefeller Center in New York City, was a $147 million conglomerate with holdings that defined American popular culture. Warner owned a bank, a cable company, a perfume label with the designer Ralph Lauren, the New York Cosmos soccer team, and three record companies whose artists included Led Zeppelin, Linda Ronstadt, Joni Mitchell, Rod Stewart, George Benson, George Harrison, Richard Pryor, and Jefferson Airplane. Warner's movie subsidiaries produced *The Exorcist*, *Blazing Saddles*, *A Star Is Born*, *All the President's Men*, and Clint Eastwood's Dirty Harry series. Warner had a publishing arm and a television subsidiary (*Welcome Back, Kotter*; *Chico and the Man*) and also owned *MAD* magazine, Looney Tunes, and DC Comics.[3]

In early 1976, Gerard had received a call from Warner's largest shareholder, Gordon Crawford, who had bought 10 percent of Warner for Capital Group. It was no coincidence that a representative from Capital Group, the original home of venture capitalist Don Valentine's Sequoia fund, was

now calling about Atari, a company with which Valentine had been working for two years. In 1975, on the strength of the *Home Pong* game and Atari's $750,000 profits, Valentine's Sequoia Capital had invested $500,000 in
Atari.[4] Valentine had also convinced three other investors to do the same:
Time Inc., Mayfield (another early Silicon Valley venture capital company), and Fidelity Investments, whose Boston representative had taken
off his jacket but nothing else when the stripped-down Bushnell had sold
him on Atari from the comfort of the hot tub.

Back in 1975, when Valentine had arrived at Atari to sign the paperwork for the $2 million funding deal, he brought several bottles of cham-
pagne, anticipating a celebration. But Bushnell and Keenan had decided
to double Atari's premoney valuation based on the company's recent successes with *Home Pong* and other games.[5] Valentine was unfazed. He put a
copy of the revised terms into his bag and headed back up the freeway, the
champagne bottles growing damp with condensation. A few days later, he
announced that he and the other partners still wanted in. Corks popped.

Now, less than a year later, Valentine hoped to sell Atari to Warner
Communications. A significant technical breakthrough at Atari's research
center in Grass Valley, 150 miles northeast of the Bay Area, meant that the
company needed another infusion of money. Bushnell and Alcorn would
visit the researchers in Grass Valley every month, but they had not moved
the group, originally established as an independent research organization
called Cyan Engineering, to Silicon Valley. In Grass Valley, the researchers
could focus on the long term, isolated from day-to-day business concerns.
It was "a very protected playground, unencumbered by the feed-me aspects" of business, according to Steve Mayer, the founder of Cyan Engineering with Larry Emmons. As Alcorn put it, "My job was to keep the
other people away from these people."[6]

Off in the still rural Grass Valley, researchers who looked more like
lumberjacks than engineers found a radical way to use a cheap new microprocessor, the 6502 built by the upstart chip manufacturer MOS Technology. (MOS Technology had launched by selling chips from a hotel room
that the founders had rented near a major industry conference.) Before
Grass Valley's breakthrough with the microprocessor, most video games
had been built around a dedicated circuit that was wired to play one game
and only that game. The *Gotcha* arcade console could play only *Gotcha*,

a maze game that Alcorn had designed. The *Pong* console sold by Sears could play only *Pong.**

But Grass Valley's new system, code-named "Stella," was something different.[7] The microprocessor inside meant that the Stella system could be programmed to play any number of games. A person who bought a Stella system could plug the box into a television and then tell the box to play different games by inserting different game cartridges.[†] This was the magic of software in action.

The Stella system, released as the VCS (Video Computer System), would go on to become one of the most successful consumer electronics products of its time, in five years selling more than 12 million units and many more cartridges.

When Warner's Gerard wrote that he had seen the future, he was referring to Stella. "I can almost remember the words," Gerard says of the call from the Capital Group. " 'Would you be interested in a very fast-growing, entertainment-based technology company?' " Warner had not acquired a company in four years, but Gerard said that he would indeed be interested.

■ ■ ■ ■

As Gerard began examining Atari, he developed doubts about acquiring it. The company had "no infrastructure," as far as he could tell, and weak financial controls. Though some divisions kept careful records, in others employees could request, authorize, prepare, and sign checks payable to themselves. Employees had also been known to receive cash advances—in anticipation of a business-related trip, for example—and then quit without returning the advance. The manufacturing line routinely had to be stopped because the plant ran out of basics such as paint, glue, and screws. The company did not require that orders be written down, which led to

---

* The Sears console played a number of games, but they were all variations on *Pong*.

† At about the same time the Grass Valley-based Atari research group developed the Stella system, one of the few African-American engineers in Silicon Valley at the time, Jerry Lawson, independently developed a similar microprocessor-based system, called the Channel F, for Fairchild in Mountain View. The Fairchild system did not have anywhere near the success of the Atari system.

regular disputes with customers regarding shipping dates, order quantities, prices, and payment terms.[8] Nearly every game was invented from
scratch with designers given total control, which meant there was no mandated standardization, even on basics such as cash doors, coin counters, or
locks for the coin-operated video games. Atari's leaders saw the company
as a complex, multifaceted organization with employees who were trusted
and empowered to be their independent, creative best. Gerard saw a cha-
otic mess.

Another worry for Gerard was the Atari team itself, which he describes
as "a bunch of guys getting whacked every day and chasing women."[*] On
one of Gerard's early visits to Atari, a group of engineers was thrilled to
show him how they had programmed a version of *Tank* to run backward.
They called it "Polish Tank." Gerard thought Alcorn was "a solid guy,"
the cautious sort who, upon realizing he had a great deal of paper wealth
after the venture capital investment in 1975, worried that "the whole thing
could still go away."[9] But Alcorn did not run the company.

While Alcorn was worrying about the permanence of his new fortune, Bushnell, raised in a conservative, middle-class Mormon family in
a small town in Utah, had gone on a spending spree. He bought a large
house and a forty-one-foot sailboat. He posed for a local paper in his
hot tub with his arm draped around a beautiful young woman. "Some
ladies feel uncomfortable with me and some don't," he told a reporter. "I
find the aura of power and money is intimidating to an awful number of
girls."[10] Gerard thought that Bushnell, "who has a hundred ideas a day
and loves them all," needed to learn to focus.[11] At times Alcorn agreed
with that assessment; he carried a special pager and instructed the engineers to ring it "only in the event Nolan got into engineering." When the
pager went off, Alcorn would rush to engineering and as he put it, "follow
behind Nolan to remind" the engineers that they worked for him, not
Bushnell.[12]

Gerard had other reasons to think twice about acquiring Atari. The future of video games was still uncertain. Pinball machines were resurging.

---

[*] Gerard exempted a few employees—most notably Steve Mayer and Al Alcorn—
from that generalization.

The video game that drew the most attention in 1976 did so for all the wrong reasons. *Death Race*, a driving game from a small Silicon Valley company called Exidy, awarded players points for running down stick fig-ures that Exidy called "gremlins" and everyone else called "people." (Upon impact, the player heard a shriek and a cross replaced the stick figure on the screen.) The National Safety Council condemned the game as "'insid-ious,' 'morbid,' 'gross, and 'sick, sick, sick.'"[13] The *New York Times* and *60 Minutes* ran stories on it. Bushnell fretted that the general public might not agree with him that blowing up tanks and airplanes in Atari games did not count as the same sort of "violence against people" as mowing down stick-figures.[14]

Despite the outcry, Gerard, who had been analyzing popular media for decades, still believed that video games would become a big business. He considered games an integral part of the human experience, with a his-tory that stretched back to markers made of stones and bones. Atari was taking that history into the new world of silicon and television.[15]

He also believed Atari, for all its problems, had virtues. The company's engineers, particularly Alcorn, Bristow, and the team in jeans and long beards who had designed Stella, were "just stunning," he recalls, "and they were astonishingly productive." Moreover, Atari was making money; Ge-rard was not sure how, given the organizational chaos, but it was.[16]

Gerard recommended to Steve Ross, the debonair founder and chair-man of Warner Communications, that Warner acquire Atari. After Ross, whose kids had loved playing *Tank* at a Disneyland arcade, agreed, he wooed Bushnell and Atari's president, Joe Keenan, regally. Ross dispatched a Warner jet to bring them to New York and a Warner limousine to convey them from Teterboro Airport to suites in the Waldorf Towers.

At dinner and in conversations at Ross's opulent apartment the fol-lowing day, Ross made his case. Warner knew how to create hits and get products to consumers; Warner's WEA group had just shipped 1.1 million copies of the Eagles' new *Hotel California* album in only three days.[17] And Warner understood how to work with artists who needed to be coddled and coaxed, a particular draw for Bushnell, who considered Atari's game designers artists. Ross also told Bushnell and Keenan that they would be happy at Warner. They could stay in California and continue to run Atari. Ross explained that he took a hands-off approach and even offered to sweeten top management's incentive structure, which already included a rich bonus pool.[18]

By the time the Atari team flew back to California on another Warner jet—Ross arranged for Clint Eastwood and Eastwood's partner, Sondra Locke, to join them—Bushnell was convinced that selling Atari to Warner was the best way to get the money to develop the Stella chip.[19] None of the other companies Atari had approached about an acquisition had been interested. Valentine was pessimistic about the prospect of an IPO. And Bushnell, who owned 49 percent of Atari stock, hoped that selling the company would enable him to take a break from it.[20] "All I could think of was 'I'm really tired, so can I get a rest?'"[21] Later he would say that he probably would not have sold Atari if he had just been able to take a vacation.[22]

■■■■

"I was in the hot tub with Nolan when he told me we had a firm offer from Warner. There was a number—and I'd get ten percent," Alcorn says. (He likely ended up with more like 6 percent.) His annual salary was $32,000. Now the stock he had thought would be worthless when Bushnell had offered it to him four years earlier was worth $1.75 million ($7.4 million in 2016 dollars).*

"Up to that point, I thought this was all a lark. Holy shit! My life was being changed."[23]

■■■■

The negotiations between Atari and Warner, held at the law offices of McCutchen Doyle in San Francisco, took months, far longer than most Warner acquisitions. Two or three times each week, Alcorn spent several hours sitting with Bushnell and Atari's two lawyers across the table from

---

* Bushnell's annual salary was $75,000; Keenan's, $40,000; Bristow's, $32,000; and Emmons and Mayer's, $30,000. At the time of the acquisition, Bushnell owned 49 percent of Atari, Alcorn owned 6.7 percent, and Keenan, 6.2 percent. Entities associated with Don Valentine (Sequoia Capital, Sequoia II, Capital Group, and Valentine's personal ownership stake) collectively owned 34 percent of Atari.

Manny Gerard and Warner's eight or nine lawyers. "It was not fun," Alcorn says. "A circus," Gerard agrees.[24]

The negotiations exposed Atari's improvisational style. Alcorn's stock grant was a handwritten note from Bushnell. The buyout of Dabney's stake in Atari had been for cash, instead of retained earnings, and so had to be done over. Atari's CFO quit in the middle of negotiations. At one point, the lawyer representing Bushnell's ex-wife called Warner's attorneys to say that Bushnell's 1974 divorce agreement had left open the possibility that she could claim half of his stock—a move that would make her a 25 percent owner of Atari. "So in the middle of this, on top of everything else, you have to renegotiate the man's divorce settlement!" Gerard recalls.[25]

But the fundamental problem that stretched the negotiations from the cool San Francisco June mornings to the surprising warmth of its September afternoons was that Warner did not trust many of the people or numbers coming from Atari.* Gerard worried that Bushnell and Keenan would quit as soon as they received money from the sale. "You are all going to become cocaine addicts and blow your brains out," he told Alcorn. Later Gerard explained, "We weren't prepared to let them completely have control of the asylum. We knew that if we let them all have a lot of money, we would all be dead."[26]

Gerard's concerns that the Atari team would abandon the company were exaggerated but not unfounded. Bushnell wanted a break, and moreover, even he feared his own fondness of excess. In college, he had taken his job as a carnival barker not only because the pay was good but also because he figured that if he was making money, that would be one time when he could not be spending it.[27]

The agreement that Atari and Warner signed in October 1976 reveals Gerard's anxieties. As an incentive for the Atari team to stick around and make the company profitable, it included a rich bonus pool: if Atari's profits reached a certain level, the top managers would split among themselves

---

* Alcorn says that a Warner attorney was fired because he was so convinced that Atari's numbers were fraudulent that he refused to participate in negotiations. "The numbers were hard to believe. Are we the mob? The mafia? . . . [Games were] a sleazy gray business area."

15 percent of any profits above that level. Moreover, Atari's senior management would receive relatively little money up front. Warner paid $28 million ($119 million in 2016 dollars) for the company, an amount that would net Bushnell roughly $10 million after taxes—the precise amount, Keenan wincingly notes, that Bushnell had once told Valentine he wanted. Of that $28 million, only $12 million was paid immediately, however, and some of that went to the venture capitalists. For the remaining $16 million, Warner issued subordinated debentures in the Atari subsidiary, not in Warner, to be paid as a fixed amount over a period of seven years.[28] In other words, if Atari succeeded, Bushnell, Keenan, Alcorn, and the other executives would split among themselves roughly $2.3 million for each of the next seven years. And if Atari proved wildly successful, senior management would receive significant bonuses on top of the regular payout.

But if Atari failed, the key employees would each receive only their portion of the initial $12 million payment; they would never see the remaining $16 million. Gerard says, "I told them they better believe their own BS if they took the deal."[29] The Atari team took the deal.

One More Year or Bust

SANDRA KURTZIG

In the waning months of 1975, Sandy Kurtzig would never have been able to imagine success on Atari's scale. After ASK failed to write a program for Hewlett-Packard's 2100 minicomputer, HP had given her a chance to develop a version of MANMAN for a different machine—but the effort was not going well. ASK's two programmers occasionally lugged their sleeping bags to Hewlett-Packard, where they figured they might end up spending the night in the low-ceilinged offices, sleeping in shifts in one of the supply closets. HP gave them free access to the flagship HP 3000 minicomputer, the machine they were writing for, but only after 6 p.m., when the HP engineers had gone home for the night.

The ASK programmers worked at HP's sprawling hundred-acre campus in Cupertino, headquarters of the company's commercial minicomputer business. Two miles south was the Rancho Rinconada neighborhood where Fawn Alvarez had grown up. Three miles west, almost a straight shot down Homestead Road, Steve Jobs and Steve Wozniak were trying to build a personal computer in Jobs's parents' garage.[*]

On this particular night, Sandy Kurtzig had joined the programmers.

---

[*] In 2010, Apple, the company Jobs and Wozniak launched, would buy HP's Cupertino campus and raze it to build Apple's new ring-shaped headquarters.

She was aware of the stakes as she watched the two young men tap away at the computer, their twentysomething faces vaguely green in the screen's reflected glow. ASK needed to write a version of MANMAN that could run on a minicomputer. The new machines were threatening the time-sharing services that provided the bulk of ASK's income. Minicomputer prices were dropping, and customers were beginning to realize that they could afford a dedicated machine of their own.[1] Writing a version of MAN-MAN for the HP 3000 would also strengthen ASK's relationship with HP. Kurtzig had blown it with the computing giant once already. She could not let that happen again. The customer who wanted the HP machine was the defense giant Boeing. Boeing's Electronic Support Division had agreed to pay ASK $50,400 for a MANMAN program on the HP 3000. The deal was more than twice the size of the Powertec job that had been canceled.[2]

Marty Browne, the rock-and-roll-loving programmer, and Roger Bottarini, a computer scientist Kurtzig had hired from UC Berkeley, had been working on the Boeing program every night for months. The plan was to write something that, with a few modifications, could be sold as a generic product for anyone buying an HP 3000 minicomputer. Browne and Bottarini were writing MANMAN as a series of text prompts, intended to be easy to follow, even for people who had never used computers. "Enter part number," the program would command, and a person at the keyboard could do just that.

Occasionally Kurtzig, who had moved herself out of programming and into a more traditional CEO role, joined Browne and Bottarini, the three of them alone at night in the HP building, aside from the security guards. "She was mostly there for moral support," Browne says. Kurtzig rarely coded, and for that Browne was privately grateful. She had written serviceable code when the company had launched, but as the MANMAN program had grown more complex and moved to new platforms, he and Bottarini found it easiest to rewrite what Browne called "Sandy code."[3] Kurtzig, meanwhile, was discovering that her strength was translating customers' needs to the technical staff.

Kurtzig used the time at HP's offices to absorb details about its corporate culture. She knew about the company's flexible work hours and traditional twice-a-day coffee breaks, complete with doughnuts, designed to promote informal exchanges among employees at all levels. She knew about the founders' famed "management by walking around," visiting employees at their desks or on the line, inquiring how work was going. Now she noticed smaller things: how, for example, every security guard's desk displayed a photo of Jimmy Treybig, who had left HP to launch Tandem Computers with backing from the young venture capital firm called Kleiner & Perkins, which had an office near Don Valentine's Sequoia Capital. The security guards feared that Treybig might return to try to steal secrets.*

Browne and Bottarini were writing good code, but Kurtzig began to worry that her company would fail again to write a program for an HP machine. Boeing kept changing the requirements for its version of MAN-MAN: demanding, for example, twenty-four alphanumeric characters for part names instead of MANMAN's standard fifteen characters. Kurtzig's primary contact at Boeing, who had signed many of his early memos to her "Love," passed along change after change, the notes becoming brusquer as Kurtzig pushed back. Browne, who remembers the entire summer of 1975 as "a nightmare," was convinced that there were more people writing change orders at Boeing than writing code for ASK.[4]

■ ■ ■ ■

Kurtzig embraced a wide range of oportunities for ASK in the years following the company's funding. ("The only long-range plan I'd had was lunch," she liked to recall years later.)[5] At a time when having more than one customer in a single industry made a young software company an expert in that field, ASK had become a manufacturing expert because Kurtzig had taken on multiple manufacturers as customers. Saying yes to Tymshare had moved ASK into time sharing, and saying yes to HP was bringing MANMAN to minicomputers.

ASK had only a half-dozen employees. With this small team, the company was either already working on, or considering working on, projects for seven different customers on five different types of minicomputers made by three different manufacturers (HP, DEC, and Data General). Computers from different manufacturers required different programs.

---

* Many people who know the highly ethical Treybig have said that the fear was absurd.

Even different models of computers made by the same company had little in common.

ASK was still selling MANMAN on the Tymshare network, and there were even a few companies still using ASK's batch-processing service operated by the punch-card ladies in Palo Alto's Town and Country Village. Tymshare and the service bureau were not cutting edge, but those businesses required little attention from ASK and together brought in enough money to fund the development work at HP. Kurtzig loathed debt.

Despite the multipronged efforts, ASK was barely profitable. The company had $143,000 in sales but only $2,900 in profits in the first six months of 1975.[6] As the year neared its end, Kurtzig decided that she would have to focus ASK for it to survive. A project for Hughes Aircraft, written for yet a third HP minicomputer, seemed headed for success. She would retain that job, continue to collect from Tymshare and the batch-processing service—and jettison everything else.

The decision to streamline meant denying the requests to write versions of MANMAN for DEC and Data General platforms. The decision would also cost Kurtzig her administrative second-in-command, who had brought in the DEC deal and resigned after she turned it down. But most significant, focusing meant pulling out of the Boeing deal that kept the ASK team awake through the night at Hewlett-Packard.

When Kurtzig told Boeing that she wanted out, the aerospace giant threatened to sue. Kurtzig countered that "Boeing Screws Small Company" might make a great newspaper headline.[7] In the end, Kurtzig extricated ASK and even negotiated a payment from Boeing that she estimates at nearly $100,000. Streamlining was painful, but she believed it was ASK's only hope for survival.

■ ■ ■ ■

In January 1976, after almost two years and two false starts, ASK delivered a functional version of MANMAN on an HP minicomputer to a customer who paid for it. Kurtzig's decision to focus all of ASK's minicomputer efforts on a single machine had yielded success. The program, written in FORTRAN with an easy-to-follow interface, would be the model on which future versions of MANMAN would be based. ASK had become one of the few software companies to transition from time sharing to minicomputers.[8]

The customer was the Industrial Products Division of Hughes Aircraft, the same company where Mike Markkula had his first engineering job. It is not surprising that two of the first three customers that wanted MANMAN on HP computers—Hughes and Boeing—were divisions of major defense contractors. Defense-based industries, with their lack of price sensitivity, had driven technological change in Silicon Valley since before the Second World War, when FMC (originally Food Machinery Corporation) had begun selling the tread technology used for orchard tractors to the US government for use in tanks.

The Hughes Industrial Products Division, based in Carlsbad, California, built laser cutters for the apparel industry and wire-bonding equipment, among other products. For $150,000, Hughes purchased an HP 21MX computer loaded with MANMAN software as well as tape drives, discs, two printers, and four terminals that the company spread among the stockroom, purchasing department, and receiving department. Employees in these departments had no previous computer experience, but by responding to the MANMAN prompts that scrolled on the screen, they could enter and retrieve up-to-the-minute information about inventory and ordering.* MANMAN "looks at the quantities we are after in our bill of materials, looks at the stock, purchase orders and work-in-process, and then comes back and tells us what we have to buy, when we have to buy it, when it should be in the stockrooms and when we can ship to the customer," explained Hughes's data processing manager.[9]

Soon Hughes reported that whereas it had once taken three full-time employees to enter stockroom transactions on paper, and three weeks for those transactions to post, with MANMAN, one person could do the same job in three hours, with instantaneous posting.[10]

Hughes, HP, and ASK jointly promoted the sale in press releases, advertising, and conference presentations. The collaboration benefited all

---

* A classic materials requirements planning (MRP) package, MANMAN consisted of six modules: inventory control, bill of materials processor, material requirements planning, processing, work in process, and management reporting/product costing.

three companies: Hughes appeared a forward-thinking business that embraced cutting-edge technology; Hewlett-Packard could announce a foray into commercial computing; and tiny ASK appeared a partner capable of working with giant companies. Kurtzig, who felt that the press release sent out by HP could be more effective, rewrote it and sent it to manufacturing trade publications. When she still did not receive the response she wanted, she sent the identical press release again.[11]

Kurtzig highlighted the Hughes sale at the 1976 American Production and Inventory Control Society show in Atlanta, where she was mistaken repeatedly for a "booth babe," a professional model hired by a company to lure visitors to its display. The mistake was understandable at a conference with a separate spouses' program (featuring a crafts fair), opened by a beautiful blonde in a *Gone with the Wind*–era flouncy dress who had been hired to greet attendees as a "Southern belle."[12]

At roughly the same time that ASK delivered the computer loaded with MANMAN software to Hughes, Kurtzig gave birth to her second son, Ken. Like his older brother, Andy, now three, Ken was born on a weekend. Kurtzig explained a few years later, mostly in jest, "I told God I had a project that had to be finished, and He'd just have to hold the baby until Saturday."[13]

■ ■ ■ ■

The only downside to the delivery of the machine to Hughes was that it left ASK without a computer. A software company could not write software without a computer.

There was hardly money to buy one. ASK had profits of less than $3,000, and the machine Kurtzig wanted, an HP 3000, cost more than twenty times that. Unwilling to burden ASK with bank debt, Kurtzig obtained a three-year, $25,000 loan from her father at 8 percent interest.[14] She also negotiated a deal with HP to buy the computer with a relatively small down payment and pay the balance with one copy of MANMAN, which retailed for $35,000.[15]

ASK installed the computer in the company's new rented office space. The low-slung building on El Camino Real in Los Altos was an easy drive from Kurtzig's home. For $714 monthly rent, Kurtzig got what she called

"1200 square feet of plush office space" and a receptionist she shared with other tenants. She had the entry painted a vibrant orange.

Kurtzig had her own office, as did Marty Browne, whom Kurtzig had named operations manager. Everyone else worked at wooden desks in a large open space, enjoying a special sense of camaraderie. "I spent several years of my life after I left ASK trying to find that again in a company," one early employee recalls.

"Sandy was basically hiring one employee per year when I got there," explains Liz Seckler, who joined ASK in 1976 as employee number six.[16] The employees were nearly all fresh graduates from Stanford or Berkeley, armed with degrees in computer science (from Berkeley) or math (from Stanford, which did not have an undergraduate computer science major until 1986).[17]

They went to lunch at the Boardwalk, a new beer-and-burger joint on El Camino resplendent with stained-glass windows, an oak bar, and a full lineup of pinball and video games. At the height of the national fascination with jogging in the 1970s, the ASK group donned their short shorts and sweatbands to run together at a local park. They visited each other's houses for dinner or on weekends. They held low-key Friday-afternoon beer busts, the pressure to leave their desks beginning to build around 4:30 p.m. "There was no real difference between night and day, whether you were working or not," recalls Marty Browne, who adds that as ASK grew and organizational structures became more formal, the group put a printer on top of the beer-filled minifridge so that they could claim that the fridge was a printer stand.

"It was an extension of the college experience," says Seckler. "It wasn't artificial fun—'Look at our Ping-Pong table and our pool table!'—it was organic; more like a dorm, where there were people a little older than you and people a little younger, and everyone was working together.

"Sandy was like the RA," she adds.[18] Kurtzig was only twenty-nine, but to a recent college graduate, she was the adult in charge that every small startup eventually needs.

To find more programmers, Kurtzig conducted on-campus interviews, which always included a programming problem that applicants had to work through on paper after they spoke with her. Her interviews were unconventionally intimate. After she told Seckler that she needed to find people who knew her better to write her recommendation letters, the two women spent much of the interview chatting about their childhoods in Southern California. Kurtzig ended Howard Klein's interview by inviting him to meet the rest of the team. She then informed Stanford's Career Placement Center that "something came up," and she would not be able to conduct the other interviews scheduled for that day. When Kurtzig thought a candidate would make a good addition to the team, she had the person come to the office to be vetted by everyone else. "Make sure you don't scare her away," she counseled the group before leaving them to interview Seckler.

"Sandy kind of fostered this collaborative environment that felt like 'If one does well, we all do well,'" explains Howard Klein, one of Kurtzig's first hires. "She enabled that environment, and we perpetuated it." Kurtzig could also take an outsized interest in her employees' personal lives. One day she called Klein into her office to take a phone call. On the other end was Kurtzig's freshman roommate, whom she thought would make a good match. She was right. They have been married since 1989.[19]

Throughout the workday, Kurtzig would throw open her office door, telephone pressed to her ear—she was almost always on the phone, it seemed—and ask someone to locate a report or answer a question. "She was zipping through all the time," Seckler says. "It was exciting to be around her."

■ ■ ■ ■

Kurtzig was not aware of it when she was working through the night at Hewlett-Packard, but the computing giant needed ASK and other small software companies.[*] HP, still focused on scientific instrumentation, computers, calculators, and other hardware products for engineers, had almost no experience writing or reselling commercial software.

In 1975, ASK entered a cooperative sales agreement with HP that

---

[*] HP allowed several small local software companies free use of its machines after-hours; the companies included Abacus, Allegro Software Development Corporation, American Management Systems, LARC Computing, Quasar, and SDG.

allowed an ASK representative to go on HP sales calls. HP relied on ASK to help teach nontechnical businesses what computers could do. It was difficult for an HP representative to show up alone at the offices of a small business and ask if the owner would like to buy a $60,000 computer, when most businesses had no idea why they would ever want a computer. But the person from ASK could explain that an HP computer loaded with MANMAN software could turn a small manufacturer into an "intelligent factory" in which anyone, regardless of technical skill or job category, could search inventory or schedule work flow with a few keystrokes.[20]

When Kurtzig proposed scrapping the sales agreement for a royalty structure in which HP would pay ASK a percentage on every system sold, she was surprised to receive a counteroffer in the mail: HP wanted to buy ASK for $1 million. Kurtzig was tempted. "I wasn't even thirty; there were plenty of other things I could do," she recalls thinking. She began to imagine what it would be like to sell the company and relax. She went to her meeting with the vice president of HP's computer division, excited but determined not to sell for less than $2 million.[21]

The meeting was a disaster, in Kurtzig's recollection. The terms of what she had thought was a firm offer never came up. Instead, there was talk of Kurtzig and her team joining HP as employees. When she demurred, someone mentioned that HP had plenty of programmers who ought to be able to write a materials requirements planning package to compete with MANMAN. Soon, Kurtzig says, the vice president who had called the meeting was irate, shouting that HP would write the program for the flagship HP 3000 computer within nine months "whether you help or not."

Kurtzig, by now yelling herself, retorted that HP would never be able to do it—and even if it somehow succeeded, ASK would get there first.

She left the meeting seething. "Damn them!" she swore. She vowed "to show them all what ASK could do."[22]

Despite Kurtzig's bravado, ASK remained precarious. Although the company had sold the version of MANMAN written for Hughes to other customers, it nonetheless ended the 1976 fiscal year with a loss of $2,664. Kurtzig had doubled her salary at the beginning of the year, but she had not taken the raise, writing herself an IOU instead. The $17,692 "due to president" at the end of the 1976 fiscal year accounted for nearly two-thirds of ASK's total liability.[23]

Kurtzig again needed to retool ASK. She drafted a formal business plan, adding accounting components to MANMAN, selling the service bureau, and repairing the relationship with Hewlett-Packard. She also assembled a four-man board of advisers that gave her some important guidance but also indicates how far outside the typical Silicon Valley culture she was. Every adviser was a small manufacturer, and three were former or current ASK customers. There were no venture capitalists, no computer manufacturers, no other software executives. The closest ASK came to a representative from the Silicon Valley tech industry was the associate from Wilson Mosher & Sonsini, the multi-faceted law firm that had worked with Tymshare and now handled ASK's legal work.

The venture capital community in Silicon Valley did not yet appreciate the importance of software. Today software accounts for more than half of venture capital investments in the Valley, but in 1977, the figure was 7 percent.[24] Venture capital was for hardware: computers, semiconductors, disk drives, and other tangible, capital-intensive businesses. In 1977, when Larry Ellison and two cofounders tried to raise funds to launch Oracle—today one of the largest companies in the United States—venture capitalists were not interested. "When they heard the investment was about software, they wouldn't even see me. In fact, their receptionists would search my briefcase to make sure I hadn't taken a current copy of *BusinessWeek* with me when I left the room," Ellison says.[25]

Kurtzig, who never used the word "software" in her 1976 business plan, would say she was in the "applications" business—in part because "software" was such an unfamiliar term that when she used it, people assumed that she sold lingerie. Software developers told rueful jokes about clueless customers who wanted to know how much the software in their computers weighed.

Even computer companies could be slow to acknowledge the value of the software that ran on their machines. Kurtzig repaired her relationship with HP, but she nonetheless had a terrible time convincing the company to designate ASK an "original equipment manufacturer," or OEM. HP sold computers at a discount to its OEMs, which then incorporated the machines into their own products. An OEM that built assembly equipment controlled by HP computers, for example, could buy the computers at a discount and then fold the cost, plus a markup, into the price of the equipment.

Kurtzig argued that the software ASK loaded onto HP's computers should count as a feature that added value to the machine in the same way that assembly equipment did. ASK should be able to buy an HP computer at a discount, load MANMAN, and then sell the combined hardware/software package to customers as a "turnkey system." Kurtzig recalls having to pull out HP's own manual and read aloud the section on OEMs to a skeptical executive before she prevailed. It would take years, but ASK would become the largest original equipment manufacturer of, and a vital partner in popularizing, the HP 3000 minicomputer.[26]

■ ■ ■ ■

As Kurtzig went over ASK's books on New Year's Day 1978—it was typical for her to work on a holiday—she was gratified to see that her efforts over the previous months had paid off. ASK had earned $33,000 in profits, a huge increase over the 1976 losses.

At the same time, she had to admit that this profit, born of what felt like round-the-clock hours in the office, amounted to scarcely more than the $24,000 salary that she paid herself. It seemed meager compensation for the strain of running her own business. She recalled, "I made a New Year's resolution: One more year or bust."[27]

No Idea How You Start a Company

NIELS REIMERS AND BOB SWANSON

In the summer of 1976, as Don Valentine was encouraging Nolan Bushnell to consider selling Atari, and the United States celebrated its bicentennial with fireworks and song, Niels Reimers tracked the country's rising fears about the recombinant DNA process he was trying to patent. Scientists' decision, made a year earlier at the Asilomar "Pandora's Box" conference, to regulate their own research, had had a perverse effect: instead of quenching fears about biotechnology, it had fueled them.

A writer for the *New York Times Magazine* warned that recombinant DNA technology, which had yet to move beyond the research stage, was more dangerous than "the A-bomb, nerve gas, biological warfare, [and] the destruction of the stratospheric ozone layer by fluorocarbon sprays." The magazine further noted that, thanks to recombinant DNA, "the spreading of experimental cancer may be confidently expected."[1] *Time* ran a cover story called "Doomsday: Tinkering with Life," almost as colorful a title as "Dr. Jekyll and Mr. Hyde and Mr. Hyde and Mr. Hyde," a warning against cloning in a Friends of the Earth newsletter.[2] A columnist for the *San Francisco Chronicle* opined that "those lovely people who gave us the atom bomb have another treat in store for us. Now they can create new forms of life, by jiggling about with genes."[3]

The Environmental Defense Fund and the Natural Resources Defense Council called for public hearings on genetic engineering. The Sierra Club adopted a resolution opposing "the creation of recombinant DNA for any purpose" outside a few carefully regulated government labs.[4]

The mayor of Cambridge, Massachusetts, asked the National Academy of Sciences to investigate whether a "strange, orange-eyed creature" or a "hairy nine foot creature," both allegedly sighted in the area, could be "in any way connected to recombinant DNA experiments."[5] (Mayor Alfred Vellucci, never a friend of Harvard—he once threatened to turn Harvard Yard into a parking lot—said, "It's nice to have freedom of the academia, but I'll be damned if we're going to get over 100,000 people in this city diseased over it.") Others worried about recombinant DNA superviruses being unleashed on crowded airlines. Or what if *E. coli* bacteria that had been genetically modified to produce alcohol for industrial purposes instead invaded the human digestive tracts, where billions of *E. coli* live naturally? We would be a race of drunks.[6]

Motivated by public outcry for government control, the National Institutes of Health issued safety guidelines requiring that some research be conducted in special facilities with air locks, the scientists garbed in outfits that resembled spacesuits. Thirteen bills aimed at regulating recombinant DNA were introduced in Congress.[7] Senator Edward Kennedy called congressional hearings, proclaiming that "Scientists must tell us what they are capable of doing, but we as members of society must decide how it should be or whether it should be applied."[8] Communities in six states, including California, debated bans on recombinant DNA research. Cambridge, Massachusetts, declared a three-month moratorium on some types of recombinant DNA research—a move that forced a leading Harvard scientist, who would later win a Nobel Prize, to move his lab to England, where his team conducted research at a secure biological warfare facility.[9]

In early 1977, demonstrators interrupted a meeting of the National Academy of Sciences. Some chanted, "We will not be cloned," while others blocked the dais with a banner that read, in huge block letters, " 'We will create the perfect race.' Adolf Hitler—1933."[10]

■ ■ ■ ■

Reimers and attorney Bert Rowland passed clippings on recombinant DNA technology—90 percent of them negative—back and forth like baseball cards. By the end of 1977, Reimers's clippings file on recombinant

DNA politics was three inches thick. Meanwhile, the controversy had prompted the Department of Health, Education, and Welfare to launch a years-long review of its policies around institutional patent agreements.[11] One historian has claimed that "recombinant DNA patenting caused so much controversy that it threatened to torpedo federal patent policy."[12] And Reimers still faced objections within his own university. Faculty members continued to complain that the patent overreached, and Stan Cohen had to defend himself against the accusations again and again.[13]

Even in the wake of the turmoil, Reimers could not contain his excitement about the technology. "I kept thinking about the potential for good," he later said. He was particularly encouraged by the thought that new medical treatments might arise from recombinant DNA.[14] He knew that although the Stanford faculty objected, the press hyped public hysteria, and patent attorneys in Washington debated, the scientific community had concluded that recombinant DNA posed no more significant risks than many other types of research.[*] The shift gave him hope.[15]

Planning for a day when Stanford would receive patent rights, he began sketching ideas about companies that might want to acquire a license. A decision to license, he noted, would come with "the potential for both income and controversy." He preferred to focus on the income.[16] Drug companies could use recombinant DNA to develop pharmaceuticals, chemotherapy medications, synthetic hormones, and vaccines that did not rely on killed or weakened viruses that could potentially cause disease. Chemical companies could create microbes to protect plants or leach metal from ores. Food and drink firms might want to use the process to speed up or improve fermentation. By July 1976, Reimers's list of companies to approach about licenses included some giants: Upjohn, Schering, Merck, Pfizer, Lilly, Dow, and DuPont.[17]

■ ■ ■ ■

In a small office in San Francisco, another man saw promise where many sensed only peril. Robert Swanson, twenty-eight years old, considered

---

[*] In 1977, several signatories of the original moratorium letter drafted a new version, never sent, to explain that they now felt less concerned about recombinant DNA.

himself an entrepreneur and venture capitalist, even though he had never
started a company and had just been fired from a four-year-old venture
capital firm, Kleiner & Perkins (today Kleiner Perkins Caufield & Byers).
Swanson was a young man who was always busy with friends and always
in a hurry. His father, an airplane maintenance crew leader, viewed life as
a series of contests to be won. He greeted Swanson after his high school
prom by asking "So, Bob, did you give her a smooch?"[18]

Swanson inherited his father's competitive spirit. As an undergraduate
chemistry major at MIT, he had accelerated his course load and talked his
way into the graduate management program, a move that had gained him
a draft deferment and bachelor's and master's degrees in five years instead
of the typical six. He was only twenty-five when New York–based Citicorp
Venture Capital, where he had worked since his graduation three years
earlier, tapped him to open an office in San Francisco. He stayed less than
a year before leaving to join Kleiner & Perkins.

Swanson's early days at Kleiner & Perkins had been promising. The
firm had been launched in 1972, at nearly the same time as Don Valentine's
Sequoia Capital, which had backed Atari. Like Valentine, Eugene Kleiner
and Tom Perkins had significant operating experience, not financial back-
grounds, and they wanted to take an active role in the companies they
funded.

The gentlemanly Kleiner, with an accent that he called continental
and others compared to Henry Kissinger's, was one of the famous "trai-
torous eight" who had launched Fairchild Semiconductor after working
for Nobel Laureate William Shockley. Kleiner had also been an original
investor in Intel.*

At fourteen, in Vienna, Austria, Kleiner had watched Nazis force his
father to open the family safe before hauling him off to jail. A police officer
who had recognized the senior Kleiner as the owner of the shoe factory
that made boots for the force had saved his life by pulling him out of a line
destined for the concentration camps. The family had fled to Belgium and
on to Spain and Portugal before settling in New York. Kleiner, who had

---

* All eight of the Fairchild cofounders were first-round investors in Intel: Noyce and
Moore as Intel's cofounders and the other six as investors.

never graduated from high school, volunteered for the US Army, explaining that he had "a personal thing to settle" with the Nazis. With help from the GI Bill, he earned an engineering degree from Brooklyn Polytechnic Institute. He also held a master's from New York University.[19]

Tom Perkins, handsome, brash, and flashy at forty-two (a decade younger than Kleiner), swaggered through the partnership's offices with a confidence born of having made Hewlett-Packard's computer division the biggest in the company. At the same time, he had launched and run an independent laser business. Gentility ran deep in Kleiner's veins, but Perkins, the first in his family to attend college, had acquired his polish through careful study, first as a scholarship student at MIT and then at the Harvard Business School, where he had listened to the legendary professor Georges Doriot's lessons about how a businessman should drink (sparingly), rise (develop knowledge and reputation before position and money), work ("worthwhile achievements take more than forty hours a week"), and even marry (a businessman's wife should read and annotate articles for her husband).[20]

Kleiner and Perkins, introduced to each other by a mutual friend, began their partnership with an $8 million investment fund, one of the largest in the young venture capital industry. Half the money came from the Pittsburgh industrialist Henry Hillman and half from other individuals, trust funds, or large institutions such as Aetna and Rockefeller University. Kleiner and Perkins, who had no interest in what Kleiner called the "invest money and hope for the best" approach, established their firm around a set of shared ideals that included an "Investor's Bill of Rights." General partners (Kleiner and Perkins) would receive profits only after the investors' capital had been returned in full. Profits would be returned to the investors, called limited partners, not reinvested in new ventures. Kleiner and Perkins could not invest as individuals in portfolio companies backed by the partnership, and one of the two men had to serve on the board of any company funded by Kleiner & Perkins. Investments would be concentrated in the West, ideally in the San Francisco Bay Area, so the partners could assist and monitor the entrepreneurs.

Eugene Kleiner had handpicked Swanson as an associate after serving with him on the board of a failing company in which both Citicorp and Kleiner & Perkins had invested.[21] Kleiner might well have recognized some version of himself in the younger man; Kleiner, too, had talked his way into an excellent education, convincing the dean of students at Brooklyn Polytechnic to take a chance on him, even without a high school diploma.

As an associate, Swanson worked alongside the senior partners to assess new investments and monitor existing ones. Kleiner & Perkins had invested in a shoe-resoling business (Tred 2, whose founder, Rory Fuerst, would found Keen shoes three decades later), a motorcycle-cum-snowmobile (Snow Job), a medical device company (Andros Analyzers), and a nearly fail-safe minicomputer firm (Tandem Computers).* Shortly before Swanson joined Kleiner & Perkins, the firm moved from 3000 Sand Hill Road, where Kleiner had had a private office before the partnership began, to the twenty-ninth floor of Two Embarcadero Center in San Francisco. The move had been Perkins's idea. Other venture capital firms, such as Don Valentine's Sequoia, were moving onto Sand Hill Road, which today is home to the largest concentration of venture capitalists in the Valley, if not the world. Perkins did not want other venture capitalists to see who he and Kleiner were taking to lunch, and, as he put it, "We didn't want to be considered part of the flock. Eagles don't flock, was our joke."[22] The new offices made for an easy commute for Swanson, who, splitting rent with a roommate, had splurged on a third-floor Pacific Heights apartment with a view of the Golden Gate Bridge.†

Swanson's promising new job at Kleiner & Perkins came to an abrupt end fewer than two years after it began, for reasons that he never quite understood. The triggering event seemed to be his interactions with a portfolio company called Cetus that built sophisticated instruments for biology labs.[23] Eugene Kleiner and Tom Perkins, worried that the company was

---

* Tandem was the company founded by Jimmy Treybig, the former Hewlett-Packard employee whose photo the security guards at HP kept on their desks during the time Kurtzig's small team was working through the night on the HP campus. Jack Loustaunou was Treybig's cofounder.

† The apartment was at 2275 Broadway, and the roommate was Swanson's Sigma Chi fraternity brother (and future Genentech CFO) Fred Middleton. Venture capitalist Brook Byers was Swanson's second roommate.

unfocused, asked Swanson to help Cetus identify products and markets worthy of sustained attention.

During one long brainstorming lunch, Nobel Laureate and University of California professor Donald Glaser, a cofounder of and consultant to the company, mentioned recombinant DNA as an arena that Cetus might explore. No one else in the meeting was interested—not the two other company founders and not Tom Perkins, who had decided to join the lunch to signal how concerned Kleiner & Perkins was.

Swanson, however, was intrigued by the laureate's suggestion. After leaving the meeting, Swanson began reading as much as he could about recombinant DNA. Soon he was excited enough to knock on Perkins's door and report, "This idea is absolutely fantastic! It is revolutionary! It will change the world! It's the most important thing I have ever heard!"[24] Recalls an early colleague of Swanson's: "He could be like a puppy, he was so enthusiastic."[25]

When Perkins suggested that Swanson join Cetus to launch a recombinant DNA effort, the young venture capitalist was thrilled. But the founders of Cetus felt otherwise. Commercializing the technology "wasn't going to happen for a while," they told Swanson, possibly another ten years.[26] Perkins recalls that the response was a bit harsher: "They said they were not interested and certainly not with Swanson."[27]

The rejection disappointed but did not worry Swanson, who assumed that he could stay at Kleiner & Perkins until another opportunity came along. He was wrong. Perkins told him that he and Kleiner "would like to have just the two of us working here." Swanson could stay on the payroll until the end of 1975, but no longer.[28] A bit bewildered by his forced departure, Swanson entertained an unfamiliar thought: "Maybe I'm not very good."[29]

But he was persistent. Several times each week as the employment clock ticked down, Swanson headed south for job interviews in Silicon Valley, zipping down the freeway in his Datsun 240Z, its $110 monthly payments looming large in his mind. He visited Hewlett-Packard's headquarters in the Stanford Industrial Park and the Intel offices in a former orchard, where Mike Markkula had worked. But Swanson, who had neither operational nor engineering experience, was not a natural fit. He even met with a Stanford professor to talk about launching a company around a new invention to concentrate radioactive waste for disposal. Through it all, and as the interviews dried up with the arrival of the winter holidays, he ruminated about recombinant DNA. "God, this seems like important stuff," he thought.[30]

"He wanted to change the world," recalls Brook Byers, who shared an apartment with Swanson and would later become a distinguished venture capitalist with a focus on biotechnology. "Bob would sit around talking about 'How do you want to spend your life?' He was twenty-eight."[31] Swanson was convinced that recombinant DNA could create a new future in which limitless cloning would make rubber, silk, medication, and fertilizer. He did not want to believe that the bacterial factories were a full decade away from commercial markets, as the Cetus executives claimed.

Swanson had never heard of Niels Reimers, Stan Cohen, or Herb Boyer. He knew nothing about molecular biology. But he loved science— he read *Scientific American* cover to cover most months—he needed a job, and he had plenty of time with little to lose. So between cheap meals on the Ping-Pong table that also served as his desk and dining table at home, he cold-called scientists who had attended the Asilomar conference on recombinant DNA almost a year earlier.

"I'm a businessman interested in recombinant DNA," he would begin. Could he ask them a few questions? Some researchers said no. Others offered vague answers to the questions that Swanson considered essential: How long until recombinant DNA could be commercialized? When could the cutting-edge science be used to mass-produce something he could sell? It would be a while, the scientists said. The technology wasn't there yet. The process was untested on a large scale.

By the time he called Herb Boyer, Swanson was officially unemployed. He had no idea that Boyer had coinvented the recombinant DNA process that so excited him. To Swanson's delight, Boyer said that recombinant DNA could probably be commercialized in the next few years. "No other scientist to whom I had spoken had been that bold," Swanson later recalled.[32]

Could I come meet with you to talk about this? Swanson asked.

Boyer demurred. He was very busy.

"I really need and want to talk with you!" Swanson insisted. Here, finally, was a top scientist who thought recombinant DNA might soon have business potential. Swanson had confidence in his own chutzpah, persistence, and business acumen. At the same time, he was a realist. He did not understand the science behind recombinant DNA—and, as his rejected phone calls had made clear, he had no credibility with biologists. Without a scientific expert, Swanson could not have a company.

For his part, Boyer had been privately thinking about commercial applications for recombinant DNA for months, ever since a pediatrician had tested Boyer's young son's growth hormone levels.[33] The boy was fine, but the doctor happened to mention that the hormone was always difficult to obtain. Boyer had thought then that if someone could isolate the gene, it ought to be possible to make vast amounts of human growth hormone using recombinant DNA. "It was a fantasy," he later recalled. "I had no idea how you would start a company."[34]

Boyer was a committed scientist—his Siamese cats were named Watson and Crick[35]—and his willingness even to consider commercial possibilities for his scientific work made him unusual in 1976. Most biologists at the time were suspicious of the corporate world. Arthur D. Levinson, who would eventually serve as CEO of Genentech and a director of both Apple and Google, recalls that as a young biochemist, if he wanted to speak to a company, he would use a pay phone down the street from his lab, rather than risk being overheard by his colleagues.[36] Brook Byers says that a scientist thinking about business "was sort of like when Bob Dylan went electric in the sixties. It was controversial. People were wondering, what's up with that, and will bad come of this?"[37]

Boyer did not have great hopes that the eager young businessman at the other end of the phone line could launch a company around recombinant DNA, but at the same time, he thought, why not talk to him? Starting a company would speed getting the technology he had pioneered to the public sphere. He could work in his lab, and at the same time, this would-be company with Swanson could offer another way to pursue practical applications. Boyer told Swanson that he could give him ten minutes on Friday afternoon.

■ ■ ■ ■

Swanson pulled into a parking spot on the UCSF campus and headed to Boyer's office. Stocky, clean-cut, and balding—a reporter would later characterize Swanson as "not a big man unless standing on his wallet"[38]—Swanson, in his good suit and pocket square, stood out on a campus bustling with casually dressed students and faculty. He also struck quite a contrast with the baby-faced Herb Boyer, who had an abundance of curls and whose sartorial choices leaned in the direction of fringed suede vests.

The pair began chatting in the lab but soon moved to a nearby bar. Three hours and a number of beers later, they had recognized that their strengths and needs were complementary. Swanson knew business but not science; Boyer, the opposite. "I can't be sure in retrospect whether it was my persuasiveness, [Boyer's] enthusiasm, or the effect of the beers," Swanson later said, "but we agreed that night to establish a legal partnership to investigate the commercial feasibility of recombinant DNA technology."[39] As Boyer put it, "You take two naïve people and put them in a room. . . . They just boost each other over the bar."[40]

Boyer, confident that a human hormone would be the logical first product, suggested that Swanson consult an atlas on protein structures to try to find the best one. It took only a little bit of reading for Swanson to home in on insulin, an obvious target for four reasons. First, the market was large: $131 million, he estimated, and likely to grow. The incidence of diabetes in the United States was increasing by 6 percent every year. Second, the existing methods of obtaining insulin for the country's 1.5 million diabetics were inefficient and prone to shortages. Every drop of the hormone was pressed from the pancreas glands of pigs and cows, and it took 8,000 pounds of pancreas gland (bought from meat suppliers such as Armour, Swift, and Oscar Mayer) to produce a pound of insulin. Third, insulin made from recombinant DNA would most likely be safer than the animal-derived product, which caused severe reactions in some people. Finally, using recombinant DNA to make insulin seemed scientifically doable. With fifty-one amino acids, insulin's structure was well understood—a critical starting point.

■ ■ ■ ■

With a partner and a product to target, Swanson asked Kleiner & Perkins to pay him a salary while he tried to start a recombinant DNA company. The founders of Tandem Computers had been granted that sort of entrepreneur-in-residence status while they incubated their company.

But Kleiner and Perkins made it clear that there was no room for Swanson.* "There was something going on between the two of them," Swanson later hypothesized, "and they wanted to work it out on their own."[41] With the exception of Tandem Computers, which looked promising, none of the companies in which the founders had invested was doing well. The fund's internal rate of return was 2.1 percent in 1973; –7.1 percent a year later; and 4.2 percent in 1975.[42] Moreover, Perkins worried that "the quantity and quality of proposals for new businesses is very poor."[43] Kleiner and Perkins had also made some significant mistakes when they launched the firm. They had taken the fund's entire $8 million up front, rather than in tranches (as is now standard), which meant that they needed to find a way to earn interest on it. They had also put the money with an arbitrageur who, Perkins later admitted, had left the fund in a "totally exposed short position of tens of millions of dollars."

Swanson, who by now was living on a $410 monthly unemployment benefit, faced a choice. The venture capitalists were not going to offer him a salaried safety net. His business partner, Herb Boyer, was willing to help, but he was not going to give up his professorship. In short, this recombinant DNA company would launch only if Swanson committed to it fully. That might mean many months without pay, with no guarantee that the company would succeed.

The prospect was frightening, but then he remembered how, when he had worked at Citicorp, the bank had fired two hundred vice presidents, some of whom had been with the company for decades, in a single day.[44] Working for someone else guaranteed nothing. He asked himself how he would feel when he was eighty-five years old if he walked away from this

---

* In his book *Valley Boy*, Tom Perkins never mentioned that Swanson had left the venture firm, much less that he was fired. Instead, Perkins drew a direct analogy between Tandem Computers and Genentech, saying that "Bob [Swanson] enabled us to use the Tandem formula again, of spinning out a venture directly from the partnership." Swanson, by contrast, described a different situation. When asked if he had talked "in some detail" with Kleiner and Perkins before launching the company, Swanson said, "Well, I told them I was going to do this. . . . Not the details." The interviewer followed up: "So you weren't going back to them as mentors and getting advice?" Swanson replied, "No. They had let me go."

chance to start a company. From that perspective, the answer was clear: "If I don't do this, I'm not going to like myself so much."[45]

Swanson's first suggestion as a committed startup CEO was laughably bad: he proposed to Boyer that they combine their first names and call the company Herbob.[46] Boyer countered with Genentech (for "genetic engineering technology").

In four years, Genentech would have one of the biggest initial public offerings in Wall Street history. But in the spring of 1976, Genentech had no lab, no offices, and no scientists. It was just Swanson, Boyer (part-time), and the confident naiveté that they shared. Had they known what was to come—that insulin would prove too complex for a first product; that the first hormone they could synthesize (somatostatin) would have such problems in the earliest stages of its development that Swanson would check himself into the emergency room due to the stress; [*] that Boyer's academic colleagues, appalled by his "selling out to industry," would ostracize him and request an investigation by the Faculty Senate;[47] that even though the investigation would find no wrongdoing, Boyer would experience several agonizing years marked by what he called "a lot of anxieties and bouts of depression";[48] that the company would sue and be sued by its partners— they might not have done it.

Swanson began with a business plan. The first goal was "to engage in the development of unique microorganisms capable of producing products that will significantly better mankind. To manufacture and market those products."[49]

Swanson and Boyer talked about what they wanted the company to feel like for its employees. Swanson wanted to offer a generous stock option plan. Boyer agreed but told him that money alone would not be enough to attract and keep top scientists. Genentech would also need to allow the researchers to publish their findings—a suggestion to which Swanson assented, with the stipulation that patent applications be filed before papers were published.

---

[*] Boyer, much more accustomed than Swanson to lab setbacks, just kept moving forward until the day somatostatin was successfully cloned and he proclaimed, "We played a cruel trick on Mother Nature."

Boyer also suggested that his coinventor Stan Cohen from Stanford join Genentech. Cohen declined. He served as a consultant to Cetus and feared that more substantial corporate involvement would taint his credibility when he spoke about the safety and promise of recombinant DNA.[50]

Swanson began visiting local investors, such as Charles Crocker, a descendant of the railroad tycoon.[51] By the end of March 1976, Swanson—now three months without a paycheck—had a few investors interested in backing him, but not enough. He began making plans to pitch to Kleiner & Perkins. It is unclear whether the firm invited him in or whether Swanson asked for the chance, but he was nervous. He feared that Perkins had a low opinion of him.[52] He needed money for Genentech, however, and he believed that Perkins's savvy and assertiveness would be assets.

On April Fool's Day 1976, Swanson, accompanied by Boyer, sat at the Kleiner & Perkins conference table, no longer a junior associate of the firm but an entrepreneur in search of capital. Kleiner & Perkins had already invested in roughly a dozen small firms, and the one that looked most promising—Tandem Computers, in which the venture firm had invested more than $1 million—was scheduled to ship its first product within the month.[53]

Swanson talked with the venture capitalists about the size of the insulin market and laid out a financing plan. After a brief six-month period during which the company would negotiate licenses from Stanford and the University of California and recruit scientists, Genentech would rent and outfit a lab and hire a microbiologist and two organic chemists to staff it. Eighteen months and a half million dollars later, Genentech would have its "desired microorganism" and could move forward as a "fully operational company" with its own manufacturing and laboratory facilities.

Boyer, using photos, spoke about the science behind recombinant DNA. Kleiner and Perkins, both with limited knowledge of biology, did not know what questions to ask, so they ran through generic, but important, ones: What are you going to do? What equipment will you need? How will you know if you've succeeded? How long will it take?

When Boyer had an answer to every question, Perkins grew excited. "The experiment might not work," he remembers thinking, "but at least they know how to do the experiment." The possibility that the experiment might not work was a problem, however. Swanson wanted more than $500,000 ($2.1 million in 2016 dollars) to hire scientists and outfit a lab for a company that would not launch if the experiment failed. There had to be a cheaper way.

As Perkins saw it, the biggest risk for Genentech was fundamental, what he called a "white-hot risk": the company might not be able to engineer the hormones it proposed to sell. As Perkins memorably asked, "Would God let you make a new life form?"[54] Without an affirmative answer to that question, the venture capitalists were not willing to invest the hundreds of thousands of dollars that Swanson wanted.

At a meeting with Swanson the next day, Perkins offered a suggestion. "We've got to figure out a way to take some of the risk out of it—something instead of me giving you all of the money, then you renting the facility, buying the equipment, and hiring the people," he said.[55] Could they keep the initial costs low by subcontracting the experiments somehow?

A few days later, Swanson came back to Perkins with a new plan developed in consultation with Boyer. Genentech would not hire scientists (or anyone else; Swanson was the company's sole employee for months), nor would it rent and outfit a laboratory. Instead, Genentech, pioneering what today is called a "virtual corporation," would be sparsely staffed. Its budget would go not to salaries or facilities but to sponsoring work at other organizations with specialized skills—in this case, academic research labs identified by Boyer. Today, this sort of outsourcing is common across many industries, with companies hiring contract programmers in India or customer support staff in the Midwest. In 1976, however, the idea was novel.

This virtual model would greatly reduce up-front risk. Only if a sponsored research team managed to engineer a human hormone—thereby affirmatively answering Perkins's question about divine cooperation—would Genentech raise more money (at a higher valuation) and try to move into large-scale production.[56] If the engineering effort did not succeed, the company would fail relatively cheaply and relatively fast.[*]

---

[*] With some modifications, Swanson and Boyer's plan would be the road map that Genentech followed in its early years. In 1976, it raised $100,000 (from K&P) to

Perkins had one more stipulation: he wanted Kleiner & Perkins to be the sole first-round investor. Swanson would need to go back to Crocker and the other interested investors and tell them that they could not be in on the first round. If Swanson did that, Kleiner & Perkins would invest $100,000—among the smallest investments in the venture capitalists' first fund. (Only two of the twelve companies invested in before Genentech had a smaller initial investment.)[57] Swanson, in return, drove a tough deal with the man who had fired him and refused his plea to work as an entrepreneur-in-residence. "He knew all my tricks," Perkins later said of Swanson, but the venture capitalist invested nonetheless.[58] Nine years later, Genentech would be a key reason why Kleiner & Perkins's $8 million fund, which had struggled so much at its launch, would be worth $165 million.[*]

Six days after pitching to Kleiner & Perkins, on April 7, 1976, Swanson and Boyer incorporated Genentech.[59] In June, the check from Kleiner & Perkins arrived. Swanson rented a tiny office in an annex of the Wells Fargo building at the corner of Sutter and Montgomery Streets in San Francisco and hired a part-time secretary. His first capital purchase for Genentech was a filing cabinet.

Boyer would later say of Genentech that he had "never dreamed that the financial rewards would amount to what they did."[60] Swanson, by contrast, dreamed big from the beginning. The second item on his list of corporate goals was "to build a major profitable corporation." A business plan refined in December 1976 promised that the company's products could one day help "to meet the world's food needs or to produce antibodies to fight viral infection. Any product produced by a living organism is essentially within the company's reach."[61]

It was an audacious prediction. But thanks in equal measure to Boyer's

---

make arrangements with universities and research teams; in 1977, the company raised another $850,000 (from K&P and others) to fund research that proved that recombinant DNA could produce a human hormone; in 1978, it raised more money to fund the production of insulin.

[*] The other big winner was Tandem. Although eight of the eighteen deals in the first Kleiner & Perkins fund were written off or marked below cost, the internal rate of return on that fund was 40 percent per annum for the first 9.5 years.

partnership and Swanson's firm conviction that he "had always been a very lucky person," Swanson was convinced that he could use an untested technology as the base for the world's first new pharmaceutical firm in many years.[62] All he needed was a license from Niels Reimers.

■ ■ ■ ■

The same month as Genentech's incorporation, Swanson sent Reimers a letter requesting exclusive rights for Genentech to use the Cohen-Boyer process to produce polypeptide hormones.* In exchange, Genentech offered Stanford 4,000 shares of stock, which Swanson estimated would account for roughly 4 percent of the company by the time it was "self-sustaining." Forever confident, Swanson ended his letter with a flourishing "We look forward to your favorable response" and a blank line on which Reimers could sign his agreement.[63]

For almost any other invention, Reimers would have been happy to sell an exclusive license. He believed that such licenses were particularly important in fields such as pharmaceuticals in which companies must invest large sums of money to develop a technology.[64] No company would make an investment if other companies could just copy the product.

But this case was different. Not only was recombinant DNA important enough to merit wide licensing, he felt, but more practically, two star members of Stanford's faculty, including coinventor Stan Cohen, were consultants to Cetus, a company that Reimers (correctly) believed would one day compete with Genentech. It would be impossible to give an exclusive license to Genentech.[65]

Reimers drove up to San Francisco to deliver the news. He remembers that Swanson "looked like he was fifteen," despite his well-cut suit. Swanson was disappointed but relieved when Reimers made it clear that no one was going to get exclusive rights to the patent. The meeting marked the beginning of a strong relationship between the two men. Swanson, who would occasionally come down to campus, kept Reimers apprised

---

* Swanson was playing a game of hypotheticals: if Stanford received the patent, Swanson wanted Genentech, and only Genentech, to be able to use the Cohen-Boyer process to make insulin and other hormones.

of everything from industry gossip to pending legislation, and Reimers made sure that Genentech was always on the list of companies receiving drafts of Reimers's evolving plans for a nonexclusive license.[66] The two men recognized that they were well matched in their vision for recombinant DNA.

■ ■ ■ ■

Recombinant DNA dominated the bicentennial summer for Reimers. In the space of four months, he spoke with the patent counsel at the Department of Health, Education, and Welfare at least seven times as the agency was deciding whether the rights to the technology could revert to Stanford.[67] After attorney Bert Rowland told Reimers that the federal patent office was not going to allow products made with recombinant DNA under the original patent application, the two men filed a follow-up application.[68]

Reimers needed to provide regular progress updates to Stanford's finance head and a few trustees. The university was in the middle of a $300 million fund-raising campaign, then the largest in the history of higher education.[69] Reimers consulted with the university's vice president for public affairs, who wrote an open letter "to those interested in Recombinant DNA" defending Stanford's move to patent.[70] Reimers fielded phone calls from Stan Cohen, who had decided that, contrary to his earlier requests, he wanted to be kept apprised of the patent's progress. Cohen told Reimers, "The complex scientific and political considerations involving recombinant DNA experiments, the moratorium on potentially biohazardous studies, etc. have made the cloning and the patent the focus of considerable attention. Thus, I am at some risk along with the University."[71] (Cohen also told Reimers that he would find it "perfectly acceptable" if the university called off the patent effort altogether.)[72] Carl Djerassi, a chemistry professor at Stanford and the developer of a key chemical base for the birth control pill, called to say that it would be at least ten years before any sort of commercial use could come from the DNA patent.

The Atomic Energy Commission, claiming that the patent "appears to be useful in the production or utilization of special nuclear material or atomic energy," requested a sworn statement from Reimers that the genetic engineering research had not been contracted with the Department of Energy. Reimers, always thinking of potential licensees, swore that it had not, and then he asked his attorney to let him know if he learned of any potential applications for recombinant DNA in atomic energy.[73]

Meanwhile, Reimers was calling the UC patent office frequently enough that Administrator Josephine Opalka at one point told a colleague not to respond to his calls. ("Let him worry a little bit," she said.)[74] Reimers could have that effect. His sense of urgency was so high, and what he was doing was so novel and controversial, that he often had to chase people down to get them to pay attention to him. "He was kind of looked at as the cantankerous one" by people outside the office, recalls his assistant Sally Hines. "He was right out there all by himself," she says, "and sometimes he was just viewed as a pain in the neck."[75] Someone else who, like Hines, worked for Reimers and respected him, agrees that Reimers had a reputation within Stanford "as a renegade. He could be a gadfly. He would repeat and repeat his position over time. He was a thorn."[76]

Reimers, who found it easier to ask forgiveness than permission, says that he "never asked the boss for a yes or a no for anything. I just did it." He adds, "I had a problem. I didn't know how to work for a boss." At some point, he recalls, a Stanford attorney came to his office and shut the door. He wanted Reimers to stop signing agreements without a prior review from Stanford's legal experts.

"When I think I need legal review, I ask for it," Reimers said amiably. "Your guys have been very helpful to me when I needed it."

"Nobody can sign things on behalf of Stanford unless there is legal review. From now on I want you to file with my office."

Reimers refused. He told the attorney, "If you can find any flaw in any of our agreements, let's meet again and talk about it." Reimers says that he never again heard about the matter.[77]

By the end of 1976, the fees for the Cohen-Boyer patent application were climbing with no guarantee that the university would ever make a penny from the invention. (The patent expenses would reach $300,000, the same amount the office brought in for an entire year for most of the 1970s.)[78] Reimers was worried about his budget. Although the office had patented twenty-one inventions, nine had brought in less than $10,000 in royalties, ten had brought in between $10,000 and $50,000, and two had brought in between $40,000 and $100,000.[79] When a consultant based in

Houston requested reimbursement of one-quarter of his travel expenses, Reimers objected, pointing out that the consultant was already in San Francisco for other reasons.[80]

At the same time that the recombinant DNA patent dominated Reimers's work, many outside of his office saw him not as the project's mastermind but as an administrative functionary who specialized in legal wrangling and licensing arcana. When Reimers tried to offer ideas about how Stanford and UC might one day use royalty money from the recombinant DNA patent—he suggested that the universities might form "a molecular biology institute, where the bulk of royalty income would be plowed back into this field"—a Stanford vice president pointedly told him to "focus on the licensing method."[81]

Every transaction with an inventor, an attorney, a government agency, a company, or the press had to be tracked by the Office of Technology Licensing staff, which included only Reimers, Sally Hines, an associate, and a part-time business school student. To maintain some order, Reimers borrowed the notion of a Friday beer bash from the startup companies that had inspired him. He introduced what he called "Henrys" or, occasionally, "Management by Beer." On Friday afternoons, Hines would bring out popcorn, pretzels, and Reimers's favorite beer, Henry Weinhard's—and the coworkers would together file the reams of paperwork that piled up every week.[82]

That Flips My Switch

█████████████ MIKE MARKKULA

One Saturday in the fall of 1976, Mike Markkula broke his Mondays-only rule for seeing entrepreneurs and drove four miles from his home in the rolling hills west of Highway 280 to a modest ranch house at 2066 Crist Drive in Los Altos. The home belonged to Paul and Clara Jobs, the parents of one of the young men who Don Valentine said were starting a computer company out of the family garage. Markkula did not have particularly high expectations for the meeting.

The garage door was already raised when Markkula arrived. Inside, "the boys," as Markkula would henceforth call Steve Jobs and Steve Wozniak, did not look like any of the entrepreneurs who had come to him for advice during his Monday office hours at his house. Shaggy-haired and scraggly bearded, twenty-one-year-old Jobs and twenty-six-year-old Wozniak were at least a decade younger. They were dressed in Levi's and casual shirts—as if they planned to repair a car, not meet with a millionaire business consultant. Jobs and Wozniak seemed like such unlikely entrepreneurs, and their garage operation seemed so sketchy, that one person who spoke with Jobs shortly before Markkula summed up his encounter in four words: "Sounds flaky. Watch out."[1]

At thirty-four, Markkula was only eight years older than Wozniak and thirteen years older than Jobs. But Markkula had little sympathy for the hippie counterculture that the young men's appearance suggested they embraced. A few years earlier, Intel cofounder Gordon Moore had said that the engineers at big technology companies "are really the revolutionaries in the world today—not the kids with the long hair and beards."[2] Markkula agreed. Early in 2016, he asserted, "I have never protested anything in my life."[3] But, he says of that day in the garage four decades ago, "What was there far overshadowed how they might be dressed."[4]

■ ■ ■ ■

Jobs and Wozniak had transformed the garage into a makeshift manufacturing line for the final assembly of the circuit boards for a computer they were calling the Apple I.* Thus far they had sold 100 boards at $500 each to the Byte Shop, a tiny new store in a strip mall in Mountain View. The Byte Shop would add a keyboard and screen and resell the computer to customers, mostly young white guys whom one early visitor described as "a handful of geeks having technical conversations with each other."[5]

Markkula, stepping carefully among the boxes of parts on the garage's cement floor, had no interest in the Apple I.† The machine required that its users possess not only a monitor, keyboard, and tape drive from which to reload every bit of software every time the machine turned on, but also to have familiarity with hexadecimal code and a soldering iron. Step 2 in the manual for the Apple I was "Type- 0 : A9 b 0 b AA b 20 b EF b EF b FF b E8 b 8A b 4C b 2 b 0 (RET)."[6]

---

* Jobs and Wozniak paid Cramer Electronics, a chip distributor, to supply chips to a company in Santa Clara that would assemble the circuit boards for the computer. Jobs, his friend Dan Kottke, and his sister Patty would plug more chips into the board, after which Wozniak would test and troubleshoot each board.

† Markkula was not impressed with the Apple I, but Apple already had its fans. The earliest fan letter, dated April 14, 1976, is a photo of a television connected to an Apple I. The capital green letters read THIS IS A PHOTOGRAPHIC RECORD OF A STATEMENT IN ENGLISH THE UNDERSIGNED HAD PREVIOUSLY COMPOSED IN 'BASIC' COMPUTER LANGUAGE FOR THE APPLE I, AN EXCEPTIONALLY NICE COMPUTER DESIGNED BY STEPHEN WOZNIAK IN ASSOCIATIONS WITH STEVEN JOBS, THE BRAINS BEHIND THE APPLE COMPUTER COMPANY, PALO ALTO, CALIFORNIA LET US HEAR IT FOR THEM !!!

Markkula was curious, however, about Jobs and Wozniak and their plans for the company. Jobs, tall and lean, said that he and Wozniak had launched their partnership, Apple Computer, on April Fool's Day, 1976.[*] Now Jobs was committed full-time to the business. Wozniak, stocky with thick glasses, had designed the Apple I. He moonlighted for Apple while working full-time as an engineer at Hewlett-Packard. Jobs read books on Eastern philosophy and vegetarianism. Wozniak had spent high school with a pinup on his bedroom wall—of a computer.[7] It was a Data General Nova minicomputer and, at nearly $20,000,[8] no more attainable for Wozniak than Farrah Fawcett-Majors, the actress whose dazzling smile and red swimsuit graced many more high school boys' bedrooms in the 1970s.[†]

Jobs and Wozniak had graduated from Cupertino's Homestead High School five years apart, and both had left college without a degree. Both had worked at Atari, Jobs for Al Alcorn, who had hired him as a technician. Wozniak, who after playing *Pong* in a bowling alley had built his own PG-rated knockoff (miss a shot, and a four-letter word appeared on the screen), had also designed a version of *Breakout* for Atari, working with Jobs straight through the night for four consecutive nights. They had both ended up with mononucleosis, and Wozniak says Jobs never paid him the full amount they had agreed on. The Apple II used a version of the same MOS 6502 microprocessor that Atari had built into its Stella system.

It was also Atari, indirectly, that brought Apple to Markkula's attention. Jobs had asked both Alcorn and Bushnell if they personally, or Atari as a company, would like to invest in Apple; both men, to their later regret, declined. Alcorn accepted an Apple computer, rather than stock, as payment for consulting he would later do for Apple. He did not think the stock would be worth anything. Either Alcorn or Bushnell pointed Jobs to Don Valentine, one of Atari's original backers. Valentine, who also heard about Jobs and Wozniak from the public relations and marketing expert

---

[*] This was the same day, coincidentally, that Bob Swanson and Herb Boyer first pitched Kleiner & Perkins.

[†] More than 5 million copies of the Farrah Fawcett poster sold in 1976 and 1977 alone. When the swimsuit was donated to the Smithsonian Institution in 2011, a curator called the poster "the best-selling poster of all time."

Regis McKenna, was not interested in investing, but he told Markkula about the company.*

When Markkula asked the garage entrepreneurs what else Apple Computer was working on, Wozniak, who was impressed by Markkula's youth, Intel pedigree, and manner—"He didn't talk like a guy who was hiding things and ripping you off"[9]—said that he had designed a new computer. He pointed to a table nearly buried beneath circuit boards, wires, tools, and parts. A keyboard and television were also discernible. As Markkula moved closer, Wozniak sat down and began typing. The screen lit up, and soon a pattern of squares appeared. The image was in color. "Watch," Wozniak said. Now a version of Atari's *Pong* game appeared on the screen. Wozniak demonstrated a few programs that he had written in a version of BASIC that he had built into the computer's memory.[10] Wozniak also showed Markkula how he had designed the computer to be flexible and expandable by including slots for users to plug in peripherals such as printers and cassette players to store data.[11]

Markkula, with his Model 33 Teletype in his home office and his familiarity with several programming languages, considered himself a fairly advanced lay user of computers. But this display astonished him. Small computers like Wozniak's typically could do one thing: display green capital letters on a black background. But here were multiple colors. Graphics. Sound. Games. A built-in programming language. Markkula found it hard to believe that he was seeing these in a computer in some guy's Los Altos garage. Such advanced features were the stuff of machines costing tens of thousands of dollars, built by teams of engineers at some of the world's most famous companies.†

---

* The scenario that seems most likely is that Alcorn told Jobs about Valentine, and then Valentine's name came up again when Jobs visited Regis McKenna to ask for help with marketing materials. Valentine says that it was McKenna who told him about Apple.

† By this time, Cromemco, another startup born of the Homebrew Computer Club (the name came from Crothers Memorial Hall at Stanford, where the founders had lived), had introduced its dazzler graphics card for the Altair, but there was no comparison.

Markkula's amazement only grew when Wozniak cleared the table to reveal the computer's circuit board. The board was the typical bright green, with a nest of wires nearly burying the plastic packages that held chips. But Markkula saw more. This was, after all, the man who had not been able to stop himself from critiquing a customer's circuit design, even though it had lost him a sale; a man who still today, after decades away from engineering work, expresses his highest level of enthusiasm with the phrase "That flips my switch."[12] Markkula knew elegant circuit engineering when he saw it. "It was spectacularly good. It was so clever and so correct," he recalls, growing excited about Wozniak's design even forty years later. "There wasn't a wasted bit anywhere, and he used a 6502 [microprocessor] because that was the cheapest. . . . I mean, it was ingenious what he had done."[13] Wozniak's anticipation of users' needs, even going so far as to think through the power supply for the peripheral devices, also impressed him.

"It was so far along, so far ahead of its time—and exactly what I would have come up with if I was going to do it myself," Markkula says of Wozniak's computer.[14]

Wozniak, thrilled by Markkula's excitement, recalls, "He talked about introducing the computer to regular people in regular homes, doing things at home like keep track of your favorite recipes or balancing your checkbook."[15] Markkula did not know exactly how people would use Wozniak's computer, but he was eager to find out. "I had been waiting for somebody to do a small computer for a long time," he explains.

Wozniak's glorious design—a computer boiled down to a single circuit board with an inexpensive microprocessor, designed to connect to an ordinary television set and with a built-in programming language—meant that the time for a small, affordable computer might have arrived.

■ ■ ■ ■

Steve Wozniak and Steve Jobs were not the only people in Silicon Valley building a personal computer. The Homebrew Computer Club, founded in March 1975, was attracting hundreds of people to evening meetings in a twenty-two-room Victorian mansion in Menlo Park that served as a counterculture-influenced children's day school and had recently gained fame as "Pine Woods Orphanage" after Disney had shot several scenes for

Escape to Witch Mountain on-site.[16] Many Homebrew members wanted to build their own computers, and all of them wanted to learn about the machines. The Homebrew aficionados, mostly men and mostly young, called themselves hobbyists: for them, building a computer fell into the same class of activity as building a model airplane. The fun was as much in the process as in the end product.

In 1975, a small Albuquerque, New Mexico–based company called MITS began selling a build-your-own-computer kit that ran on an Intel 8080 microprocessor. Although quite rudimentary—the computers were programmed by flipping switches, and their output was not text on a screen or printer but a series of flashing lights—the MITS Altair offered the processing power of a $20,000 minicomputer at a cost of only $395 (unassembled). MITS never claimed that the Altair was a personal computer; the company called the machine a "minicomputer kit." As one historian describes it, "The Altair 8800 often did not work when the enthusiast had constructed it; and even if it did work, it did not do anything very useful."[17] At the second meeting of the Homebrew Computer Club, someone programed an Altair to play the Beatles' "Fool on the Hill" through a transistor radio's speakers.[18] That sort of playful innovation was typical.

The Homebrew spirit ran mostly in an idealistic, antiestablishment direction. One of the club's founders, Fred Moore, had served a two-year prison sentence for violating the Selective Service law.[19] Lee Felsenstein, who opened every meeting by welcoming people to "the Homebrew Computer Club, which does not exist," had worked for the radical *Berkeley Barb*, which had advocated the building of People's Park.[20] "Everyone in the Homebrew Computer Club envisioned computers as a benefit to humanity—a tool that would lead to social justice," explains Wozniak, a regular attendee.

"There was a strong feeling that we were subversives. We were subverting the way the giant corporations had run things," recalls another early Homebrew participant. "I was amazed that we could continue to meet without people arriving with bayonets to arrest the lot of us."[21]

Although many members claimed to distrust business, the Homebrew Computer Club, along with similar clubs popping up around the country, was a cauldron of entrepreneurship. Jobs and Wozniak had shown off the

Apple I at a Homebrew meeting, and Wozniak demonstrated the Apple II, the machine that had so impressed Markkula, at Homebrew meetings throughout the computer's development. In January 1976, twenty-one-year-old Bill Gates published his "Open Letter to Hobbyists" in the Homebrew newsletter, excoriating people who were copying software rather than buying it. Members of hobby computing clubs such as Homebrew and the Southern California Computer Society launched roughly a dozen companies, most of which did not look too different from Apple's early incarnation: a technical expert and a would-be business type selling a few computers or peripherals out of their garages or bedrooms.* After meetings, the Homebrew hobbyist-entrepreneurs would adjourn to the Oasis bar on El Camino Real. Seated around deeply carved wooden tables lit by the neon glow of beer advertisements on the walls, the young innovators would help one another with design problems. The race to solutions was intense, the glory of having the first or best idea outweighing any potential concerns about helping a competitor.

Homebrew-style hobbyist-entrepreneurs typically had technical skills, ambition, and drive—but no money, little concern for how less technically proficient people might use the machines, and no idea how to build a business. Meanwhile, a different set of people interested in building small computers had a nearly opposite set of skills and shortcomings. In 1975 and 1976, at one major firm after another, in companies that made semiconductors and companies that made minicomputers, a few employees were trying to convince someone in the operation to build an inexpensive computer for consumers. They found the task almost impossible.† Wozniak had offered the Apple I design to Hewlett-Packard—twice.[22] HP

---

* Among the best known of those companies: IMSAI, Cromemco, Processor Technology Corporation, North Star Computers, and Southwest Technical Products Corporation. Hobbyist-entrepreneurs launched some two hundred companies between 1975 and 1977.

† Among established computer makers, only Commodore made an early move into personal computing, and this only after Wozniak and Jobs showed the Apple II to a Commodore employee, Chuck Peddle. Peddle had also sold Wozniak the MOS Technology 6502 microprocessor when Peddle was working for MOS Technology.

had declined because the machine was not designed to serve its core market, scientists and engineers who needed reliable and preassembled off-the-shelf equipment. HP also worried about quality control. The Apple I connected to a customer-provided television set, and HP worried that the display might work better or worse on certain models.

At Intel, several marketing executives, including Markkula's former boss Ed Gelbach and Bill Davidow, who would go on to become a prominent venture capitalist, tried to convince senior management to sell Intel's microprocessor development systems as personal computers.* Cofounder and president Gordon Moore killed the idea.

At National Semiconductor, Gene Carter, who had shared the closet-sized office with Mike Markkula when they were at Fairchild, tried to persuade National to build small computers. Working with an engineer who would go on to do essential work on the Tandy/Radio Shack TRS-80, Carter had put together a business plan recommending that National create a separate division to design, manufacture, and sell a small computer. The company decided that, at the time, there was no market.[23]

Semiconductor companies such as Intel and National worried that if they began selling personal computers, a product that used their microprocessors, they would at best compete with their own customers and at worst lose money. Intel had already tried selling digital watches built around its chips, and National had done the same with calculators; both efforts had been expensive failures. "The technology is going to move faster than we can be sensible in applying it," one National executive admitted when asked about moving into consumer markets.[24] Large computer companies such as Hewlett-Packard and DEC, meanwhile, were caught in what Harvard Business School professor Clayton Christensen has called the "innovator's dilemma": they had committed resources to an existing market (minicomputers) and could not see, or did not want to see, an emerging market for smaller, cheaper, less powerful machines that could one day destroy the established market. Xerox, where Bob Taylor's group and the

---

* Development systems, designed to make it easy for customers to debug software written for the microprocessor, could be programmed to simulate any number of environments, from controlling a lathe to running a cash register.

systems science lab had designed the accessible, user-friendly Alto, was similarly stymied.

In essence, companies with money and experience did not want to pursue the personal computer business, and the hobbyists who did want to pursue it lacked capital, business knowledge, or both. The retired Mike Markkula, who consulted only on Mondays, had the distance, the experience, and the capital to bridge the divide.

I've Never Seen a Man Type That Fast
⬛⬛⬛⬛⬛ BOB TAYLOR

It had taken years for Bob Taylor to reach this moment in November 1977. The Boca Raton Hotel and Club, steps from the Atlantic, gleamed in front of him on a lovely fall morning. Inside milled some three hundred executives and their wives, flown in from all over the world on first-class tickets to spend four days in the sun at the Xerox World Conference. They had slept in luxury rooms, listened to Henry Kissinger opine on the Soviet Union, and whiled away free hours at all-expenses-paid deep-sea fishing trips, tennis lessons, cocktail parties, and casino nights.[1] Now, on the last morning of the last day, they had assembled for the highlight of the conference: Futures Day, an invitation-only glimpse at "the shape of tomorrow" via a demonstration of the Alto personal computer system.[2]

Every night for the past three evenings, champagne glasses had clinked and orchestras had played. But the days had told a different, uglier story. In 1977, Xerox was in trouble. The World Conference was a last-ditch multimillion-dollar attempt at a morale booster.[3] Three decades of uninterrupted growth had crashed to a halt. In 1975, the company had taken an $84.4 million write-off on the Scientific Data Systems computer company it had bought for $918 million and that Taylor had warned against. At around the same time, Japanese copier companies began encroaching on Xerox's markets, sending Xerox's share price from $179 in 1972 to $50 on the first day of the conference five years later.[4] While their wives attended fashion shows, the executives sat through somber presentations by Xerox's top brass. CEO C. Peter McColough called for "three or four hundred million dollars of expense reductions," while President David Kearns said he did not want the company growing "by one single person" in the next year.[5]

The conference organizing committee, hoping to end on a positive note, had budgeted $220,000 for the final-day look at PARC's computing breakthroughs.[6] For Taylor, the chance to demonstrate PARC's work culminated an effort that he had begun two years earlier. The Alto system by that time included most of the elements that the executives were about to see in Florida, including the graphical user interface, mouse, and network. In Palo Alto, inside PARC, the demand for Altos had reached the point that people were coming in at night to use them. There was even talk of instituting sign-up sheets for each machine.[7] But few at Xerox corporate headquarters seemed to care or notice.[8] Efforts by Taylor's boss, Jerry Elkind, to convince Xerox to reorient its computer strategy away from big machines in favor of "an Alto-like personal computer system" had yielded no result.[*]

Once the Alto system was working well, Taylor decided to close what he, with apologies for military overtones, called an "information systems gap" between PARC and the rest of Xerox. He wanted Xerox employees outside of PARC to use Altos during their regular workdays. "We are all concerned about transfer and the lack of in-place recipients of our systems," he wrote. He wanted a "user experiment" to determine how well the Alto and its software worked when "actually used by people."[9] To that end, in 1976, PARC made a short film about the Alto, which Taylor

---

[*] In 1974, Elkind wrote a remarkable letter to Xerox leadership, drafted with assistance from Taylor, Lampson, Thacker, and Bill Gunning (then head of the systems science lab). The personal computer, Elkind wrote, is "an idea whose time has arrived. Technology now makes it possible to produce practical and affordable personal machines." He also explained that the Alto's microcode "allows us to use the same basic hardware for many different applications in much the same way that Hewlett Packard has been able to provide several different calculator models from the same hardware." He suggested that Xerox could pursue seven possible applications for the Alto ("super calculators" for specific user communities; a custom message system for the military; and systems for accounting, computer research, personnel management, stenotype, and publishing).

narrated while sitting on the corner of a desk, pipe in hand, his tie as wide as his head. The Alto would be transformational, he explained, eliminating much of the "drudgery of office work" and freeing office workers "to attend to higher-level functions so necessary to a human's estimate of his own worth."[10] His push worked. By the time of the Boca Raton conference in 1977, some four hundred Altos had been installed at Xerox.

But none of the new users was a senior executive. The November morning in Boca Raton would be PARC's best chance to introduce the Alto to the men who would determine whether the machine would stay a curiosity within the company or become a real product in the wider world.

Forty-two people, many of them from Taylor's lab at PARC, had been working full-time on a forty-minute Hollywood-meets-Engelbart's-Mother-of-All-Demos demonstration for the conference's final Futures Day presentation. Chuck Geschke from Taylor's lab served as second-in-command to John Ellenby, who was responsible for the effort. Geschke, who would go on to cofound the software giant Adobe with fellow PARC alumnus John Warnock, says he only began to see himself as a manager after Taylor told him that he had potential.[11]

The PARC team commissioned a musical score and hired a professional narrator and lighting designers with movie credits to their names. After a rehearsal on a Hollywood sound stage, they had $1.6 million in equipment—a dozen Altos, five printers, twenty-five keyboards (two of them with Japanese characters), servers, tens of thousands of feet of cable, a small studio's worth of video and multiplexing equipment, pounds of documentation, and dozens of mice, tools, repair parts, and power supplies—loaded onto a pair of DC-10s and flown to Boca Raton. There the PARC group would hold four more rehearsals over three days.[12]

And now, here in Florida, Futures Day had arrived. Ellenby and Geschke, along with their team, were running the final checks and wishing each other luck. The executives were assembling.

There was only one problem with Futures Day from Taylor's perspective: he was not allowed into the room.

■ ■ ■ ■

The Futures Day presentation required a special pass—Xerox had hired extra security to enforce the rule—and Taylor did not have one.[13] The pass was a hostage in a battle of wills between Taylor and George Pake, the head of PARC who had hired Taylor only after telling him that without a PhD, he was not qualified to run the computer science lab. The relationship between the two men had deteriorated to such dysfunction that Pake had not cleared Taylor to receive a pass to Futures Day, and Taylor had refused to ask Pake for one. "I didn't want to give him the opportunity to say no," Taylor recalls.[14]

The larger issue at play was a fundamental disagreement about the nature of PARC itself. Taylor believed that the computing groups were the only worthwhile part of the research center, and Pake thought that Taylor had no idea what he was talking about. In Taylor's early years at PARC, he had believed that Pake would one day "come to his senses" and devote the bulk of PARC's budget to building small, user-friendly, networked computers. Taylor asked at least three different eminent computer scientists to come to Palo Alto and explain to Pake how revolutionary and important the work in the computer science lab was. "He never did understand," Taylor says, adding that Pake put an Alto on his desk only after "he was just sort of intimidated into doing it."[15]

For nearly every PARC budget cycle, Taylor lobbied Pake to take funds away from the other labs and give them to the computer science lab. By Taylor's accounting, the computer science lab was responsible for 80 percent of PARC's accomplishments but received only 18 percent of PARC's budget.[16] When Xerox instituted austerity measures in 1975 and Pake cut an equal amount of funding from each lab's budget, Taylor, who thought that cuts should reflect each individual lab's importance, was enraged.

But Pake wanted to tread carefully. Believing that it was the job of a research center to cast a wide net, he hesitated to eliminate work that might later prove valuable to the company. In 1977, he said that a fear that Xerox might not "receive its money's worth" from PARC had "haunted me since I came out here to start PARC."[17] Taylor, meanwhile, thought the wide-net argument was an excuse to cover up Pake's real reason for not targeting cuts: the mild-mannered director was afraid of angering people by singling out certain projects or labs as more worthy than others.*

---

* Bill Spencer, who did not agree with Taylor on much, similarly described Pake, whom Spencer liked very much, as overly concerned with pleasing everyone.

The battle between Taylor and Pake might have remained a not-atypical managerial push-pull—if Taylor had not openly derided the other research groups at PARC. Taylor reserved particular disdain for the physicists in the general sciences lab.* Pake was a physicist. Taylor enjoyed jokes with punch lines about the real-world stupidity of physicists. During a talk to the general sciences lab, he informed the assembled physicists that one of their wives had told the wife of a researcher in his lab that "a computer scientist isn't really a scientist." He then added, "We've got mathematicians in CSL [the computer science lab] who can run rings around any mathematician you've got"—a startling display of aggression, based on a second-hand rumor, hurled at a group that had invited him to speak.[18]

Every performance review that Taylor received glowed with praise, but many also commented on his behavior toward other labs: "Bob should be more receptive to people from other Xerox groups and to their work" (1972); "He tends to regard other research domains as intrinsically less important. This attitude, expressed both implicitly and explicitly . . . works to Bob's disadvantage because people are consciously wary of him" (1980); "I would very much like to see Bob show a more generous attitude toward the work of others" (1981); "I have urged, and continue to urge, Bob to 'cool it'" (1982).[19] Pake did not write these comments—there was almost always a layer of management between the two men—but he did have to deal with the fallout of Taylor's partisanship, fielding complaints from other labs and wrestling over budgets every year. Taylor says that Pake responded, in typically passive form, by ignoring him. "He came into my office twice in thirteen years."

Even now, at Futures Day, Taylor was causing trouble for Pake. Taylor had spent the past year managing the computer science lab while his boss, Jerry Elkind, was on a temporary assignment with corporate engineering on the East Coast. When Elkind's return was imminent, a few months before Futures Day, the research staff of the computer science lab had begun

---

* Not everyone in the general sciences lab was condemned. Taylor admired the work of the inventor of the laser printer, Gary Starkweather. Chuck Thacker, the Alto's hardware mastermind, says that Taylor's denigration of the physicists at PARC was baseless.

lobbying Pake to find a different job for Elkind and make Taylor the permanent head of the lab. Taylor says he had nothing to do with instigating the coup, which seems to be the case.*

Pake was not eager to give Taylor full charge of the lab. He was wary of Taylor's influence over his researchers, once allegedly likening Taylor to the cult leader Jim Jones, who in 1978 convinced more than nine hundred followers to commit suicide by drinking Kool-Aid laced with cyanide.[20] Pake responded to the researchers' request by removing Taylor from lab management entirely and naming him a member of Pake's own technical staff.

By the time the Boca Raton conference began, Taylor had been reporting to Pake for only a few months, but there were already problems. Pake might well have been plotting the move he would make reluctantly in January, after Futures Day, when he would put Elkind in charge of efforts to introduce the Alto into universities and some government agencies. He would name Taylor head of the computer science lab.

■ ■ ■ ■

But for now, Taylor, not invited to Futures Day, prowled a loading dock behind the Boca Raton hotel, trying to figure out how to break in. The moment marked a new nadir in the Pake/Taylor battle of wills. Taylor knew that he could have walked into Futures Day without an invitation—plenty of people would vouch for him—but that was not his way. If Pake did not want him there officially, he was not going in officially.

He was looking for a back entrance when he spied, near the loading dock, a group of men with camera equipment and lights waiting to enter the building. After a few minutes, the door opened. Taylor followed the workers though. "They went upstairs to a balcony to mount their lights to shine down on the stage," Taylor recalls. "I went up there with them."

---

* Thacker, Lampson, and Elkind agree that Taylor was not involved with the coup, and a May 1977 email also makes it sound as though Taylor had not been consulted about the planned lobbying (though it does say, "I think Bob would like to be invited enthusiastically to be the manager").The email's author, Bob Sproull, wrote that the researchers wanted the change "because Bob is 'good,' and not because Jerry is 'bad'"

Ensconced in the balcony, Taylor crouched behind a spotlight and looked down. He saw Pake sitting with Bert Sutherland, the head of the systems science lab that worked closely with Taylor's. ("Sutherland got an invitation," Taylor noted.) He turned away and waited for the show to begin.[21]

■ ■ ■

The house lights went down. A film appeared on the screen. "Here is our future, the modern office. Our opportunity," a voice intoned as the camera scanned past earth-tone fabric wall art hung over an earth-tone sofa in the PARC lobby. "Beneath the chrome and coordinated colors lurk huge problems, for this office is little changed in generations."

The film went on for several minutes, flashing to statues from Greek antiquity; the Robert Frost quote about the road less taken; and images of a rising sun, paper airplanes morphing into jetliners, scientists looking through scopes, men in jeans sitting in beanbag chairs, and a group of children playing *Pong* around a television set. ("Are today's children telling us something? Are they telling us they are ready for tomorrow?" the narrator asks.)

A new voice then boomed into the hall: "The shape of tomorrow may be here today. Yes, welcome to the all-Xerox office system we call Alto." With that the demonstration began.

The forty-two people in Florida from PARC, along with a separate team back in Palo Alto connected via television, showed how a computer could edit documents, draw bar charts, toggle between software programs, and pull up documents and drawings from stored memory. The presenters manipulated mice, highlighted text on screen, worked with people on other Altos, completed expense forms electronically, forwarded them for processing, typed in foreign characters, sent emails, and printed documents.[22] A narrator assured the viewers, "Does it seem complicated? We can assure you it's not. It is what Xerox calls a *friendly* system. In field trials, an experienced typist became proficient within hours, and even beginners learn in a day or two." Since more than four hundred Altos were in use around the company, the presenters explained to the executives, all of the apparent magic was working for Xerox every day.

It had taken fewer than four years for the system—a complex collection

of hardware, software, networks, printers, and servers—to blossom into
reality from the vision of the man now watching surreptitiously from behind the spotlight.

For the executives who had never used or seen an Alto, the demonstration had to have been eye-opening. Outside of research labs, computers came in two flavors: big and hobbyist. Both were the purview of
specialists. Although Sandy Kurtzig and others were developing software
for people who had never before used computers, minicomputers were
generally tended by experts and used only in very limited ways by non-
experts (responding to prompts on a screen, for example). The new hobbyist machines interested only hackers such as those at the Homebrew
Computer Club, who were happy to flip switches or type in long strings
of characters in order to hear a tinny rendition of a Beatles song played
through a transistor radio.

Few people at PARC, including Taylor, knew about or paid attention
to the Apple II, which had been introduced six months before Xerox's
Boca Raton conference. While the Apple II marked a significant step beyond hobby machines toward a more user-friendly computer, it lacked the
Alto's graphical user interface, mouse, ease of use, and network capabilities. Even five years after the Apple II's introduction, lay users complained
that it took many hours to figure out simply how to begin to use a personal
computer. What does it mean when the screen prompts you to "load a
file"? What is a "boot diskette"?[23] When Fawn Alvarez used a computer—
an Apple II—for the first time at ROLM, she could not figure out how to
turn it on, and once it was on, she was afraid to turn it off, for fear that "it
might lose its information."[24]

The Alto represented a different class of machine. The hobby computers were modeled on the big computers, but the Alto was modeled on the
Licklider-Taylor vision of interactivity and ease of use. The Alto, intended
for use by everyday office workers, was also networked to other computers, which made file storage, email, and printing possible.[25]

Xerox president David Kearns would later call PARC's presentation
at Futures Day a "technological extravaganza," saying that "people told
themselves that they had seen the future of our technology, and it was
impressive."[26]

But when Taylor emerged from his hidey-hole behind the lights and visited the room where a number of Altos had been arranged for use by the executives, he did not see the enthusiasm that Kearns described. After greeting George Pake with a hearty hello that, Taylor says with relish, infuriated him, Taylor noticed that it was the wives, not the Xerox executives, sitting in front of the Altos, typing away and using the mice. The husbands, unimpressed, stood around the perimeter of the room with their arms crossed.* Later someone told Taylor that he had heard one executive ask another what he thought of the demonstration. The second responded, "I've never seen a man type that fast." He had missed the point.

Xerox would go on to try to commercialize a successor to the Alto, so it is not accurate to say that the company had no enthusiasm for the technology presented at Futures Day. But the reaction that Taylor witnessed among the assembled executives, something between indifference and incomprehension, is understandable. Aside from the introduction of the Xerox copier, the fundamental technologies of the business office— electric lights, typewriters, and telephones—had not changed in decades. Why should people embrace this radical change now?

Moreover, most people in Xerox's executive suites focused on copiers that handled information and documents after they had been created. By contrast, the PARC researchers were pushing for a move backward in the process: to the *creation* of information and documents. The computer seemed even more foreign because the California upstarts were insisting that work in the office of the future would be centered on screens. The prediction was ominous for a copier company that made most of its profit by selling paper. Taylor did not know it, but strategic planning sessions among the highest-level executives often debated whether copiers could become obsolete. Some executives wondered "whether PARC technologies were not only speculative, but also potentially subversive to future Xerox profitability."[27]

The PARC team tried to address the concern at Futures Day. Again and again, in the film accompanying the demonstration, the narrator was scarcely able to say something bad about paper. Speaking of "office problems," he offered, "The prime symptom of the problem—Dare we

---

* Chuck Geschke has an identical memory.

say it?—is the medium by which information is transmitted and stored. The villain—Can we face the truth?—is, yes, paper." The narrator assured viewers that the technologies on display represented not a radical break for Xerox but a logical next step. He promised, "A paperless office? Not at all. But paper will have a new value. Paper will have significance. Emphasis." Time-lapse photography showed parts being swapped out of a copier to transform it into a laser printer. "Familiar machines, modified for electronic printing," the narrator soothed over pulsing music.

Given Xerox's fear of the paperless office, perhaps it is not surprising that the one PARC technology that the company would profitably bring to market, the laser printer, is the only one that directly consumed paper.

There may also have been something to Taylor's observation that it was women, not men, in front of the Alto computers after the formal demonstration. In corporate America in the 1970s, typing was the job of secretaries, and secretaries were women. As late as 1980, computer manufacturers worried that the machine might never be adopted in offices because, as one computer marketing manager put it, managers "regard a keyboard as something that doesn't suit their status."[28] The researchers in Taylor's lab all knew how to type because it was impossible to code without typing, but few men in the United States were typists.[*]

Jerry Elkind, the boss Taylor recruited at Pake's insistence, thinks the bias against typing worked against the Alto and its successor machines even after the Boca Raton demonstration. The flagship software program on the Alto was a word processing program called Bravo. (Its writer, Charles Simonyi, would later move to Microsoft, where Bravo became the design influence for Microsoft Word.) Word processing is, of course, a typing-intensive undertaking. Moreover, in the late 1970s and early 1980s,

---

[*] A 1977 *Datamation* article on the "automated office" promised that the "clerical stigma" associated with a keyboard wouldn't be a problem in the future because "In the electronic office, managers will be keyboarding, too." The article, which summarized Citibank's experience with twelve pairs of minicomputers linked over dial-up telephone lines to connect senior managers and their secretaries, noted, "Our secretaries have shown far more flexibility and adventurousness in using the system than our managers have."

a document produced by a word processor was not qualitatively different from one that was typed. It was easier to edit the word-processed document, and some fonts and special characters were more easily available, but in the end, the computer and the typewriter produced similar documents.

By contrast, the "killer app" software that ultimately catapulted the personal computer into commercial markets was the spreadsheet.[29] Spreadsheets required only keying in numbers, something men regularly did on adding machines and desktop calculators. And the spreadsheet provided an enormous benefit over earlier calculation tools: change a single number, and the change would ramify through the entire series of calculations. In the past, changing a value had meant erasing and recalculating everything. Whereas word processing required a big cultural shift (learning to type like a woman) for a small practical advantage, spreadsheets required essentially no cultural shift for a huge payoff. Butler Lampson has said that the people in the PARC lab built the software they needed. They needed a word processor to communicate with one another. They did not need spreadsheets. The Alto optimized the wrong application.

■ ■ ■

By 1977, some researchers at other Xerox-sponsored labs in other parts of the country were jealous of PARC's being held up as the company's great hope.[30] David Kearns, the then president of Xerox, has written of "a brutal clash of cultures . . . the West Coast systems gang pitted against the East Coast copier-duplicator gang." Many of the East Coast executives had come from IBM and Ford, and some saw Taylor's group as "people who spent their time coming up with sophisticated ideas that never made any money for the corporation." Meanwhile, Taylor's group, in Kearns's words, "regarded the copier people as the past, a group of stodgy individuals completely out of touch with the future path of the world."[31] Some executives at Xerox singled Taylor out by name as a key source of friction. "I talked with Bob Taylor often and it was apparent that he had a deep contempt for all persons from Rochester and Stamford," recalled one.[32]

The rift between Taylor and the senior management of Xerox had opened the moment Taylor arrived in Palo Alto and told George Pake that Xerox had bought the wrong computer company. The lab was less than two years old when a Xerox executive visited a Dealer meeting, only to be pelted with questions as to "whether or not people at the corporate level were listening in a responsive way to what we at PARC have to say." (The answer: "a qualified yes.")[33]

The split deepened after *Rolling Stone* published Stewart Brand and Annie Leibovitz's 1972 article that praised the PARC computer scientists' "bent away from hugeness and centrality, toward the small and the personal, toward putting maximum computer power in the hands of every individual who wants it." Brand recalled that the "East Coast headquarters embarked on a major flap about unauthorized information, photos, four-letter words, and the scurrilous *Rolling Stone.*" Back at PARC, Taylor and other employees were subject to new security rules, and Taylor was called out in his annual performance review for the "stumble."[34]

Some people in Taylor's lab smoked pot or used harder drugs outside of work. Some practiced transcendental meditation in their offices.[35] Alan Kay stocked the PARC library with every book from the Whole Earth Truck Store, and another researcher volunteered with a San Francisco–based nonprofit that aimed to establish public computing terminals around the Bay Area.[36]

Yet the computer science lab was far from a hotbed of hippie activity or even beliefs. Xerox did not allow alcohol at its facilities, so even as the workers at Atari were smoking pot on-site and other small companies began holding beer bashes on their campuses, PARC researchers took tea breaks. Taylor and many researchers wore collared shirts and slacks and worked typical business hours. Chuck Thacker calls the core of the lab "very straight folks" and notes that although many of them had been at Berkeley during the Free Speech Movement and People's Park protests, they had worked in unmarked buildings and "tried to stay under the radar because we would not have liked to have our data center burned down."[37] PARC researchers who coded through the night and wore casual clothes did so not because they were making a political statement but because they came from an academic computer science background in which such behavior was the norm.

Many in the lab, Taylor included, disliked top-down, centralized authority. But the roots of personalized interactive computing stretched back much further than Watergate or Vietnam, to people such as Licklider, who were more concerned with making computing useful for the average person than "sticking it to the Man." Xerox PARC was an elite research institution. Most of its staff had very little in common with the hackers at the Homebrew Computer Club.

The differences were hard to perceive, however, peering at PARC not from the San Francisco Bay Area, but from the executive floor of Xerox headquarters three thousand miles away. According to the former head of computing for Xerox, PARC was "the Berkeley campus, but one up." His offered evidence: "People jog. There is tofu in the cafeteria." Though he conceded that at PARC, "the girls are slim and drawn-out, and there are some real nice girls there," he nonetheless deplored the overriding philosophy of the place, which he described as "computer lib."[38]

PARC was a small part of the giant machine that was Xerox. Former president David Kearns's 330-page book about the company devoted only a dozen pages to PARC.[39] That this enfant terrible, brilliant and not ready for polite society, was so often a source of corporate aggravation did not bode well for the computer science lab or for Bob Taylor.

There Are No Standards Yet

MIKE MARKKULA

Even by the time Mike Markkula visited Steve Jobs and Steve Wozniak in the garage in the fall of 1976, Apple was a profitable, albeit very small and very amateur, operation. The circuit boards sold to the Byte Shop for $500 each cost Apple about $220 to assemble.[1] Before Markkula, however, Apple was a business by only the loosest definition. The family bedrooms and garage were rent free. The sales force was Jobs and Wozniak driving around to electronics stores and asking the owners if they wanted to sell Apple computers.[2] The only two people being paid for their labor were Jobs's sister and a friend, Dan Kottke, who earned $1 per board and $4 per hour, respectively, for their work. Jobs and Wozniak had come up with the $666.66 retail price for the Apple I by adding 30 percent to the $500 they were charging the Byte Shop and rounding so that the price would contain repeating digits—something that Wozniak enjoyed seeing.[3]

Markkula's note card commitment had been to help promising entrepreneurs in any way he could one day per week. Standing in Jobs's garage, Markkula knew that Wozniak's Apple II computer was a magnificent answer to the hopes of anyone who had ever longed to own a machine. What he did not know, however, was whether a company could be built around that machine. He gave Jobs and Wozniak the same advice that he had shared with other aspiring entrepreneurs: write a business plan. Figure out your supply costs, the size of the market, the distribution paths.

He thinks he even suggested that since it was impossible to estimate the potential size of a nonexistent market (for personal computers), the number of telephones in US households might provide a good starting point.

Over the coming weeks, as the autumn weather crisped, Jobs (and occasionally Wozniak) would drive to Markkula's new house, a larger home a few blocks from his old one. The young men met Markkula in the small cabana he had built in his backyard near the pool. Wozniak was wowed: "He had a beautiful house in the hills overlooking the lights of Cupertino, this gorgeous view, amazing wife, the whole package."[4] At the end of every meeting, Markkula assigned homework: think through who the competition might be, what a reasonable profit might be, how you might staff the company, how fast you would want it to grow. Each of those factors would form a component of the plan that would tell Jobs and Wozniak if they could build a viable business.[5]

Every meeting, Jobs returned without having done the work.

As the weeks passed, Markkula realized that Jobs and Wozniak were never going to write a business plan. How could they? Wozniak had his job at Hewlett-Packard and no interest in starting a company. Had it been up to him, he would have given away his computer designs or sold them at cost.[6] Jobs was ferociously interested in launching a business, but in the fall of 1976, that meant trying to deliver the boards that the Byte Shop had ordered and then using that income to buy supplies to build more boards. Twenty-one years old and with fifteen months' experience in the corporate world (all of it working for Atari as a technician), Jobs could not have known how to answer the questions and make the estimates that Markkula requested.[7]

The only way Markkula was going to see a business plan, he realized, was to write it himself. He had never done that for the other entrepreneurs he had advised. No one else had risen to that level of importance in his assessment.[8] Writing a business plan for Apple would bleed his commitment beyond his Mondays-only regimen for thinking about business.

But Wozniak and his computer were just too good. Jobs, whom Markkula had already begun thinking of as "a diamond in the rough," was essential, too. "He was a really smart kid," Markkula recalls.[9] Jobs had been the first to see the business possibilities in Woz's original computer, and he had evidenced a tenacious audacity in promoting the machine. For the Apple I, he had convinced a number of chip manufacturers to extend Apple thirty days' credit and then had turned around and demanded immediate payment for the Apple I boards on delivery—the same boot-strapping move that Bushnell had adopted at Atari's launch. Jobs called potential customers multiple times in a row, badgering them until they picked up the phone.

In mid-November, Markkula sat down at his desk to write a business plan for Apple. Typing so quickly that he misspelled "business" as "buis-ness," he defined the company's major objectives and markets, recom-mending that Apple "rifle-shot the hobby market as the 1st stepping stone to the major market." He defined pricing strategies, writing that "the basic computer must be more economical than a dedicated system in specific applications, even though all features of the apple [sic] are not used." He decided that Apple should sell peripheral products (monitors, cassette re-corders for loading software, interface boards for printers and teletypes), as well as computers—and to expect that profits from the peripherals would equal those from computers.

The more Markkula worked, the more intrigued he grew. The com-pany could realistically expect 20 percent pretax profits, he calculated, enough to support an R&D program that would make "significant tech-nological contributions to the home computing industry" and serve as a base for future products. Time to market was essential. "It is extremely im-portant for Apple to be the first recognized leader in the Home Computer marketplace," he wrote. When he calculated that Apple could grow to $500 million in annual sales in ten years—faster than any other company he could think of—he felt a jolt of excitement.[10]

Markkula thought he was writing a business plan to convince some-one else to invest in the company, an effort that had already proven chal-lenging. Not only had Atari and Don Valentine declined, but so, too, had the owner of the first retail computer store in New York City, to whom Jobs had offered 10 percent of the company for $10,000. (As the would-be investor, Stan Veit, later recalled, "Looking at this long-haired hippie and his friends, I thought, 'You would be the last person in the world I would trust with my ten grand.'")[11] Jobs's attempt to interest the computer maker Commodore in Apple had failed as well, though the company soon intro-duced its own rival personal computer. Even the investment banker who would eventually take Apple public, Bill Hambrecht, declined to invest in the first private funding round for the company because the same internal team that had advised him (correctly) that CB radios were a passing fad said the same thing about personal computers.[12]

But Markkula had two advantages that other potential backers lacked: he had seen the Apple II, and he had run the numbers, even if the early plan was, as he later described it, "a view from 50,000 feet."[13] Drafting the business plan for Apple, he experienced that wonderful moment when the desires of his heart and the dicta of his mind meshed. He loved the Apple II, and every calculation he ran said that the computer could be the basis of a wildly successful business.

Markkula could imagine only one significant potential problem for the young company: marketing. People had to be taught to want a computer at a time when only businesses, universities, and governments used the machines. Television and movies typically depicted computers as terrors. In *2001: A Space Odyssey* (1968), a computer kills an astronaut. In *Demon Seed* (1977), an artificial intelligence program builds a robot that impregnates a woman. Just months before Markkula met Jobs and Wozniak, a US senator had asked if it were true that computers "will soon be able to secretly interpret a person's brainwaves."[14]

Steve Jobs had great potential as an evangelist, Markkula could tell. But reaching the size and scale that Markkula anticipated for Apple would require much more than the passion and charisma of a starry-eyed twenty-one-year-old. Apple would need a marketing expert who understood logistics and how to coordinate planning, forecasting, sales, and customer service; someone who could bridge the needs of middle-class suburban families to the tinkerings of Wozniak, Jobs, and the hippies at Homebrew. Markkula knew the perfect person for the job: himself. "I knew there was not another person who had the foggiest idea how to market a personal computer," he says.[15]

Apple's potential excited Markkula, but he did not necessarily want to be excited. After the first thrill at the numbers in his high-level business plan, he had tried to hold himself back. "That's not what you retired for," he reminded himself. [16] He was enjoying his newly relaxed life. "We had finally gotten into a rhythm of having me around the house that was really pleasant," he says.[17]

By contrast, running marketing for Apple would require enormous effort and time. It would mean trying to launch not just a product or a company but also an industry, working with one cofounder who did not want to be a founder—when Markkula told Wozniak he would have to quit Hewlett-Packard to launch Apple full-time, Wozniak almost chose to stay at Hewlett-Packard[*]—and another whose passion for the product regularly outran his diplomacy. Joining Apple meant that Markkula would have to say good-bye to his woodworking and volunteer work, his ski trips and tennis dates and, at least for a while, good-bye to the unfinished items on his note card.

But then he started thinking about what it would mean to launch a company that his plan said would grow faster than any he had ever heard of, a company, moreover, that would be, as he put it, "my own, with my own set of values and my own way of managing." He wanted to do it.

"Gosh, the opportunity here is just too great," he told his wife, Linda. "We've got to try." She had seen how interested he was in the little computer and the two young men. She knew that he could not imagine any other situation that felt so perfectly in sync with his skills and interests; it was as if Apple had a Markkula-shaped hole that needed to be filled for the company to succeed. Markkula was also confident, he told Linda, that even though they were comfortable now, if Apple succeeded as he expected, "it would change our lives economically."

The couple had what Markkula calls "some really serious discussions about whether or not I should go back to work." In the end, she told him that if he wanted to join Apple, he should. They agreed that he would work for no more than four years. He had spent that long at Hughes, Fairchild, and Intel. After four years at each of those companies, he had felt he "had the thing running out just like a sewing machine." It always got boring. He would spend long enough at Apple to launch it well and then return to his quiet retirement.

---

[*] Wozniak recalls that Markkula told him, "You go to a company when you want to turn an idea into money." Wozniak also says that he was ultimately convinced to cofound Apple by a friend who told him that as a cofounder, he could refuse to move into management and instead remain an engineer as long as he wished.

Markkula had started out trying to convince an imaginary investor that Apple's business was sound. Instead, he had convinced himself. He would provide the money Apple needed.

Apple Computer incorporated on January 3, 1977. Jobs and Wozniak each contributed one-half interest (valued at $5,308) in their partnership. In exchange, each received 26 percent of the new company. For an equivalent 26 percent stake, Markkula paid $91,000 (35 cents per share).[18] The investment was substantial—about $400,000 in 2016 dollars—but Markkula, who was wealthy thanks to his Intel stock and investments in more than two dozen wildcat oil wells in Louisiana, estimates that it was less than 10 percent of his net worth.[*]

More important than the direct investment, in Markkula's mind, was his decision to personally guarantee a $250,000 line of credit for Apple from Bank of America, where he had a contact from his work at Intel. Markkula told the banker that Apple would be profitable within the year, at which point the bank should remove Markkula's guarantee. Obtaining a line of credit immediately was unconventional, but Markkula was thinking about the long term. Establishing a banking relationship early on would give Apple the approvals and credit it would later need to grow.

Markkula joined Apple as chairman of the board and head of marketing. Jobs and Wozniak were happy to have him in charge. They had known since the moment they had formed their partnership that they would benefit from the assistance of someone with more business experience.[†] As

---

[*] Markkula bought a condominium at Northstar ski resort in Lake Tahoe from a Louisiana oilman who also invited him to take a quarter-interest in a well that had not yet been drilled. The well hit ("big time," according to Markkula), and he reinvested the money in other wells for a smaller fractional interest.

[†] The original partnership included a third partner, Ron Wayne, whom Jobs knew from Atari. (Jobs and Wozniak each owned 45 percent of the partnership, and Wayne had the other 10 percent.) Wayne was decades older and a stickler for paperwork; at Atari, he wrote the documents specifying how engineering change notices were numbered and how printed circuit boards were identified. For Apple, Wayne pulled together the partnership papers and filed them with the Santa Clara Registry Office. Ten days later, he dissolved his participation. Years before, he had designed a slot machine and gone to work for a company that had promised to market it. The

Jobs later explained (a bit confusingly, since Markkula's money was the only money being offered to Apple), "We didn't necessarily want Markkula's money—we wanted Markkula."[19] Jobs added, "Woz and I decided we'd rather have 50 percent of something than all of nothing."[20] Without Markkula, there would have been no company.

■ ■ ■ ■

Markkula began recruiting. His first offer was to Mike Scott, one of the two men with whom he had shared the closet-sized office at Fairchild. Markkula and Scott shared a birthday, February 11, and every year had lunch to celebrate. Markkula had followed Scott's career since leaving Fairchild for Intel. In 1972, Scott had moved to National Semiconductor to work for the third Fairchild office mate, Gene Carter, as a marketing manager. Scott was then promoted to direct hybrid operations, a job in which he oversaw every aspect of the hybrid circuits business, including manufacturing.*

At their birthday lunch in 1977, Markkula told Scott that he wanted him to run manufacturing at Apple and to serve as the company's president. Blunt, blazingly smart, organized, and uninterested in office politics, Scott was not a glamorous hire. But like Markkula, he was a planner. Markkula was confident that with Scott as president, Apple would never deal with supply shortages, order-processing problems, or other not-flashy interruptions that could hamstring or destroy a rapidly growing company. "Good management doesn't mean coping effectively with problems," Markkula once said. "It means running things so you don't have many serious problems."[21] By that definition, Scott, who once likened a startup to a never-ending chess game where "the challenge is to put together a whole system that works without being minded," was an ideal president.[22]

Scott brought one other asset: he was tough. Markkula took as his

company had failed in what Wayne describes as "a grubby battleground of corporate dirty tricks and greed." Wayne feared the same fate for Apple—not because he distrusted Jobs and Wozniak but because he had doubts about the Byte Shop.

* Hybrid circuits combine digital and analog electronics.

own model Robert Noyce, the aw-shucks Intel cofounder nicknamed "Dr. Nice." Scott modeled his management style on that of Charlie Sporck, the towering National Semiconductor CEO who was both beloved and so emotionally volatile that he had once broken his own wrist banging on a table for emphasis. Noyce and Sporck had made a powerful team at Fairchild (and at Intel, Noyce had found another Sporck-style manager in Andy Grove). There was every reason to expect that Scott would be an excellent complement to Markkula.[23] Moreover, as Markkula puts it, "a key part of the job description [was] to keep Steve [Jobs] off my back,"[24] a task that required a firm hand.* Markkula liked Jobs and also knew that he could be relentless about matters such as the precise shade of tan or the curve of the corners of the Apple II case. Markkula was happy that Jobs, like Markkula himself, cared so much about the details and appearances of Apple's products. But he did not want to manage him.

Instead, Markkula wanted to focus his attention on bigger questions. His first step was to develop an even more detailed business plan for Apple. "To have a real plan, to really build a business, you need to know what parts to order. And who are you going to get them from? How much are they going to cost? What kind of terms?"

To help answer those questions, Markkula hired a consultant, thirty-two-year-old John Hall. A finance expert who had worked at an education startup and helped to draft the financial section of ROLM's business plan, Hall now served as international group controller for Syntex, a Palo Alto–based pharmaceutical company best-known for its pioneering birth control pill.[25] Markkula and Hall, who took a vacation from Syntex, spent a focused two weeks in one of the open spaces at Apple's new offices, talking and planning, with Hall taking notes that he would type on Markkula's IBM Selectric typewriter. Markkula wrote the overview and marketing section, paying special attention to competition, the size of the market, and Apple's target customers: hobbyists, small offices (dentists, attorneys, and the like), and people who might some day want a computer in their

---

* Scott once said of Jobs, "The question between Steve and me was who could be most stubborn, and I was pretty good at that.... He needed to be sat on, and he sure did not like that." Jobs said, "I never yelled at anyone more than I yelled at Scotty."

homes. Hall wrote the finance section, penciling in numbers on long, green worksheets gridded with faint lines. Scott, who had begun helping with the plan even before officially joining Apple, estimated a bill of materials and manufacturing costs. From time to time, Jobs or Wozniak would join the conversation. Jobs, who was particularly interested, provided details of contracts with parts suppliers and also helped Markkula and Hall edit and polish the final draft. Markkula had the final say over the plan's contents.[26]

In the end, the group produced a thirty-page document that laid out strategy and detailed budgets for products, manufacturing, marketing, head count, and finances. Apple's targets seemed aggressive to Hall, who, despite helping to write the plan, found it almost impossible to believe that a company that had sold roughly five hundred computers in six months would be selling $13.5 million worth in two years' time. ("Hell, I didn't have any idea of what a personal computer was or what it was going to be," Hall says. "I was thinking, 'Are people really going to buy this? What are they going to do?'") Hall urged Markkula to be more conservative and plan the financials around the expectation that Apple would achieve only 70 percent of the sales that Markkula was predicting. Markkula refused. "He wanted the goals out there to drive people toward them," Hall says.[*] As a compromise, Markkula finalized the plan with two alternative sets of financial data: one assuming that Apple met its most aggressive growth targets, the other assuming 70 percent.[27] That decision turned out to be wise: Apple's 1978 sales were $7.8 million, about 60 percent of the most aggressive expectations.

After the plan was finished, Markkula asked Hall to join Apple as CFO, but Hall thought the salary was too low and the stock option package

---

[*] Throughout his tenure at Apple, Markkula continued to set goals that one person characterized as "ridiculously aggressive." Employees sometimes joked that the straitlaced Markkula—he calls marijuana "funny cigarettes"—rolled himself a joint and toked up before doing any forecast. "His goals were like somebody saying that we are going to build a ship that can break the speed of light when he hadn't broken the sound barrier," Trip Hawkins, who was at Apple from 1978 to 1982, says. "He was such an optimist, and he wanted to inspire you."

insufficient. He also worried that he would not get along well with Scott over the long term. He asked to receive his $4,000 consulting fee in stock, but Markkula told him that he was saving the stock for employees. "He never made a counteroffer. I always wondered about that," Hall says. He left Apple with cordial handshakes all around and soon began a new job at Intel, later moving to Colorado with a startup called Cadnetix, where he was CFO. He has seen Markkula only twice since, and he never personally invested in Apple, but he has tracked the company's progress with pride, happy knowing, as he puts it, that he "had a little part to do with it."

The business plan called for Apple to target the education market after capturing the hobby market, and in 1977, Markkula was already thinking about how to do so. He thought back a dozen years to when, as a twenty-something engineer at Hughes, he had noticed that the only oscilloscopes in the company were made by Tektronix. When he had asked why, he had learned that most of the scientists and engineers had used Tek scopes as undergraduates and graduate students and had little interest in trying anything different. Markkula wanted to capture that type of loyalty for Apple.

He did in 1979, when Apple launched the Apple Education Foundation, setting aside $250,000 (and later far more money) to support and advance new methods and techniques of learning through the use of small computers. The Apple Education Fund would help teach a generation of students about computers—and ideally yield future sales for Apple. "Upwardly-mobile parents and students in the future are expected to buy . . . the brand that was used in school," a confidential Apple document explained.[28]

Within two years of the Apple Education Foundation's launch, more educational software had been developed for the Apple II than for any other personal computer. In 1982, under a program called "The Kids Can't Wait," spearheaded by Steve Jobs, Apple donated some nine thousand Apple IIs worth $21 million to schools in California.[*] That donation, of

---

[*] In a confidential document, Apple assured its dealers that the Kids Can't Wait donations "are providing California dealers with buying-source contacts and follow-up business." The California donations were a scaled-back version of Jobs's original vision, which died in the US Senate, to give every school in the country a computer in exchange for a tax write-off.

course, led to tax write-offs for Apple, even more educational software
being written for the machine, and more students who would leave for
college and, as Markkula put it, "say 'I want an Apple II'" when it came
time to buy a computer.[29] By 1983, after the foundation had made grants
totaling more than $750,000, 73 percent of high schools and 84 percent of
colleges that owned computers owned Apples. The next highest competitors had only 43 percent and 48 percent penetration.[30]

In 1977, when he was fleshing out Apple's business plan and hiring Mike Scott, even Markkula, the master planner, might not have predicted the success of the Apple Education Foundation. But he already
knew that he needed the mental space to think about those sorts of
long-term endgames. Having Scott in charge of day-to-day management
would free him to undertake the strategic thinking and planning that
Apple needed.

After Jobs and Wozniak interviewed Scott and approved his hiring,
Scott and Markkula agreed to a novel reporting arrangement. Since Markkula was board chairman, Scott reported to him; but since Markkula also
served as vice president of marketing, he reported to Scott, who was president. Markkula thought the intertwined relationships sent an important
message that organizational charts and titles were not going to matter at
Apple. By any other measure, however—who hired whom, who had more
stock (Markkula had four times the amount Scott did), who held ultimate
authority—Markkula was in charge.[31]

One of Markkula's first moves was to have the company buy back as
many Apple I machines as he could find. "There were some reliability issues," he says. "I didn't want to let the company get started on a bad taste,
and so we gobbled up all the Apple Is."[32] That endeavor was part of a larger
effort, spearheaded by Markkula, to make Apple look as large and sophisticated as possible. His biggest concern was that a giant firm such as Texas
Instruments might enter the market before Apple had developed a foothold. "If we weren't big enough, we'd get squished," he says.[33] "It was go for
broke. We had to dominate the business or go bankrupt trying."[34]

To help, Markkula contacted Regis McKenna, the semiconductor veteran whose eponymous marketing and public relations firm had been key
to Intel's successful introduction of the microprocessor. When Jobs had
first approached the McKenna agency about helping to promote the Apple

I, he had been sent off with a referral to a printer who could typeset the brochure and instructions to come back after Apple's first ad, written by Jobs, had run.[35] Now, with Apple's finances more stable, Jobs and Wozniak returned to the McKenna agency's offices to give a formal presentation. When the cofounders arrived toting a small portable color TV, a cassette tape recorder, and a wooden box filled with wires and circuit boards, many at the agency "were collectively mystified," recalled an account executive. "But since these kids were obviously smart, and appeared convinced that this box would revolutionize the world, we listened."[36] McKenna was impressed with Jobs and Wozniak early. His first notes, from even before the company incorporated, highlight the appeal of Apple's origin story— McKenna jotted down that Wozniak had sold his calculator and Jobs his van for seed money. McKenna also pinpointed a practical reason that the company looked promising. "There are no standards as yet, and Apple has opportunity to set standards," he wrote.[37]

In much the same way that McKenna had helped Intel explain what a microprocessor was, so now did his firm turn its attention to helping Apple develop a brand and explain the personal computer. The agency developed a rainbow-striped logo of an apple, the designer crafting it with a bite missing to provide scale, so it was clear that it was an apple and not, say, a cherry.[38] The McKenna agency also produced a glossy brochure printed on heavy stock, with a Red Delicious apple on the front beneath the phrase "Simplicity is the ultimate sophistication." At $10,830, the work was costly, but Markkula placed a high priority on first impressions—so high that Jobs would later say that it was Markkula who taught him to do the same.[39]

■ ■ ■ ■

The quality of Apple's efforts was especially striking given that the event inspiring them was the first West Coast Computer Faire at the Civic Auditorium in San Francisco, a weekend-long conference advertised on mustard yellow hand-illustrated flyers headlined "SAN FRANCISCO BAY AREA—WHERE IT ALL STARTED—FINALLY GETS ITS ACT TOGETHER." But the Faire organizers thought as many as ten thousand people might attend,[40] so Apple paid for a booth one visitor described as "big, and flashy, and right in front."[41] Markkula knew that the Faire would mark not only the

debut of the Apple II and the company itself, but also that of a rival machine: the Commodore PET.

Markkula coached Jobs and Wozniak ahead of the Faire's opening day.
They needed to dress nicely and trim their beards, he said.* He had them
practice their pitches for the Apple II and walked them through what he
saw as the computer's most attractive features. He learned only years later
that despite his attempts to orient the young Apple cofounders to serious
business endeavors, Wozniak had devoted considerable energy during the
Faire to an elaborate prank that involved printing and distributing eight
thousand flyers about a fake computer. Wozniak later reminisced, "Thank
god Steve and Mike didn't find out I'd done this. Mike, at least, would've
said, 'No, don't do pranks. Don't do jokes. They give the wrong image to
the company.' "[42]

By weekend's end, more than twelve thousand people had attended
the Faire, many of them not hobbyists but onlookers curious to know
more about computers. The visitors were an ideal audience for the Apple
II, the only machine on the floor that a reporter described as so easy to use
that "you don't really have to know anything beyond how to hit a switch."
(The Commodore PET, by contrast, was more of an overgrown calculator, with tiny calculatorlike buttons instead of a keyboard and no capacity
to expand to accommodate printers or drives.)[43] When a reporter asked
Markkula, who was staffing the booth on Sunday morning, to describe the
market that Apple was pursuing—"somebody that wants to do programming, play games, or what?"—Markkula replied, "All of the above and
more. We really want to be *the* computer company, not the small-business
computer company or something else—just the personal computer company!" He explained that the Apple II would be ready to ship in a few
weeks and that the $1,298 price (more than $5,100 in 2016) included two
game paddles and a padded carrying case.[44]

■ ■ ■ ■

---

* Even a year after the Faire, Apple had to issue memos telling those on booth duty
  that "The dress code is suit coat and tie, Please. You are representing Apple, and it is
  essential that we convey an image of professionalism. We are in the big leagues and
  must present the proper image."

Douglas Engelbart presenting during the "Mother of All Demos" in San Francisco, December 1968.

Protests at People's Park in Berkeley turn violent, May 1969.

If she can only cook as well as Honeywell can compute.

Her souffles are supreme, her meal planning a challenge? She's what the Honeywell people had in mind when they devised our Kitchen Computer. She'll learn to program it with a cross-reference to her favorite recipes by N-M's own Helen Corbitt. Then by simply pushing a few buttons obtain a complete menu organized around the entrée. And if she pales at reckoning her lunch tab, she can program it to balance the family checkbook. **84A** 10,600.00 complete with two week programming course **84B Fed with Corbitt data:** the original Helen Corbitt cookbook with over 1,000 recipes 5.00 (.75) **84C** Her Potluck, 375 of our famed Zodiac restaurant's best kept secret recipes 3.95 (.75) Epicure **84D Her tabard apron,** one-size, ours alone by Garden House in multi-pastel provincial cotton 28.00 (.90) Trophy Room

Honeywell's $10,600 kitchen computer in the 1969 Neiman-Marcus Christmas Catalog.

Al Alcorn's original
Atari badge.

COMPUTER
SPACE

NA-2010

Flyer for *Computer Space*, the first
video game from Atari. The company
was called Syzygy at the time.

Fawn Alvarez on her first day of work on the ROLM assembly line, Halloween 1975.

Bob Taylor in Palo Alto, enjoying his beloved 1967 Corvette Stingray.

Atari founders Ted Dabney and Nolan Bushnell with Larry Emmons and Al Alcorn.

The Byte Shop in Mountain View, the first store to carry an Apple computer, 1976.

Niels Reimers at
the Stanford Office
of Technology
Licensing, 1974.

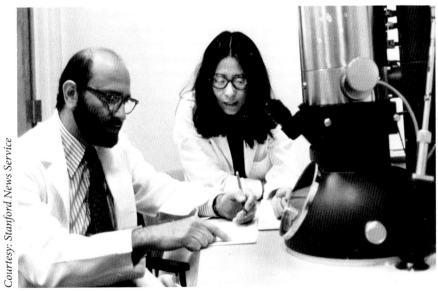

Stanford professor Stanley Cohen and Annie Chang, June 1974. The work in
Cohen's lab and in Herb Boyer's lab at the University of California, San Francisco,
inspired Niels Reimers to file for a patent on the recombinant DNA process on
behalf of the two universities.

Genentech cofounders Herb Boyer and Bob Swanson shortly after they launched the company.

"WE WILL CREATE THE PERFECT RACE"

ADOLF HITLER - 1933

Protestors interrupt a National Academy of Sciences forum in March 1977. The forum discussed the merits and dangers of recombinant DNA research.

The main beanbag meeting room at Xerox PARC. Left to right: Jim Mitchell, Ed Fiala, Terry Roberts, Vikki Parrish, Wes Clark, and Ed Taft.

Children enjoying an Alto computer at PARC. 1977. The machine was never marketed to the public. On the screen is the Smalltalk "Shazam" animation system (done in 1975 by Ron Baecker and using Steve Purcell's 2.5D graphics system) that had been modified by twelve-year-old Susan Hamet.

One morning shortly after the Faire, Gene Carter, the third man who had worked in the tiny Fairchild cubicle with Scott and Markkula ten years before, walked into the lobby at Apple's new headquarters at 20863 Stevens Creek Boulevard in Cupertino.[45] The visit was Carter's second; the first time, Markkula and Scott had given him a tour of the small twenty-by-sixty-foot space. Now Carter walked past Scott's, Jobs's, and Wozniak's offices, which ringed an open area, to Markkula's tidy corner office. It was almost laughable in its sparseness; a white metal desk with a laminate top, four matching metal vertical hanging file drawers, and a small bookshelf. Behind the desk, a hook on the wall for Markkula's coat hung beneath an open copy of the Apple II brochure prepared for the West Coast Computer Faire and tacked to the wall.[46] A coffeemaker stood on a not-particularly-stable-looking table.

"I want to come to work here," Carter said. He had seen the Apple II at the West Coast Computer Faire. The machine and its marketing materials had impressed him.

"Are you crazy?" Markkula asked. Carter had done well at National and did not need to work at a startup. Moreover, Markkula was going to be handling marketing, also Carter's area of expertise.

Carter was undaunted. He had long believed in the promise of small computers. In 1976, after his attempt to convince National Semiconductor to build a small computer had failed, he had quit and asked the venture capitalist Don Valentine (his former boss at both Fairchild and National) for money to start a company. He had come to ask Markkula for a job only because Valentine's response to his request for funding had been to ask, "Well, do you know what your buddies Scott and Markkula are doing?"

Carter handed Markkula a sheet of paper. On it, he had proposed a job and compensation for himself. "I want to run sales," he told Markkula. "Whatever you put out there for marketing, I will follow." Carter had never worked in sales, much less run a sales organization, but he was tired of pulling together marketing plans and materials that a sales force ignored. At Fairchild, he had taken the same classes as the salesmen, learning how to call on customers, manage accounts, and master the basics of the seven-point sale. He figured he knew the process.

"I want ten percent of the company," he added.

Markkula grimaced. "I only paid eight percent for the president. Why would I pay a sales guy ten percent?"*

"Well, I know how much work it's going to be, and that's what I want," Carter said.

Markkula offered Carter much less stock. Carter was adamant: 10 percent or no deal. Markkula looked at his friend. Then he wadded up Carter's proposal and threw it into the trash can near the far corner of his desk.

"I knew he wasn't kidding, and I knew what I wanted," Carter recalled years later. He drove home, disappointed.[47]

Toward the end of lunch with his wife, Carter leaned across the kitchen table. "I *really* want to do this," he said. Even if he could not get as much stock as he hoped for, he wanted to work for Apple.

If he felt that way, she said, he should tell Mike. Carter climbed back into his car and drove again to Apple. He walked into Markkula's office. "You've got a deal," he said.

Markkula told Carter to start in August, when Apple would have enough money to pay him. Carter countered that he would begin work immediately—for no pay. "I wanted to be sure that someone else didn't get *my* job," he explains. "I just showed up every day and worked. I copied tapes. I tested systems. I wrote datasheets, whatever it took." He worked for three months before receiving his badge and first paycheck in August 1977.

Scott and Carter, along with Markkula—three men in their mid- to late thirties who had spent the past dozen years as middle managers in semiconductor companies—formed the core of Apple's earliest business operations. Joining them was a CFO, Ken Zerbe, hired from another semiconductor firm, American Microsystems. Most histories of Apple focus on the young, long-haired, Levi's-wearing technologists who worked with Jobs and Wozniak and looked so different from the typical buttoned-up,

---

* According to Apple's IPO prospectus, Scott purchased 1.28 million shares of stock at a penny per share a few months after he was hired. (He later purchased another 1.92 million shares at 9 cents each.) Gene Carter purchased 160,000 options at 9 cents per share a few months after his hire; he also held another 620,000 shares at the IPO.

IBM-type technology worker of the 1970s.[*] But in many ways, Markkula's task was to make Apple an exemplary technology company according to the "typical" measures. It needed to be profitable, manufacture reliable products in predictable volumes, and deliver them through stable sales distribution channels. Markkula looked to managers whom he knew and trusted in the semiconductor industry to make that happen. "We all knew the semiconductor business inside out and backwards," he explains.[48] They had already worked through manufacturing disasters, supply problems, failed processes, and design glitches.

They had no worries about hurting one another's feelings or sugar-coating their opinions. What mattered was moving forward. "There was always a sense of urgency from all of us," Carter says. "We had enough common ground or enough business friendship that we were able to scream at one another and get away with it, close the door and just scream at one another."[49]

Although the team Markkula assembled in Apple's first months was experienced, each man was new to his specific job. Markkula, who had never served on a corporate board, was board chairman. President Mike Scott had never run a company; as he explained years later, "I approached Apple as a chance to see if all the things I had learned about management would actually work."[50] Sales head Gene Carter had never worked in sales.[51] Cofounder Steve Jobs, the vice president of administration, had only administered a company (Apple) with fewer workers than could be counted on one hand. The only people who seemed obvious choices for their jobs were Steve Wozniak, who ran engineering, and Mike Scott, in his capacity as manufacturing head.

But it worked. Occasionally lost in the stories of Jobs's bare feet and Wozniak's practical joking is the reality of how hard and effectively the young men and their older semiconductor colleagues worked together. The birth of Apple's floppy disk drive almost exactly a year after the company incorporated, and six months after it began shipping Apple IIs, offers

---

[*] Rod Holt, an outstanding Atari engineer (and eventual Apple Fellow) brought in by Jobs to do the switching power supply, was older and more experienced than Jobs or Wozniak.

a fine illustration. The Apple II was selling well, but Markkula was convinced it would never make the leap to a broad consumer market unless
there were some way to make it load software faster. Asked to recall his
early strategy for the consumer market, Markkula said, "Software, software, software." The company was only ten months old when it formed
a formal "software bank" with plans to encourage users to develop programs by awarding them gift certificates for Apple products.[52] Without
easy-to-use, ready-to-load software, the computer had no value for most
people. By the fall of 1977, Apple had about two dozen software programs,
mostly games and a few educational programs, in the works. But loading
even something as simple as a game of Hangman from the Apple II's cassette drive could take many minutes. A hobbyist might be willing to wait
that long, but most consumers would not.

In December 1977, Markkula told Wozniak and Jobs that he wanted a
floppy disk drive for the Apple II. IBM had invented the floppy disk drive,
about one hundred times faster than cassette drives, six years before at its
San Jose research center. Markkula knew that eight-inch disks and drives
were in use in some specialized IBM products and a few Altair-style machines.

Wozniak had never used a floppy disk, but he was confident he could
build a drive.[53] He asked Markkula: If the disk drive were ready in time,
would Apple pay for him to fly to Las Vegas for the Consumer Electronics
show in January? Although computers were so new that the show planners
did not have a special area designated for them and put Apple's booth with
the digital watch manufacturers, Wozniak did not mind.[54] He had never
seen Las Vegas. Markkula agreed that if the drive was ready, Apple would
pay his way.

Although Markkula set the objective, it was Jobs who knew how to
jump-start Wozniak's work. Much like Nolan Bushnell with Al Alcorn at
Atari, Jobs could channel Wozniak's genius in a direction that would interest a wide range of people. Apple was Wozniak's second startup. Years
before, for a partnership called Computer Conversor that he had started
with a man named Alex Kamradt, Wozniak had built a cheap teletype system to connect to the Arpanet. The business had failed in part because
once Wozniak had built a terminal that worked for his purposes (which
meant that he knew how to fix it every time it broke), he wanted to move

on. Kamradt could not convince him to continue to work on the product until it stopped breaking. "The genius is nothing unless you can get it out of him," Kamradt later explained. "I couldn't."[55] Jobs could. It had been Jobs who suggested certain technical improvements to the Apple II, and when Wozniak had said that implementing those improvements would be too expensive, it was Jobs who cold-called chip distributors and convinced them to give him a few chips for free so that Wozniak could do the work.[*]

Jobs brought Wozniak one of the newest disk drives, made by a manufacturer called Shugart Associates.[†] Jobs also gave Wozniak manuals and schematics, knowing that he would want to understand everything about disk drives.

In a feat of engineering as impressive as his Apple II—Wozniak calls it "my most incredible experience at Apple and the finest job I did"[56]—Wozniak, working with a high school student named Randy Wigginton, designed and built a prototype of a functional 5¼-inch disk drive for the Apple II.[57] In only two weeks, he designed a controller with about one-tenth the typical number of chips, moving much of the work previously done by dozens of chips into a few lines of software.[58] The accomplishment earned Wozniak his first trip to Las Vegas, Wigginton by his side.

Jobs took Wozniak's design to Shugart Associates, where one executive recalls his surprise upon learning that his next meeting was with "this guy . . . with holes in both knees [and] the most dark, intense eyes." Shugart became Apple's original supplier of parts for the drives.[59]

Markkula, meanwhile, decided to price the drive at only 10 percent over cost, even though the market was bearing significantly higher prices.[60] No longer was he anticipating, as he had in the company's first

---

[*] Wozniak on Jobs: "One day he asked me, 'Why don't you use these new 16-pin dynamic RAMs?' I had looked at them in my work at Hewlett-Packard, but they were new and I couldn't afford any parts that didn't come my way almost free. I'm a little bit shy, and I didn't know any of the reps, but Steve just called them up and talked them into giving us samples. I jumped on it. I thought it was a great part because you could replace 32 chips on a board with just eight."

[†] Al Shugart, the company's founder, had invented the floppy disk drive while at IBM. Xerox acquired Shugart Associates in 1977.

business plan, that peripheral devices would be an important source of revenue for Apple. "I didn't care if I made a profit on that damn thing or not. I just wanted everybody to have one because it was the key to doing something worthwhile," he says. "An Apple II without Disk II [Apple's disk drive] was useless."[61] On the day the drives were to ship, Markkula had every employee report to the warehouse, where they, at times joined by family members, worked late into the night to pack the devices.[62]

Wozniak was the inventor, but Jobs was the link between Wozniak's creations and the outside world. Only a year after Apple launched, at a trade show with hundreds of vendors, Jobs caught the attention of a reporter from *The New Yorker*, who quoted his promise that "for the first time, people can actually buy a computer for the price of a good stereo, interact with it, and find out what it is all about." He claimed that Apple was "the largest personal computer company in the world."[63] Markkula was another layer around Jobs and Wozniak, reading the broader market and devising how to sell to it—skills that Jobs would later develop to dazzling effect.

At the same time, Carter was building a sales team and a distribution network.[64] He refused to hire sales reps from IBM "because they never had to fight for an order." He wanted salesmen who already "knew what it was like to be the underdog."[65] Many of the 180 authorized Apple dealers in 1978, all of whom would service the computer in addition to selling it, were entrepreneurs themselves—former electronics representatives so new to the retail business that Carter developed a newsletter and step-by-step "Dealer Start-Up Cookbook" for them.* The cookbook included everything from model business plans, to forms to certify items in stock, to basic advice such as running demonstrations in color rather than black and white ("The Sizzle Sells the Steak"). The cookbook also recommended using double-sided tape on the back of posters, rather than splaying a strip across the corners that customers could see.

"Don't be afraid to ask for an order or commitment!" Carter exhorted,

---

* Carter, who estimates that many of the dealers had a net worth of less than $50,000, worked with Apple's CFO, Ken Zerbe, to approve provisional shipments to dealers with questionable credit history.

while also urging "technical types" to refrain from "point[ing] out the limitations of the machine rather than highlighting the benefits." He asked, "Has a car dealer ever told you your car may be recalled because of some defect? Or that it won't go 0 to 60 mph in 5.6 seconds?" Of course not. "Just remember that your customer has no basis for judging, generally, and will accept your criticism as a reason not to buy. The limitation for you may well be something he will never be able to use or want to use. He looks to you for professional advice."[66]

Across the country there were nearly four hundred dedicated computer stores, with names such as Byte Shop, Micro Store, Computer Mart, and Digital Deli. Another four hundred stores, ranging from stereo and TV repair shops to small department stores, included among their electronics offerings small computers made by Apple, Commodore, Tandy, IMSAI, and others. Entrepreneurs were willing to try selling computers, but major retailers were not yet ready. Sears announced that it would sell computers and then backed away. Macy's had intended sales for Christmas 1977 but changed plans when the demo model did not work.[67]

As the small computers became more popular, top computer scientists tried to determine how the machines they called "small-scale gadgets" or "very small things" related to their own work at universities and research centers. "What's going on doesn't fit the type of computer science and computer technology I'm familiar with," admitted Bell Labs scientist V. A. Vyssotsky in 1978. "I don't understand what's going on out there, and I think it's important. I suspect that it will get away and become a new technology on its own unless we can understand it and come to terms with it." Ed Feigenbaum of Stanford said that he did not know how to help personal computer makers avoid the mistakes that the academics had already made. "It's really a critical problem because that hobbyist market is reinventing everything bad along with all the good things they are doing," he said.[68]

■ ■ ■ ■

Back at Apple, Mike Scott oversaw production. Anticipating rapid growth, he installed computerized management information and order-processing systems geared to a much larger organization, a move that Markkula applauded.[69] Scott also developed a system of standard production modules that could be moved and duplicated within and across various facilities, making it easy to grow or modify a production line very quickly.[70] (Apple assembled its computers from chips, boards, cases and other parts, mostly from suppliers who built them to the company's specifications.) Scott also wrote the in-house reference manual for the Apple II, the "red book," so called for its fire engine red cover.[*] He orchestrated two moves in rapid succession, first to the small offices on Stevens Creek Boulevard and from there, a few months later, to a twenty-thousand-square-foot facility at 10260 Bandley Drive in Cupertino. The Bandley building, built for Apple, was split into four quadrants: engineering, administration/marketing, manufacturing, and a large empty space that a drawing of the floor plan from January 1978 jokingly labeled "tennis courts?" Within two months, the would-be courts were warehouse space. Four months later, Apple had to rent two more buildings for its growing operations.[71]

Markkula wrote one of Apple's first software programs, a $20 package to balance a checkbook called, fittingly enough, Checkbook 1977. (Never eager for attention, he wrote it under the pseudonym "Johnny Appleseed.") He also played an almost paternal role for several of Apple's young employees. Electronic Arts founder Trip Hawkins, who came to Apple in 1978 and was another of Markkula's "diamonds in the rough," says, "Scotty, with his get-out-of-my-way style, had an operations-based, hard-ass mind-set. He crisply executed every day. But Mike was good at encouraging, at keeping Steve [Jobs] and all of us going and believing. He conveyed the feeling that we were important, that the work was important, that he believed in us." Hawkins recalls that a few months after joining Apple, he visited a venture capitalist to propose launching his own computer business. When he got back to Apple after the meeting, Markkula met him at the door. "What are you doing, Trip?" he asked. "It's really inappropriate that you did that. You're going to have a great career at Apple. Now get back to work." Hawkins returned to his office, stunned. "How the hell did he find out that we were even doing this?" he wondered. "And now he is promising me a great career here?" Decades later, he asks a rhetorical question: "How can you not be motivated by somebody like that?"[72]

---

[*] Chris Espinosa, while a freshman in college, would soon use the red book as the basis of Apple's first published technical manual.

As Apple neared the end of its first year of operation, much of Markkula's work focused on raising money and building a board of directors. Markkula wanted expert company builders on Apple's board, and the only way to get them was to allow them to invest. By fall of 1977, Apple was no longer the questionable operation with two seedy-looking guys in a garage. The Apple II had begun shipping in May, and already the company was profitable. In 1977, Apple had $49,000 in net retained earnings on $756,000 revenues and robust European sales that accounted for nearly 25 percent of the company's profits.[73] The company had twenty-four employees, a business plan spun into a seventy-page private offering memorandum, and sales on track to grow tenfold in the next year—as well as first-mover advantage in a personal computer market that Apple predicted would reach $290 million by 1980.[74] (The market would, in fact, be more than twice that size.)[75] Markkula's leadership was a strong draw, as well. Peter O. Crisp, the managing partner of an early investor, Venrock (the Rockefeller family's venture capital firm that was also a first-round investor in Intel), says that all Venrock knew about Apple "was that they were honest, and Markkula recommended them."[76]

In the fall of 1977, Apple sold 150,000 shares at $3 each, most of them to people Markkula knew from the semiconductor industry and wanted on the board. Arthur Rock, the venture capitalist behind Fairchild, Intel, Teledyne, and Scientific Data Systems, had been impressed by Markkula at Intel board meetings. Rock invested in Apple and advised the company informally for three years before joining the board.[77] The venture capitalist Don Valentine, who had rejected Jobs's original request for money, now invested and joined the board. So, too, did Hank Smith, the same Hank Smith whom Markkula had hired years before to write the technical documentation for Intel's microprocessor and who now worked for Venrock. Jobs and Scott were also on the board. At one memorable meeting, when Jobs removed his shoes, Markkula rebuked him with a stern "You are excused until you can behave like a board member." Jobs, who had not spent much time in boardrooms, put his shoes back on.[78] Atari might hold meetings in hot tubs, but a company run by Mike Markkula followed *Robert's Rules of Order*.

■ ■ ■ ■

The memorandum for the private offering in 1977 offers a wonderful glimpse into Markkula's mind as Apple was just beginning to accelerate. An early draft listed fifteen "benefits to be obtained from operating [one's] own, personal, home computer," including "personal pleasure and enjoyment," "reduced pollution," "better financial decisions," "more free time for any purpose, leisure or profit," "better educational opportunity," entertainment, fire and theft protection, and "personal comfort," as well as an "improved standard of living."[79] The document reviewed small computers recently introduced by Commodore and Tandy/Radio Shack that sold for roughly one-third the price of an Apple II. The memorandum also warned that even more formidable competitors—Atari, Texas Instruments, and RCA—would likely enter the market within a year. One line from a draft of the offering memo that did not survive the editing process referenced "Apple's overall strategy for attacking (developing?)" the personal computer market.[80]

The memorandum also captured the challenge that Apple faced at the end of 1977 and the reason why Markkula attached so much importance to marketing. "Apple's limited experience with existing products indicates that a *minimum* of two hours of one-on-one discussion is needed to convince 'Joe Average' that he needs a computer," Markkula wrote. "Due to a general lack of knowledge of the benefits offered by the computer, most potential customers of 1980 do not have even the slightest desire to purchase one today."[81]

TRIUMPH
1979–1981

"The 1980s are going to be the Golden Age of Electronics." So predicted Ben Rosen, an analyst Fortune *called "personal computing's greatest visionary power broker," in 1979.* * *"The Energetic, Exuberant, Exciting, Electronic Eighties," Rosen averred, would bring a second industrial revolution, one based on technical breakthroughs that would increase productivity and raise living standards.[1] Another argument for the "smokeless" high-tech industries, persuasive at the time but later proven tragically false, was that the new industries were nonpolluting.*

The promise of a bright electronics-based future offered a welcome contrast to the reality facing the United States in 1979. American hostages had been kidnapped, blindfolded, and paraded in front of television cameras in Iran. The United States was facing double-digit inflation. Some 1.5 million American manufacturing workers had lost their jobs in the past decade, and Japanese competition had decimated the US steel, automobile, and television industries.[2]

Electronics and Silicon Valley came to represent a new era in which Americans could once again dominate. Silicon Valley had added 50,000 new

---

* Rosen would soon act on his optimism. He cofounded Sevin Rosen Funds in 1981 and went on to invest in Citrix, Compaq Computer, Cypress Semiconductor, Electronic Arts, Lotus Development Corporation, and Silicon Graphics.

jobs in each of the past three years. The region now accounted for one out of every five new jobs in the United States.[3] High-technology employment in the San Francisco Bay Area grew by 77 percent between 1974 and 1980, and per capita personal income growth in Santa Clara County outpaced that of the rest of California by more than 10 percent.[4] With unemployment in the Valley hovering around 4 percent, the classified advertising section of the San Jose Mercury News swelled to ninety-three pages.[5] The Wall Street Journal wrote of a "latter day Gold Rush."[6]

New industry groups such as the National Venture Capital Association (founded in 1973) and the Semiconductor Industry Association (founded in 1977) sent some of Silicon Valley's most recognizable entrepreneurs to Washington to testify about the importance of venture capital, microchips, biotechnology, personal computers, and software.* The lobbying efforts paid off. In 1978, Congress slashed the capital gains rate from 49 percent to 28 percent and eased the "prudent man" rules that had restricted pension funds' abilities to invest with venture capitalists.† Within a year, venture capitalists had more than a half-billion new dollars to invest, and pension funds, universities, and other institutions across the country had begun experimenting with placing small amounts of money with venture capital firms.[7]

The Wall Street Journal editorial board wrote that a "counter-revolution taking place in American politics" was leaving big business "powerless."[8]

---

* The suggestion to send entrepreneurs, rather than paid lobbyists, came from Regis McKenna. Among those who went to Washington were Andy Grove, Steve Jobs, Robert Noyce, Tom Perkins, and Bob Swanson. The first electronics industry association was the American Electronics Association (originally known as the West Coast Electronic Manufacturers Association), founded by David Packard in 1943.

† Key individuals who worked behind the scenes to promote the capital gains tax reduction and "prudent man" revisions included William F. Ballhaus of Beckman Instruments, Roger Kennedy of the Ford Foundation, and venture capitalists David Morgenthaler and Reid Dennis. Ed Zschau, an entrepreneur who led the capital gains lobbying efforts on behalf of the AEA, went on to serve in Congress as the Republican representative from California's Twelfth District from 1983 to 1987. In 1986, Silicon Valley further proved its political influence as the United States imposed 100 percent tariffs on $300 million worth of imports from Japanese microchip firms.

California governor Jerry Brown established and chaired a California Commission on Industrial Innovation; members included David Packard, Steve Jobs, and Charlie Sporck of National Semiconductor, as well as the dean of the Stanford business school.[9] California's 1982 budget supported establishment of computer centers and a software clearinghouse, and also set aside $25 million for math, science, and technical education. In Washington, some Democrats, led by Gary Hart, Tim Wirth, and Michael Dukakis, were such outspoken advocates for entrepreneurial high-tech industry that they were collectively known as "Atari Democrats." For the first time in decades, the White House hosted a conference on small business. President Reagan established a Commission on Industrial Competitiveness chaired by Hewlett-Packard chairman John Young.

The groundwork had been laid for today's tight connections between Silicon Valley and Washington.[10] The Silicon Valley story—a tale of risk-taking upstarts whose willingness to explore new territories brings progress and huge financial rewards—had updated a beloved American story for the high-tech era.

"Every time I drive down and think about what's happening in this valley," said one San Francisco venture capitalist in 1980, "I have an orgasm."[11]

Looks Like $100 Million to Me!

NIELS REIMERS AND BOB SWANSON

Venture capitalist Tom Perkins was trying to convince Genentech co-founders Bob Swanson and Herb Boyer to take the company public. Three-year-old Genentech had successfully used the recombinant DNA technique to synthesize human insulin and now supplied small batches of the hormone to large pharmaceutical companies such as Eli Lilly. These companies would manufacture the hormone in volume and sell it.

Genentech was profitable, though barely. It employed more than one hundred full-time employees (more than one-third of them with PhDs), who worked in a warehouse in South San Francisco. Rents were low and activists scarce in the self-proclaimed "Industrial City." For a while, Genentech's neighbor was a distributor of soft porn.

Perkins, Swanson, and Boyer agreed that Genentech needed cash to build manufacturing facilities. The company's tepid efforts to raise money by selling itself to Lilly or Johnson & Johnson had failed. Perkins believed that a public offering would give Genentech the opportunity to set the template for this new industry, recently christened "biotech."[*] A successful IPO would also provide a profitable exit for his venture capital firm.

---

[*] Fred Middleton says that the word "biotech" was coined by E. F. Hutton analyst Nelson Schneider for a conference in October 1979. Earlier, people had spoken of "genetic engineering" or "gene splicing."

The industry had not entirely shaken the mad-scientist reputation that had shadowed its earliest years, but overall, the public had become less fearful and more admiring of biotechnology. Genentech had hired Regis McKenna to help explain and promote the new technology, and whether due to his influence or simply the passage of time, the company had been the subject of glowing stories in publications that included *Time*, *BusinessWeek*, the *Wall Street Journal*, and the *New York Times*. Headlines proclaimed, " 'Astonishing' Report on Gene Research," "Gene-Splicing Field Is Swiftly Approaching the Commercial Stage," and "The Bold Entrepreneurs of Gene Engineering."[1] A few large companies that could imagine using genetic engineering in their products had even invested in several young biotech companies.[*]

Perkins believed that Genentech had everything it needed to go public: profits, a product, big-name customers, and public approval. Moreover, he warned Swanson and Boyer, if another, less conservatively managed company went public first, it could dim Genentech's prospects later. Genentech needed to move quickly to define for the public what it meant to be a biotech company.

Swanson disagreed. He thought that Genentech was not ready. Profits were tiny—only slightly more than $100,000 on roughly $3 million in revenues—and came not from selling products to patients or hospitals but from supplier agreements with the big pharmaceutical companies. Swanson did not believe that a barely profitable business without product revenue was a good candidate for a public offering. It did not help that the University of California was suing Genentech for alleged theft of genetic research materials.[2][†] Moreover, the US Supreme Court had not yet determined if novel life-forms—the heart of Genentech's intellectual property and products—could be protected with patents.[‡] There was too much uncertainty.

---

[*] In the fall of 1979, the chemical company Lubrizol paid $10 million for 15 percent of Genentech. In 1977, Standard Oil of Indiana paid $10 million for 23 percent of Cetus; Standard Oil of California bought another 26 percent for $13 million the following year.

[†] Shortly before the IPO filing, Genentech made a $350,000 payment to the University of California.

[‡] In the landmark *Diamond v. Chakrabarty* decision that came down shortly before Genentech's IPO, the US Supreme Court decided that such life-forms were patentable.

Swanson did not want to add to his list of concerns the attention and reporting requirements that would come with public ownership. He was already running Genentech in a near-constant state of anxiety, "worried most of the time," as he put it. Often, he said, he felt as if the stress was "chewing on my heart."[3]

Swanson said he wanted to wait a year. Perkins hurled his pencil onto the table. "That's the most ridiculous thing I ever heard!" he yelled, "Why don't I just sell all my stock and resign from the board?"[4]

Perkins asked Boyer to break the tie. The scientist demurred, saying only, "I always vote with my friends."

Chairman and CEO battled on. Finally, Perkins, playing on Swanson's competitiveness, asked him how he would feel if Cetus, the company that had declined to hire him to launch a recombinant DNA effort back in 1974, became the first biotech company to go public. That decided it.

■ ■ ■ ■

The next step was to find investment banks to underwrite the offering. Eugene Kleiner asked Alfred "Bud" Coyle, a partner at New York's Blyth Eastman Paine Webber, to come out of retirement for the Genentech offering. Here, as had been and would be the case so many other times in Silicon Valley, connections forged among an earlier generation of entrepreneurs supported the following generation's success. Three decades earlier, Coyle, along with his then junior associate, Arthur Rock, had brokered the deal that had launched Fairchild Semiconductor, of which Kleiner was one of eight cofounders. The Fairchild deal—funding eight young unknown scientists and engineers who had decided to leave their Nobel Prize–winning boss and launch a competitor company—had been revolutionary. Kleiner knew that Coyle was a risk taker.

The Genentech offering was similarly radical. "It was science, it was hardly profitable, and it had essentially no revenue," says Brook Byers, Swanson's former apartment mate who joined Kleiner & Perkins (now Kleiner Perkins Caufield & Byers) in 1972. Genentech, attempting to be the first new pharmaceutical business launched in many years, would compete against established giants. Without Coyle behind the offering, Byers muses, "no New York investment banker would have paid any attention to it."[5]

Perkins suggested that the West Coast investment bank be Hambrecht & Quist, an operation he knew well and whose principals he considered "Boy Scouts" among the cynical moneymakers of investment banking.[*][6] Bill Hambrecht and George Quist had launched their partnership a dozen years earlier, in 1968. At the time, they believed that West Coast venture capitalists did not know how to take small high-technology companies public and East Coast bankers, who knew how, did not want to take the risk. Hambrecht and Quist embraced what was then an unusual business model: their business served as both a venture capital firm that made early investments in risky young companies and as an investment bank that made money by taking young companies public.

Long Island native Hambrecht, amazed by how easy it was to try something new at the western edge of the continent—the new firm raised $1 million from four prominent San Francisco families in a single afternoon— credited California's "pioneer traditions."[7] "Nobody ever asked me, 'Well, why do you think you could do it?' We just said we could!" Hambrecht says. "We found out by doing some of these—yes, there was a market for it."[8]

Bill Hambrecht may be alone among wildly successful financiers in having chosen his college (Princeton) because Norman Thomas, the six-time presidential candidate from the Socialist Party of America and a hero to Hambrecht's grandfather, was an alum.[9] Bill Hambrecht would gain fame in 2004 when he employed an "auction model" to make it easier for the general public to buy into the Google public offering. But he had been trying to increase public access to high-potential companies since early in his career. The year prior to launching his partnership with George Quist, he had visited an entrepreneur, Bill Farinon, who wanted Hambrecht (then with the investment firm F. I. du Pont & Co.) to take his electronics company public.[†] Hambrecht had explained that the business was too small. To go public, companies needed to hit certain targets: $1

---

[*] It was Hambrecht & Quist that turned down a first-round investment opportunity in Apple because they thought the personal computer was another CB radio-like fad.

[†] Farinon's company, coincidentally, was the first one to use ASK's MANMAN software.

million profits, $10 million in revenues, five years of profitability. Those were the same metrics that had made Swanson hesitant to take Genentech public.

Farinon demanded, "Who set up those rules?"

"I don't know. They're just the rules," Hambrecht said,

Farinon was indignant. "Those goddamn rules. They were set up for you guys so you could make a lot of money! They have nothing to do with me." He pressed Hambrecht. "Do you think I have a good company here? Would you put in some of your own money?"

Hambrecht admitted that he would love to buy stock in the company.

Farinon looked triumphant. "If you're willing to buy stock for yourself, why aren't you willing to sell it to the public?"[10] That challenge became the credo of Hambrecht & Quist.

By the time Perkins called Hambrecht to see if he might be interested in taking Genentech public, Hambrecht & Quist, after struggling through several rough years, was a success. The bank had four senior managing partners and twelve general partners.[11] Hambrecht reviewed Genentech's financials and agreed to join a tour of the company's facility. He listened to the explanations of the cooling and warming rooms, visited the wet labs with scientists in white coats, and peered at the state-of-the-art fermenter with multiple dials and colored valves and pipes. After the tour, Perkins asked Hambrecht, a history major, what he thought of the operation. Hambrecht paused, looked around, and said, "Well, it looks like $100 million to me!"[12] He wanted his firm's venture arm to invest in the next round of funding, and the banking side to bring Genentech public.

■ ■ ■ ■

In September 1980, Swanson began a four-day, six-city European "road show" tour in preparation for the company's initial public offering, slated for the next month. Accompanying him across the continent were cofounder Boyer and board chair Perkins, as well as Genentech's CFO and three bankers. From Paris through Geneva and Zurich and on to Edinburgh, Glasgow, and London, the men moved as a pack. Black cars ferried them from airports, to breakfasts with pension fund managers, to one-on-one meetings with representatives from major trusts, to lunch presentations before rooms of insurance fund managers, to afternoon appointments and the occasional dinner—and back to the airport.

The presentations had been refined and repeated to the point that any member of the group could have given any part of them. But the men stuck with the script. CFO Fred Middleton ran through the financials, pointing out that Genentech had been the first to synthesize human insulin and was profitable. Boyer talked about the science behind recombinant DNA, using a string of brightly colored children's plastic pop beads as props. He would connect the beads end to end to represent a plasmid and then pop the circle open to insert new beads that represented DNA from a different source. To demonstrate inserting the plasmid into *E. coli* bacteria, he placed the mixed ring of beads into a clear plastic box.[13]

"People just listened and gaped," recalled Middleton of the road show presentations. "Every time we asked for questions, there weren't any. People didn't know what to ask. There were no experts. There were no analysts. Everyone was just amazed."[14] Nearly anyone who heard about the science behind Genentech's business walked away with the same thought: I'm not sure exactly what this is, but it feels huge. Already the company had found a way to synthesize life-sustaining human insulin in a test tube. What would come next?

Swanson spoke of the company's origins and promise. He was a natural, poised and enthusiastic, the model of a successful executive at only thirty-two. Over the past three years, he had raised more than $1 million in three rounds of private financing. Within the company, he knew how to wheedle, badger and cajole the scientific staff, armed with "just enough [science] to be annoying," as one Genentech researcher put it. Genentech's scientists hardly lacked motivation, however. Dave Goeddel, who headed the successful push to be the first to make human insulin, liked to say, "You either came in first or you might as well be last." He designed a T-shirt adorned with the Harley-Davidson logo, a DNA double-helix, and the phrase "Clone or Die."

The road show blur left Swanson "wrung out like a wet dishrag," in his own description. At day's end, some of the others repaired to the hotel bar or lobby to sink into the plush seats and enjoy a cigar. But not Swanson. For him, the trip was about much more than bankers and investors and balance sheets. There was someone else along on the road show, someone who did not even work for Genentech. And Swanson felt that he owed her an apology. He had never meant for it to happen , but the trip was doubling as his honeymoon.[*]

He had married less than a week before the road show had begun. He and his new wife, Judy, had enjoyed a single weekend together before the six other executives joined them for a honeymoon that Swanson characterized as "Snow White and the Seven Dwarves." Judy Swanson tried to squeeze quick tourist look-sees into the two-hour blocks while her new husband pitched to potential investors. When the group spent several days in London, she rented a car and drove to the Lake District. "What's a nice American girl like you doing alone?" she was asked.

"I'm on my honeymoon," she replied.[15]

■ ■ ■ ■

On October 14, 1980, Genentech began trading on the NASDAQ exchange under the symbol GENE. In the first twenty minutes, the opening price of $35 per share zoomed to $89—the fastest first-day gain in Wall Street history. ("George Quist is the one who really made that happen," theorizes venture capitalist Brook Byers. "I think he just picked up the phone and called all of the people he knew in New York and said, 'You ought to buy this.'")[16] When the closing bell rang, the price had settled a bit, and Genentech was valued at $532 million ($1.6 billion in 2015 dollars). Swanson and Boyer's original $500 investments were each worth about $65 million ($199 million in 2015 dollars), as was the stake that Kleiner & Perkins bought for $100,000 in the first round and an additional $100,000 later.[†]

At least one scientist, seeing Boyer as a turncoat, not a pioneer, remarked that the windfall might not be good for Boyer's scientific career, since business still retained an unsavory reputation in scientific circles.[17] But many more disagreed. A dozen biotech companies, all of them with

---

[*] The SEC, worried about a possible violation of the "quiet period" before the offering, had forced Genentech to delay its IPO, which pushed it into the weeks after Swanson's wedding.

[†] Kleiner & Perkins owned roughly 15 percent of Genentech at the IPO, about the same as Swanson and Boyer.

scientists in key roles, went public in the two years after Genentech's IPO.[18] Recombinant DNA, one chronicler has theorized, "pushed genetics from the realm of science into the realm of technology."[19] Venture capitalists knew how to build companies around technology.

The morning of Genentech's IPO, Perkins phoned Bob Swanson back in California and roused him from a sound sleep. "Bob," he said to the drowsy young entrepreneur he had once fired, "You're the richest man I know."[20]

■ ■ ■ ■

On the same day Genentech went public, Paul Berg, the Stanford biochemist who had organized the Asilomar conference on the safety of recombinant DNA and who had criticized the Cohen-Boyer patent for "claiming everything," was awarded the Nobel Prize in Chemistry. The Nobel Committee cited him for his "fundamental studies of the biochemistry of nucleic acids, with particular regard to recombinant-DNA." Berg shared the prize with Frederick Sanger of the MRC Laboratory of Molecular Biology in England and Walter Gilbert, the Harvard researcher whose team had worked in a British biological warfare facility during Cambridge's 1976 moratorium.

At no point in the Nobel proceedings was Stan Cohen or Herb Boyer's fundamental work mentioned.[21] Some observers claim that Cohen and Boyer had merely "reduced to practice" the breakthrough work by the prizewinners and others, such as Janet Mertz and Peter Lobban. It is also possible that Cohen and Boyer's acquiescing to Reimers's suggestion to patent the recombinant DNA process cost Cohen and Boyer the prize.[22] If successful, the patent application would mean that Cohen and Boyer would receive credit as "inventors," as well as substantial financial remuneration, for research that built on other scientists' work (as all science, including that of all the Nobel laureates, does).

"The Nobel Prize issue bothered me for a long while, to be honest about it," Cohen admits, "but it's something I've gotten over."[23] (Boyer, who says that he is happy and "never would have expected to do what I have done," has also said that he was disappointed by the Nobel announcement.)[24] Asked if it had been difficult to learn, on the same day that the prize was announced, that the cofounders of the company he had declined to cofound were each now worth tens of millions of dollars, Cohen says, "It was a delightful surprise that the business world had recognized that the valuation of a company devoting itself to this area could be so great."[25] Cohen, who began accepting his inventor's share of royalties on recombinant DNA in 1986 and donated them to charity, never measured success in terms of prizes or dollars.[26] "My turn-on comes from research," he explains. "Evolution has given us TB and malaria and cancer. All of medical science is altering what evolution has given us." He also adds, "I tried to do what was right."[27]

■ ■ ■ ■

Two months after the Nobel Prize announcement, on December 2, 1980, attorney Bert Rowland called Niels Reimers at Stanford's Office of Technology Licensing to say that after six years, the recombinant DNA patent was going to be issued. "Method patent allowed!" Reimers wrote in a triumphant memo to the file that by this time filled several drawers. He printed the new patent number at the bottom of the memo: 4,237,224.[28] That patent and related ones would be the source of nearly $255 million in income to Stanford and the University of California in the coming years.

Reimers had been expecting the patent to be issued for months, but there's nothing like official paperwork to kick off a celebration in a licensing office.* The Henrys celebration that Friday was jubilant.

A proud Stanford vice president told the *San Jose Mercury*, "This is the first time in my memory that a whole field of science has broken open to commercial application in ways that are both obvious and obviously profitable."[29]

At about the same time, Congress passed the Bayh-Dole Act. The act essentially codified in law the objective Reimers had pursued for five years: the right of universities to claim ownership to an invention that

---

* In March, Rowland had called it "95% definite patent will issue by end of year," and in June, Reimers was feeling confident enough to drop a line to Bob Beyers, who had pointed him to the original *Times* article about the DNA breakthrough, thanking him for giving Stanford "the chance at obtaining what could be the most significant patent property Stanford (and UC—sigh) will ever have."

had been funded by Federal research grants. Reimers had pushed hard for the bill, meeting with senators and members of the California House delegation. Passage of Bayh-Dole was a high priority for him. He wanted all universities automatically to have the same rights Stanford had won piecemeal.

■ ■ ■ ■

The patent in place, Reimers turned to refining the terms of the recombinant DNA license. The beauty of the Cohen-Boyer process—its simplicity—also meant that as the patent application sat in limbo for six years, scientists around the world had begun using the process. At the time the patent issued, the National Institutes of Health was spending $91.5 million to back 717 recombinant DNA projects; the National Science Foundation was spending $15 million on 184 grants; and the Department of Agriculture was supporting $5 million in research.[30] Academic and pure research institutions could continue their work without buying a license, but now that the patent had been issued, any company employing the technique needed to buy a license from the Office of Technology Licensing. That included Genentech, which used the process in developing its hormones.

Reimers had to design a license that would accomplish two hard-to-reconcile ends: bring in money to Stanford and the University of California and, at the same time, be priced fairly enough that companies would be willing to buy a license, rather than challenge the validity of the broad patent in court.[31]

For years, he had played with different models for terms. In the end, he adopted a shrewd approach that played on corporations' instinctive dislike of uncertainty. Licenses that offered cheap terms would be available only to those who signed up within four months. After that . . . Reimers got creative. He decided not to offer any information about the terms that would be available after the four-month window closed.[32] ("I don't know where that idea came from; I guess in here," he says, tapping his head.) In essence, he was telling companies that they could take a deal now or take their chances later.[33]

Reimers sent drafts of the licensing terms to roughly a half-dozen companies, including Genentech.[34] This gave important companies a first look, and their feedback gave Reimers a better understanding of what mattered to potential licensees.[*] Soon after receiving the draft terms, Genentech's Bob Swanson, who had been in touch with Reimers since the month the company had launched, began petitioning for a custom license. Reimers refused. There would be only one deal. It could not be customized. To make the point, he had the final licensing paperwork bound into little pamphlets that looked more like finished books than legal documents that could be changed by scratching out and initialing corrections.[35]

As much for publicity as for sales, Reimers placed advertisements in *Science*, *Nature*, and the *Wall Street Journal* during the first week of August 1981. The notice, which read almost like a birth announcement— "Stanford University, in cooperation with the University of California, announces that patent licenses are now available"—thanked the sponsoring organizations; laid out the terms of the license; warned obliquely that after December 15, terms "will be less favorable to the licensee"; and ended with the promise that the universities would use the royalty revenue for "educational and research purposes which, in turn, may result in yet other scientific advances for public use and benefit."[36]

Back on the Stanford campus, Sally Hines, Reimers's longtime assistant, taped a large piece of paper to the front of her desk. She announced that she would keep a running tally of companies that signed up for a license. Reimers sent Andy Barnes, a business school student hired on a one-year contract, around the United States and to Europe and Japan to encourage companies to buy licenses.[†] In Japan, where the government had named development of recombinant DNA processes a national goal and Silicon Valley was widely admired, Barnes says that he was treated "like a rock star" with media events and even a small group of paparazzi trailing him.[37] But in New York and Chicago, company representatives told Barnes that the patent was too broad to be enforceable.[38] In Europe,

---

[*] Draft licensing agreements went to Eli Lilly and Company, Genentech, Hoffmann-La Roche, Schering-Plough, SmithKline & French, and Upjohn. Cetus and Monsanto may also have seen a draft.

[†] Although Stanford did not hold an international patent, many foreign companies paid for licenses.

companies expressed concern about locking themselves into a royalty rate that might later prove too high.[39]

Barnes recalls, "The plan was a little audacious. Everywhere the response was the same: 'This should be in the public domain. How can Stanford be doing this?' "[40]

■ ■ ■ ■

Reimers hoped that Genentech would be the first company to sign a licensing agreement. "Their company had the name. They were not an old-line firm like SmithKline or Bristol Myers. They were clearly making a splash in the field," he recalls.[41] Soon, however, he learned that he could not be choosy. A Japanese company was the first to sign an agreement—Sally Hines proudly noted it with a large "1" on her butcher-paper tally—but that was it for more than a month. By mid-September, with only three months until the window closed, only one other company had signed up.[42]

Stanford president Donald Kennedy began getting antsy. "I'm concerned about the last stages of licensing on Cohen-Boyer," he wrote to a senior administrator in an early use of the campuswide email system. "Neils [sic] does not seem to have thought thru a firm strategy for how we approach the preannounced deadline, or what we do about rates afterward."[43] Kennedy was also concerned that Reimers, without consulting a superior, had set aside $200,000—more than half the total income brought in by the Office of Technology Licensing in the previous year—to contest potential challenges to the patent.[44] Reimers wanted this amount set aside not only to defend the patent but also to "be sure we get the money from Cal."[45]

Privately telling himself that he would be happy if thirty companies bought licenses, Reimers prepared his office for an onslaught, even arranging to have a database program set up on Hines's new computer to keep track.[46] Seven years earlier, the patent application had been typed on a typewriter. Now a computer would calculate its value.

■ ■ ■ ■

License contracts trickled in. The tally on Hines's desk moved into double digits. Then, on the morning of December 15—the day the licensing window closed—Hines noticed something unusual. Delivery trucks, one after the other, were pulling into the circular drive in front of Encina Hall, the drivers hopping out to bring envelopes into the Office of Technology Licensing. "We were just like kids with our big eyes," Hines recalled. "So many trucks were there, and they were all for us."[47]

As the day neared its close, the paper on Hines's desk was a mess of numbers and cross-outs. Seventy-two companies had signed up for a license.

Reimers, at his desk, reviewed the names of the new licensees, comparing them to his list of targets. Every one he wanted was there—except Genentech.[48] Then, at the last possible moment, Genentech's contract arrived. Seventy-three contracts, representing $730,000 in immediate income and unknown amounts in the future, had come in.

Reimers, his satisfaction evident, reported the success to Stanford president Don Kennedy.[49]

Sitting in a Kiddie Seat

AL ALCORN

"I think I really did it this time," Nolan Bushnell told Al Alcorn. It was a few weeks before Christmas 1978, and the old friends were sharing a drink in the bar of a New York hotel. The expensive libations, the polished mahogany and brass along the bar, and the twenty-four-hour room service and beds so comfortable they swallowed a man like sleep itself—Alcorn knew that Bushnell had grown accustomed to it all. Gone was the wide-eyed Utah boy who could be wowed by a ride on a private jet. Bushnell had his own plane now (as did Alcorn, Keenan, and a number of other early Atari employees). Bushnell's house was as nice as this hotel.

Alcorn had spent the day at an event introducing Atari's personal computers and proprietary software that included games as well as tax preparation and record-keeping programs. The marketing for the computers, called the Atari 400 and 800, took a direct lunge at Apple's customers, featuring a brochure striped in a rainbow pattern reminiscent of the Apple logo and emblazoned with the phrase "Computers for People." Atari computers, an official press release noted, were "a natural evolution of Atari's technological expertise, planning, and ongoing consumer research."[1]

Bushnell had been across town at a budget meeting with Manny Gerard and other top Warner executives. He and Alcorn had connected at the bar at the end of the day.

Bushnell took a deep swallow. "After that meeting, I think they are going to fire me," he said.[2]

■ ■ ■ ■

The meeting agenda had been standard enough: Finalize Atari's 1979 budget. Over the past three weeks, divisions throughout Warner had met with members of the office of the president, as well as CEO and chairman Steve Ross. Typically, the head of each division presented a budget that was then discussed in an open forum around the conference table in the main boardroom at Warner's Rockefeller Center headquarters. Surprises were unusual. The division's staff had agreed on the budget in advance, and the Warner leadership took a loose approach, asking questions but allowing the division heads to take the lead.

Something different had happened when Bushnell unfolded his six-foot, four-inch frame from his leather chair and started to speak. As he told Alcorn, "I began calling them dumb shits."[3]

The friction between the leaders of Atari and Warner had begun almost at the moment the acquisition papers were signed. Shortly after, Bushnell had taken a much-needed vacation, and then another. He had remarried. He had bought a 13,000-square-foot, thirty-seven-room mansion on sixteen acres in the foothills of the Santa Cruz Mountains—previous owners included the Folger family, of the eponymous coffee company[*]—and begun remodeling projects.[4] Bushnell's absences had become problematic enough that other employees would joke about them. Often Joe Keenan, the president of Atari, would be gone, too. "After we sold the company we were significantly less motivated to bust our humps," Keenan admitted.[5]

Meanwhile, the microprocessor-based cartridge game once code-named Stella and now called the VCS was not proving the "fucking huge" success that Warner's Manny Gerard had hoped. In the first year, chip suppliers could not meet delivery schedules. The next year, the supply side

---

[*] The Folger family had sold the estate after twenty-five-year-old Abigail Folger, granddaughter of the Folgers founder who had built the house, was murdered in Los Angeles by members of the Manson family in 1969.

was under control, but 300,000 game systems went unsold. At the same time, competitors caught up to Atari and began selling their own cartridge systems.[6] Warner had invested some $120 million in Atari, and still the division was losing money.[7]

Gerard decided to hire an executive consultant to spend six months at Atari and recommend changes. He says, "It was time for some adult supervision."[8] That's when the problems really started.

■ ■ ■ ■

The consultant, Ray Kassar, had spent twenty-eight years at Burlington Industries, one of the fifty largest companies in the country. He had run Burlington's sizable home furnishing division, which he preferred to call "home fashions."[9] Recently passed over for the chairmanship of the company, he had accepted the Atari consulting job mostly for the guaranteed income.[10] He had never played a video game and, as a lifelong easterner, had no desire to live long term in California.

But Kassar knew how to sell into the living rooms that Warner hoped one day would have a VCS game system connected to every television. At Burlington, he had helped convince people throughout the United States to make redecorating and updating their homes an ongoing activity.

Near the end of his consultancy, Kassar reported back to Warner that the Atari subsidiary was a mess. Under Bushnell and Keenan's leadership, he said, Atari "wasn't about the business, the quality of the product, the returns, the advertising, the marketing—none of those things."[11] He recommended a return to business basics, coupled with the right marketing and advertising push for the VCS system. With such changes, Kassar believed, the Atari division as successful as Gerard had hoped.

Kassar's recommendations could not have deviated further from Bushnell's vision for Atari. Bushnell believed that "Atari's reason for everything it did was innovation."[12] That priority came from Bushnell's experience with arcade games that were popular for only a few months. To stay profitable and ahead of copycats, he was always eager to try a new idea from Atari engineers. "An engineer doesn't always come in a body that can talk," he once explained. "But they're not shitheads. You've got to have enough faith in them to say, 'I don't know what you're talking about; here's some money, go show me.'"[13] When the engineers said that they could develop a better version of the VCS video game cartridge system—Super Stella—Bushnell was happy to let them try, even if the original VCS was not selling well. If they said they could build more powerful computers than the ones Atari was selling, he told them to get to work.[14] His encouragement was a powerful motivator, particularly in tandem with Alcorn and Bristow's insistence that only the most promising ideas be pursued.

Bushnell had expansive plans for Atari. He wanted to open restaurants with arcade games and use animatronic actors, like those in Walt Disney World's Hall of Presidents, to entertain kids waiting for dinner. Bushnell had even come up with a mascot for the restaurants, a large rat called Ricky. He kept the Ricky Rat costume in his office.

Ray Kassar came from a different background and thought differently about business. The home furnishings market did not hinge on developing new products. A towel or bed does not change much from one year to the next. Instead, he had succeeded by selling old products in new ways. Bushnell's near-compulsive pursuit of the next great new idea, in Kassar's opinion, was misguided. He felt that Atari needed to spend the money to market the products it already had, especially the VCS game system.

Bushnell, who claims that Kassar or another representative from Warner once told him to "innovate kind of like you did last year—none of this new stuff,"[15] also had a financial incentive to pursue the development of new products over the marketing of old ones. He and Alcorn, along with a few top Atari employees, participated in a bonus plan that entitled them, as a group, to 15 percent of Atari's profits.*

The struggle between Bushnell and Kassar was a generational clash between one of the newest industries in the world—semiconductor-based electronics—and one of the oldest: textiles. It was also a regional battle. Bushnell had spent his life in the West; Kassar had spent his in New York, Rhode Island, and Massachusetts. As Kassar explained, "We're all more serious in the East. You have a job and you do it the best you can and, you know, it's not a playground."[16] Bushnell's description of work at Atari is rather different: "You didn't worry that much about legal liability, people

---

* Bushnell says, "Warner would say, 'Well, hey. We're all in the same family.' Bullshit. Our bonus program was set up on that [profit]."

getting drunk. Almost everybody went home with everybody else. Sexual harassment, or things like that, wasn't an issue."[17]

Bushnell hosted naked hot tub parties, loved to tweak authority, and, according to Kassar, once wore a T-shirt emblazoned "I Love to Fuck" to the office. Kassar enjoyed antiquing and claimed to have secured both a job at Burlington Industries and a scholarship to the Harvard Business School after charming the heads of each organization in a single holiday weekend.* Perhaps inspired by the silk business his family had owned in Syria, Kassar felt best in a beautifully tailored suit.[18]

By the 1978 meeting that Bushnell said might get him fired, the budget, likely reflecting Kassar's influence, called for Warner to invest millions of dollars in advertising the VCS. Bushnell's objections to the plan were at the heart of his "shitheads" speech at the meeting, though what precisely he said is a matter of debate. Gerard says that Bushnell told Ross, "Sell off your inventory; it's all over for the 2600," a prediction for the VCS game system (commonly nicknamed "the 2600") that proved wrong. Bushnell claims that he told Warner to drop the price of the console, not to abandon it.[19] Keenan further recalls Bushnell telling Warner that the computer effort was doomed without a $600 million investment, a warning that proved prophetic. Gerard has no such recollection. Bushnell also objected to some of Warner's plans for Atari's small pinball division and wanted to spend more on product development.[20]

Whatever the specific areas of friction, there was only one question in the end. Who would control Atari, Bushnell or a Kassar-like executive chosen by Warner?

Three days after Christmas 1978, Bushnell had his answer. Warner assigned him to "other duties within Atari."[21]

A week later, Warner announced that Atari had a new president and CEO: Ray Kassar.[22]

■ ■ ■ ■

* Kassar says he got his start after spending a holiday break at the Palm Beach villa owned by Spencer Love, the founder of Burlington, who also happened to be the father of Kassar's roommate at Brown. The dean of the Harvard Business School was also a guest, and by the time Kassar returned to school, the tycoon had offered him a job and the dean had offered him a scholarship.

By any metric used on Wall Street, Kassar did a fabulous job in his first years at Atari. The company that was losing money when he arrived in 1978 accounted for nearly two-thirds of Warner's $227 million profit three years later.* Kassar's decision to boost advertising budgets helped fuel that growth, as did his critical move to sell video games not only at Christmas but year-round. In 1980 and 1981, Kassar received $10 million in bonuses as well as Warner-covered rent on a luxury apartment in New York's Trump Tower.[23]

Though Atari's coin-operated arcade games also did well (thanks in large measure to *Asteroids*, which shipped 70,000 units in 1980 alone), the biggest profits came from the VCS cartridge system and its twenty-seven games—particularly *Space Invaders*. Each cartridge sold for roughly $30, with an 89 percent gross margin for Atari, according to Gerard.[24] The VCS accounted for almost a quarter-billion dollars' worth of sales in 1981.[25] A few other companies, most prominently Mattel, made cartridge systems to compete with the VCS, but 80 percent of the 4 million video game systems in American homes in 1981 were made by Atari.[26]

Within Silicon Valley, Atari, which had grown to nearly ten thousand employees and fifty buildings, became a prestigious employer for computer programmers and game designers.[27] Saying you wrote games for Atari, one programmer noted, "didn't get you laid, but it was seen as cool."[28] Programmers had complete control over every aspect of a game, from design and rules to graphics and sound effects. Some programmers started developing the games on paper before typing code into the hexadecimal language the VCS system could understand. To test the game, the programmer would load the code onto an eight-inch floppy disc and carry it into the main lab space for debugging on the development system that all the programmers shared.

The debugging process provided the first hint of a game's potential popularity. Some games were so good that other programmers could not stay away. They would cluster around a screen in the main lab space, test-playing the game, offering suggestions and shortcuts. A really good

---

* The figures are for the consumer electronics division, which was almost entirely Atari. Figures for 1982 were 50 percent of sales and 62 percent of profits.

game was one that the programmers insisted be restarted as soon as play had ended. "Self-modulating feedback," one person called it. Every game was the product of an individual mind, refined by a team of players.

The work was creative and arduous. Programmers developed original ideas for games and also found inspiration while playing coin-op arcade games (some of the most popular cartridges were versions of arcade titles). Warren Robinett modeled the wildly popular *Adventure* game on a similar text-based game developed at Stanford's Artificial Intelligence Laboratory, where several members of Bob Taylor's Xerox PARC team had once worked.[29]

The unwritten expectation was that a game should go from concept to production in six months, a challenging task made no easier by the technical quirks of programming for the Atari VCS. "It's one of the most difficult hardware systems to write for that I've ever seen," says Howard Scott Warshaw, who programmed the *Raiders of the Lost Ark* and *E.T. the Extra-Terrestrial* cartridges. "To program a game on it was a game in and of itself."[30] Changing something on the screen required rewriting the code to redraw the entire screen. The technical limitations of the VCS that were beginning to cramp graphics and complexity also led to creative programming.[31]

"I lived inside these games while I was working on them," says Carla Meninsky, who programmed *Dodge 'Em* and *Warlords* and was one of the only women programmers at Atari. "People would be sleeping on their desks, under their computers, on the floor. These were your friends. You lived with these people.

"The light changed outside, but nothing changed inside."[32]

To deal with the pressure, the programmers ran pranks, from floating a frog-shaped balloon over the building to stealing the sign from the Apple facility nearby. When Warshaw was writing the *Raiders of the Lost Ark* cartridge, he sported a beat-up hat and carried a whip around the office. Programmers wore jeans and T-shirts and worked long, odd hours, shoving doorstops under their doors from the inside so they could sleep without interruption. An MRB—"marijuana review board"—committee was "very much part of the creative process" and met on the roof when its members felt the need. One designer took to climbing the walls of a particularly narrow corridor. Pressing one foot against each wall, he would

scooch himself up several feet above the ground and then begin scoot-
ing toward the other end of the corridor. He once clocked his head on a
ceiling-mounted smoke detector.[33]

If the game developers' offices retained elements of the old free-
spirited Atari, other areas changed after Ray Kassar became CEO. Kassar,
who had moved to San Francisco and was chauffeured to Silicon Valley
every day, installed a formal executive dining room with waiters in black
tie.[34] The programmers derided the CEO and his business-focused executives as empty-headed "beautiful people" who had never played a video
game, much less loved one. Even outsiders began to wonder if Atari had
"too many pinstripes and not enough flannel shirts and long hair."[35] "Marketing thought the programmers were lazy; the programmers thought
marketing was stupid," says one programmer. "We didn't like them; they
didn't like us."[36] A product manager hired in 1981 says that he, like many
of his colleagues "on the staff side," had never played a video game. When
asked about the engineers, he said, "I didn't deal too much with those
people."[37] Tensions ran so high that some employees refused to speak to
others and would send memos via overnight mail—even if the recipient
sat in the next cubicle over.

Al Alcorn watched with mounting dismay as the antagonism between
engineering and marketing, techies and suits, led several important early
employees to leave. When two of the three division heads at Atari quit,
Kassar hired a paper products marketing executive from American Can
Company to run the computer division and a vice president at a perfume
maker to run the home video game division.[38] A few months after Bushnell was removed from the presidency, Joe Keenan, Bushnell's onetime
co-CEO, left as well.[*]

Bushnell had paid Warner for the rights to his ideas about a restaurant
with arcade games. With Keenan's help and financing from Don Valentine,

---

[*] Bushnell and Keenan legally remained Warner employees. When Warner acquired
Atari, the parent company reserved the right to remove key employees from authority but keep them on the payroll until 1983, seven years from the acquisition. While
in this well-paid limbo—"on the beach" was the Warner term for it—Bushnell and
Keenan could not work in any capacity that competed with Atari.

Bushnell launched a line of Pizza Time Theatre restaurants, with Ricky Rat, now called Chuck E. Cheese, as mascot. A few months later, Joe Decuir, an instrumental member of the engineering team for both the VCS and Atari's computers, left to launch his own engineering company.

Next to go, still within a few months of Kassar's hiring, were four of Atari's top VCS programmers. Among them, David Crane, Larry Kaplan, Al Miller, and Bob Whitehead had written some of the cartridge system's most popular titles; by Crane's estimate, their cartridges accounted for 60 percent of all cartridge sales and at least $50 million of Atari's sales in 1979. The four men had also helped write the operating system for Atari's computers. Their combined annual salaries were less than $200,000.

The programmers had asked Ray Kassar for a pay raise or a bonus, as well as recognition as the games' authors on the cartridges. (Already some designers had taken to hiding their initials as "Easter eggs" in secret rooms that players could discover in the games themselves.) Kassar allegedly responded that the game programmer was no more essential to the company's success than was the line worker who put the cartridge in a box.[39]

The group of four programmers decided to leave, but they had no idea how to start their own company. They asked Joe Decuir, recently of Atari himself, how he had launched his engineering consultancy. He sent them to Larry Sonsini, the young attorney who had handled the Tymshare IPO and soon would begin working on Genentech's. Sonsini, who continued to counsel the programmers, also introduced them to the venture capitalist Bill Draper, of Sutter Hill Ventures.* Draper agreed to invest after he took home a copy of a bridge game and found the experience of playing it "magical."[40]

With Sonsini and Draper's help, the four programmers launched a company called Activision. Soon Activision was publishing game cartridges compatible with the VCS (and later with competitors' systems as well) and fighting off a lawsuit by Atari alleging theft of trade secrets.

---

* Draper's father cofounded Draper, Gaither & Anderson, the first venture capital firm in Silicon Valley to embrace the limited partnership model. Draper's son, Tim Draper, is a founding partner at the venture capital firm Draper Fisher Jurvetson (DFJ).

Activision profiled its programmers in the manuals that came with its cartridges and would go on to create a number of bestselling games, including *Kaboom!*, *Pitfall!*, and *River Raid.** Today Activision, in a different incarnation, is best known for its *Call of Duty* and *World of Warcraft* game series; the company's 2016 market capitalization was $30 billion.

In 1981, Atari's marketing vice president and a group of programmers, including the creators of *Asteroids* and the *Space Invaders* cartridge, started yet another competitor, Imagic.[41]

Since nearly all of Atari's profits on the VCS system came from the cartridges, the company could not afford to have its programmers become competitors. Within weeks of the Activision team's departure, Atari began offering programmers a bonus of roughly $10,000 for each title. That bonus was on top of salaries starting at about $20,000, a figure that programmer Carla Meninsky recalls felt like an enormous sum when she first applied at Atari, so absurd that she practiced saying "twenty thousand dollars" in front of her bedroom mirror so she could request it with a straight face at her hiring interview.[42] Salaries and bonuses for experienced programmers could go much higher. Despite these efforts to encourage game designers' productivity, the number of cartridges released by Atari dropped by half the year after the Activision founders left.[43]

Even with the higher pay, many on the engineering side felt that Kassar and the managers he hired did not appreciate their ideas or their work. Kassar gave an interview in which he called the technical minds behind the games "superstars" but also "high-strung prima donnas." Many programmers felt the jab was a closer approximation of Kassar's real feelings.[44]

■ ■ ■ ■

Alcorn understood why people he admired left Atari. Every week he and several other senior technical employees met with Kassar. Initially happy

---

* Some Activision games overlapped with Atari's offerings. Both companies offered *Checkers*, for example, and the competing programmers (Carol Shaw at Atari and Al Miller at Activision) unknowingly consulted with the same professor at Stanford, Arthur Samuel, who was working on a computer version of the game.

to participate, Alcorn realized after a few meetings that nothing came of them. Kassar would nod his head and even authorize people to start research, but no project ever seemed to move beyond a prototype.

"It was like sitting in a kiddie seat in the back of the car, turning a toy steering wheel," Alcorn said.[45] Kassar was willing to give his most senior and technically knowledgeable employees an illusion of influence, but he wanted them in the back seat playing with their toys while he figured out where to take the company. The technical team had a private name for their sessions with Kassar: the "limp-dick meetings."[46]

Alcorn resolved to earn Kassar's respect through technical ingenuity. With Bushnell no longer around to goad him into trying things that seemed impossible, he photocopied a list of "Idea Killers" to avoid: "Too radical." "Contrary to policy." "It won't work." "The boss won't like it." "What's the potential profit?" "Can you guarantee that it will work?" He kept the list, which opened with neurosurgery pioneer Wilfred Trotter's warning that "we have begun to argue against a new idea even before it has been completely stated," with his most important work papers.[47]

At a meeting in 1980, Alcorn proposed that Atari build a cartridge-type system like the VCS—but the games would include three-dimensional holograms. Holograms were an exciting prospect at the time. In the blockbuster 1977 movie *Star Wars: A New Hope*, the droid R2-D2 projects a three-dimensional hologram of Princess Leia asking for help, and two characters play Holochess, in which three-dimensional chess figures hover over a circular chessboard.[48]

Alcorn was less interested in reproducing *Star Wars* fantasies than in an excuse to research holographic technology, which he thought looked "hard" and "fun," with "technical challenges that were just fabulous."[49] Of course, he provided a different justification to Kassar, who measured success in dollars, not technical accomplishment. He said that the hologram-enhanced game system could sell for perhaps half the price of the VCS system and thereby open a new market for Atari.[*]

---

[*] Alcorn would lower the price by embedding the game software in the console itself, rather than in the cartridges (as with the VCS). A cartridge would contain only a hologram and a key that told the base unit which game to play.

Kassar told Alcorn to talk to marketing. The marketing manager told him to write a business plan.

"We didn't need a business plan for *Pong* or the VCS," Alcorn replied. He needed one now.[50]

Once marketing approved Alcorn's plan, he began assembling a team. Several of the people he recruited told him that no matter how good the product was, Kassar would never release it. Kassar cared only about the VCS, they said.

Alcorn was unconcerned. "I'm Al Alcorn," he reminded them. His work had helped launch the company in 1972 and save it in 1975. That had to be worth something.

It took only a few months for his team to develop a prototype of the holographic game system. Called Cosmos, it was about the size of a thick paperback book, with its own tiny screen. Cosmos could play eight cartridge games, each of which loaded a hologram into the system to serve as a backdrop with the illusion of three-dimensional playing fields. In the *Super-Man* game, for example, Superman flew past and between buildings that seemed to pop out from the screen slightly. Cosmos was no *Star Wars* Holochess, but *Adweek* called the game system "the most dramatic technical video development since color TV."[51]

Alcorn took the prototype to the 1981 Consumer Electronics Show and says he received some eight thousand orders for the system. Atari's marketing vice president told the press that Atari had budgeted "in excess of $1 million" for the Cosmos introduction.[52]

When Alcorn returned to Atari and reported on this progress at the next meeting with Kassar, the CEO seemed unimpressed. "He sat there with his lips pursed together," Alcorn says. "Then, after a minute, he just shook his head."

Atari never sold the Cosmos game.[*] Alcorn saw Kassar's decision as proof of a cultural shift at Atari that he describes in four words: "All creativity had ceased!"[53] The team behind the original VCS system had designed a follow-on system, but it was not released until the end of 1982,

---

[*] Somewhere between 250 and 1,000 game systems had been at least partially built when the Cosmos project was killed; none was sold.

and many members of the technical staff were unhappy with it. Next-generation computing efforts met similar frustrations.* Steve Bristow, one of Atari's most senior engineers, spent years on a networking project, complete with advanced telephone systems (some with video capability) that would combine voice and data and perhaps even make it possible for gamers to play against each other over the telephone lines. The system, called Atari-Tel, was announced but never released.[54]

Kassar was simultaneously afraid to say no to engineering and afraid to release a new product that was not as successful as the VCS. He allowed Atari engineers to pursue new ideas—but only to a certain point.

In Alcorn's mind, the problem boiled down to risk aversion. "When we were young at Atari, every year we risked the whole company on new products. If the VCS had failed, or *Home Pong* had burned up, we'd have killed the company," he says. Bristow concurs: "The thinking moved from 'We don't know what we can do, so let's give it a try' to 'You need to prove why you need to do this new product.'" Decades later, Alcorn still cannot understand why, at a time when Atari's revenues were so high that it would have made no discernible difference to the bottom line if the Cosmos hologram system failed, Kassar was unwilling to risk its release. Kassar and the rest of the Warner executive team, in Alcorn's estimation, were paralyzed by "the fear of failure": "They weren't Silicon Valley, they weren't start-up guys, they were not risk takers—so nothing came out!"[55]

Even more important than Kassar's East Coast roots, however, was his lack of technical background. He was a marketing expert who believed, correctly, that much of the success of the VCS had come from the marketing push behind it. He somehow failed to recognize that Atari had a product to market only because its engineers had been given the freedom to develop it. Kassar seemingly had only one way to evaluate a new idea: he asked if it would make as much money as the VCS. That was a flawed

---

* Although Atari's computers initially sold relatively well, largely on the strength of the company's superb reputation in video games, the machines' appeal soon dwindled. Atari computers lacked expansion slots and suffered from a shortage of software as well as confused marketing that positioned them as a hybrid game system and personal computer.

metric, as Alcorn knew. No new product could, from the moment of its release, compete with the VCS, one of the most successful consumer electronic products of its decade. Moreover, no product is a hit forever; for evidence, Atari needed to look no farther than a few miles down the road. Around the time Alcorn was lobbying for the Cosmos game, 75 percent of Hewlett-Packard's sales revenue came from products that had not existed five years earlier.[56] No good comes to a consumer electronics company without follow-on ideas. As Alcorn likes to say, "If you don't obsolete your own products, someone else will."

After Kassar's final dismissal of the Cosmos hologram game, Alcorn tried to follow up with a private conversation. It was impossible. "Been trying to contact [Kassar] by telephone for the last few days and have had very poor results," he noted in a memo to himself. [57]

The lack of response was itself a response. "They didn't need me anymore," he explains.

Alcorn, the onetime careful adventurer, no longer needed Bushnell to push him to pursue radical change. He quit.[*]

---

[*] Alcorn took the same "on the beach" arrangement Bushnell and Keenan had taken in 1979: he officially remained an Atari employee but did no work for the company.

Can You Imagine Your Grandmother Using One?

BOB TAYLOR

Bob Taylor was sitting in his Xerox PARC office in early 1980, his pipe on its stand near the Alto on his desk, when Larry Tesler walked in. Strictly speaking, Tesler was not in Taylor's group—he worked in the systems science lab—but he had been an integral part of the Alto team. He was also a regular visitor to Taylor's office, so Taylor expected nothing out of the ordinary when Tesler shut the door behind him.

"I'm leaving Xerox," Tesler said with little preamble. "I'm going to Apple."

Apple? The company making small computers? Taylor had known Tesler was interested in small machines, but why would anyone leave PARC for Apple? PARC offered nearly unlimited resources, access to powerful computers, an easy connection to the world-class Stanford Computer Science Department, good pay, colleagues who were among the greatest minds in the field, and a free choice of work projects. Taylor was confident that PARC was the top place in the world for computer scientists.

Apple, by contrast, had . . . what? Sales and customers to worry about. Time-to-market pressures. A technical team without formal academic computer science training: amateurs trying to cobble together machines. A publicity-hungry cofounder who had dropped out of college and who allegedly interfered with technical decisions.

What could Apple possibly offer over Xerox PARC? Taylor wanted to know.

People had been asking Tesler some version of that question for years, if not about Apple, then about the class of computers that Apple built. Tesler had been dabbling in the world of small computers almost since its inception. Fred Moore, a cofounder of the Homebrew Computer Club, had been Tesler's neighbor and had invited him to the club's first meeting. When Moore had told Tesler that "everybody is going to build their own computer," Tesler had found the idea absurd. Most people can't put together a couch, he told Moore. He skipped the meeting.

By chance, however, Homebrew's second meeting was held at the progressive, hippie-rooted school that Tesler's daughter attended.[1] Tesler, a trustee, was just leaving when he noticed a stream of young men making their way from the parking lot, carrying boxes of wires, boards, and assorted electronic gadgetry.

It's worth a look, he thought, following the men into the wooden shed that served as the school's auditorium.

"It was literally guys showing off lights flashing," he recalls. "I wasn't interested. I was interested in things that were easy to use."[2]

His interest sparked a year later, however, when he attended another Homebrew meeting. More people had joined the group, now gathering in an auditorium at the Stanford Linear Accelerator. By the end of the evening, he had to admit that he was impressed—less by the computers than by the rate of progress they represented. Machines that had once displayed results only in a series of flashing lights were now putting letters and numbers on screens. People were exchanging cassette tapes loaded with software and sharing programming tips.

Tesler bought a Sol computer from a pair of hobbyists calling themselves Processor Technology. He also acquired a PET computer from Commodore.* He tinkered with the computers the way a Formula 1 mechanic might fiddle with a go-kart. What could the little guys do? He experimented with coding for the machines and had soon written an

---

* Tesler says he did not buy an Apple II because he wanted lowercase letters and the early Apple IIs used only caps.

accounting program for his finances and a painting program for the students at his daughter's school.

He also took the machines to PARC. A few researchers quietly outed themselves to Tesler as hobby machine enthusiasts, but the general reaction was either lack of interest or contempt. *Those* were supposed to be computers? There was no mouse, no network, no way to print, and no way to send anything to anyone else. Whereas PARC had email, users of the small computers would type notes, photograph the screen, take the film for processing, and then mail the photo of the screen to the note's intended recipient.* Many people at PARC and other research centers used the same word to describe the small computers, a word that would have infuriated Markkula, who saw the Apple II as a powerful tool: toy. Small computers were toys.[3] "You couldn't do real programming. You couldn't collaborate—there was no network. There were few applications, no FORTRAN or C-compiler," recalls one computer expert. "You couldn't do more than one thing at a time, couldn't store large amounts of data. What could you do with it?"[4] Taylor could imagine that the machines might help students wanting an inexpensive way to learn programming, but he thought the general public would have little patience for computers like the ones Apple was building.[5] "Can you imagine your grandmother using one?" he asked Tesler.

Tesler conceded that the machines he lugged into PARC were not computers in the same mold as the Alto. They were nowhere near as easy to use, nowhere near as elegant to look at, and nowhere near as useful for communication. Yet people could buy them. In the last three years, Apple, Atari, Commodore, Radio Shack, and Texas Instruments had begun shipping personal computers. Nearly 200,000 had been sold.[6]

By contrast, in 1980, eight years after work on the Alto system had begun, and three years after the Futures Day demo that Taylor had watched from behind a spotlight, it was still impossible to buy an Alto computer. Apple's Trip Hawkins recalls visiting the Xerox booth at the National

---

* Intel cofounder Robert Noyce's son, an early adopter of computer technology (and a teen at the time), did precisely this, send a note to his grandmother. Apple's first fan letter was a photograph of a screen.

Office Machine Dealers Association conference in 1979. Hawkins, like many people at Apple, had heard about the Alto. (By then, the company had given a number of demos.) When Hawkins spied one of the computers in the booth's back corner, behind a display of Xerox's dedicated word processing machines, he asked to see it.

"Do you have any plans to commercialize that?" Hawkins asked an older man whose badge identified him as a senior vice president in charge of office products.

The man turned to look at the Alto, which was not even plugged in. "No."

Hawkins said, "I see you don't have the mouse with it. Why is that?"

"Because I don't like it."

"Why not?"

"Because it can roll off the table."[7]

Some 1,200 Altos were in use inside Xerox in the late-1970s, but only a handful had made their way beyond the company's walls. Congress used a few loaner Altos and printers (also provided by Xerox) for committee reports and other large documents. President Jimmy Carter's assistant for information management, who had an Alto installed in his own office and another in his assistant's, used the machines to track projects and maintain lists. He called the Alto "the most powerful piece of equipment I've ever worked with." Altos were also being real-world-tested at select universities and a few companies.

Most of these "probe sites" were pleased with the machines on loan, but Xerox did not commercialize the Alto. More than a year after the probes began, the director of technical services for the Senate was still waiting to see whether Xerox was "finally going to follow up and give us all the benefits of a production model."[8]

It never did become possible to buy an Alto. By the time Tesler sat down in Taylor's office, Xerox had decided to bring a different computer, the Star, to market at $16,000 ($45,000 in 2016 dollars). The Star was a complex, impressive, and beautiful "executive workstation," developed under the guidance of former PARC researchers with little input from the marketing or sales people at Xerox.

Tesler saw no evidence that the Star, which seemed to have every feature imagined by a computer science researcher shoved into it, would be released any time soon. By the time the computer hit the market, he predicted (correctly), the small machines would already have a foothold, and the Star would be altogether too much—more power than almost anyone would need at a price higher than most people would be willing to pay.

Tesler could not help but contrast the situation at PARC with what he had seen of Apple. By 1980, he had an inside view. His first extended encounter with Apple came in December 1979, when he, along with Adele Goldberg (also from Alan Kay's group in PARC's systems science lab), gave two demonstrations of the Alto to a group from Apple that included Steve Jobs.[9]

Six months before the demos, in June 1979, Xerox had purchased 100,000 shares of Apple for $1.05 million in Apple's second round of private investment.* An investment in Apple made sense for Xerox because if the smaller company did well, Xerox could make a good deal of money, and if Apple failed, Xerox would be well positioned to acquire it.† Apple, by this time, was seen as such a potentially lucrative investment that Markkula could hand-select investors. He allowed Xerox to invest because he thought that Xerox retail stores, which sold printers and copiers, could potentially also serve as sales outlets for Apple computers.[10]

When several Apple engineers requested to see a demo of the Alto, Xerox was happy to oblige. Goldberg and Tesler's first demo to Apple was a standard presentation that did not include the most impressive parts of the Alto's capabilities. When Jobs learned that Apple had not seen the machine's full abilities, he demanded a more extensive demonstration.

---

* In this private investment round, Apple sold 720,000 shares at 10.5 cents per share. Xerox bought more shares than any other investor. With stock splits, Xerox's stake was 800,000 shares at Apple's IPO, representing 1.6 percent of Apple. Xerox sold 10 percent of its Apple shares at the IPO.

† Tesler speculates that Xerox, which had a unionized workforce, hoped that Apple, which had no unions, could become a lower-cost, high-volume manufacturer for Xerox computers. He recalls Roy Lahr, the manager of advanced development for Xerox, joking that Xerox spent more to build the plastic case for a computer than Apple spent on its entire machine.

Goldberg refused. She said she would give the confidential demo only under a direct order from senior Xerox management. Jobs made calls, Goldberg received the order, and Jobs and the Apple engineers got their first glimpse of the Alto's graphical user interface and mouse, its icons and menus, and the ease with which it could be reconfigured to meet different users' needs. Throughout the second demonstration, Jobs was alternately jumping around, yelling in excitement, and standing so close that Tesler could feel his breath on his neck.

The Apple engineers impressed Tesler. They understood what was magical and beautiful about the Alto, and they asked the right questions. They might not have had the pedigrees of the PARC researchers—Tesler himself did not have a PhD—but when it came to the inner workings of a computer, they were savvy. When Jobs called Tesler after the demo and asked him to come to Apple to work on the Lisa computer, Tesler agreed.[11]

Bob Taylor learned of the Alto demos for the Apple engineers only later. He assigns some significance to their having taken place while he was out of town. "I never would have let him in," he says of Jobs, "and if he was already in, I would have thrown him out." It seems unlikely that Taylor, who was not even in the systems science lab that gave the demos, could have done anything to stop them. Goldberg had already pushed back as hard as possible.

Tesler was not the first person Taylor had watched leave research at PARC for work in the burgeoning high-technology industry. Bob Metcalfe, one of the inventors of Ethernet, had left a year earlier to start 3Com, a networking company.[*] Within two years, Chuck Geschke and John Warnock would depart to launch Adobe; Charles Simonyi would leave to become employee number forty at Microsoft, where he would lead the technical development of Microsoft Word and Excel; and David Liddle and Don Massaro would start a personal computer company called Metaphor Computer Systems.[12]

Most people left for the same reason Tesler did: a sense of frustration

---

[*] HP acquired 3Com in 2010 for $2.7 billion. Metcalfe had left PARC briefly before 1979 but had returned.

bordering on futility at Xerox. "An engineer lives for one thing—to build something that millions of people use," explains Chuck Geschke, who adds that "money was not an object."[13] (Nor was money a significant factor in Tesler's departure; he had to ask what a stock option was when Apple offered it to him.) What was the point of being the best in the world if your ideas never reached anyone?

In the end, Apple, IBM, and companies such as Adobe and Microsoft commercialized many of the innovations from PARC's computer science lab and systems science lab. Critics claim that Xerox could have owned personal computing and a significant chunk of the software industry, if only the company had not "fumbled the future."[14] Xerox's defense is that it netted more money from the laser printer developed at PARC than was spent on the entire research center—a very real win that should over-shadow any theoretical might-have-been losses.[15]

By Taylor's accounting, "winning" in computing had nothing to do with money or commercial success. For Taylor, winning meant realizing his and Licklider's vision by building a truly interactive computer that facilitated communication. By that measure, PARC had already won by the late 1970s. Altos were being used in many parts of Xerox, and Taylor's lab had already moved on to new contests: faster machines, smaller ma-chines, ever-more-interactive machines that would understand speech commands. The people who mattered to Taylor—the world's leading academic computer scientists, not venture capitalists, shareholders, the buying public, or the press—knew he had won. Chaired professors at MIT, Stanford, and Carnegie Mellon were impressed by PARC's work and stunned by the caliber of Taylor's lab. In 1983, Stanford computer science professor Donald Knuth called the computer science lab "the greatest by far team of computer scientists ever assembled in one orga-nization."[16]

If Xerox was too shortsighted to turn top work into products, if the general public was so undiscerning as to want to buy a toy and call it a computer, that didn't mean that Taylor's team had lost. What did those people know?

The mind-set of Taylor and other leaders in the lab was fundamen-tally different from the traditional Silicon Valley way of thinking. In 1984, Butler Lampson, the technical leader of the computer science lab, saw the

Apple Macintosh, the computer that had been at least partially inspired by the Alto demo Tesler and Goldberg had given to the Apple group five years earlier. Lampson's reaction was not outrage at having been copied or irritation that someone else had made money from PARC's innovations. "I thought the Mac was fantastic," he recalls decades later. "It was just great that Mac came along."[17]

When the earliest researchers left for commercial enterprises, Taylor was hurt. ("He took it personally," Tesler says.) But as people continued to leave, and almost always for the same reason, Taylor's pain became rage. His fury was not directed against the people leaving—"He understood," says Geschke—but against Xerox. By being so slow and unresponsive, Xerox was chasing his team away. As Taylor later put it, "I was of two minds. I understood their frustration because I was frustrated, too. But I wasn't ready to give up." Taylor had created the computer science lab. He was proud of its work and venerated its researchers. To turn his back on it all would be a huge personal and professional loss. In 1980, he had considered leaving for Hewlett-Packard, which had offered him more money, a better title, and the opportunity to take some of his team with him. But Taylor had not liked the man who would have been his boss at HP. He had stayed with the devil he knew. "I suffered from a failure to see reality," he now says.

PARC's splendid isolation from the entrepreneurial ferment in the valley below its glassed windows is hard to believe now. In many ways, the PARC team and Silicon Valley entrepreneurs moved in the same general direction: toward smaller, faster, more personal and user-friendly devices built in casual work environments. They did so, however, along parallel paths, PARC's originating in academia while much of the Valley was influenced by countercultural ideas. When Geschke and Warnock left PARC in 1982 to launch Adobe, they were so unaware of the entrepreneurial ecosystem around them that they called Warnock's thesis adviser, Dave Evans, in Utah for advice. Evans put them in touch with the venture capitalist Bill Hambrecht, whose office was only a short distance from Xerox PARC.

Driving home after Hambrecht agreed to fund them, Geschke turned to Warnock: "Do you realize we're going to start a business? Have you ever taken a business course?"[18] Warnock had not. They stopped by

Kepler's bookstore in Menlo Park and bought a how-to-start-a business book.[*]

■ ■ ■ ■

As the 1970s drew to an end, a virtual drawbridge between PARC and Silicon Valley began to lower. To Taylor's dismay, it seemed that traffic on the bridge ran primarily in one direction: from PARC to the Valley. As the venture capitalist Don Valentine put it, "We drank at the Xerox PARC well deeply and often."

---

[*] Geschke does not recall the book's title but says, "There was a chapter called 'Market Gap Analysis.' The thesis was that if you are going to start a business or a new product, look for a place where there's a need but no presence of competition. Invent the solution, take it to market, and you'll have 100% market share. Shame on you if you can't keep it. Basically we've done that [at Adobe] with everything we've done. . . . That chapter was worth its weight in gold."

Young Maniacs

███████████████ MIKE MARKKULA

From the moment Mike Markkula funded Apple and demanded the highest-quality marketing, publicity, and advertising campaigns, he had worked to make the young firm appear credible and professional, a stylishly creative company that built tools, not toys. "People DO judge a book by its cover, a company by its representatives, and a product's quality by the quality of its collateral materials," he wrote in December 1979. "Our image is the combined result of everything the customer sees, hears, or feels from Apple."[*] To bolster Apple's image, Markkula also brokered alliances and branding agreements with major companies. The telecom giant ITT distributed computers in Europe. Bell and Howell, a company that

---

[*] Markkula wrote in "The Apple Marketing Philosophy: Empathy, Focus, Impute" (December 1979) that he wanted Apple marketing to run on three underlying principles: (1) Empathy: "If we have empathy for our customers and dealers, we will truly understand their needs better than any other company." (2) Focus: "In order to do a good job of those things that we decide to do, we must eliminate all of the unimportant opportunities, select from the remainder only those that we have the resources to do well, and concentrate our efforts on them." (3) Impute: "We may have the best product, the highest quality, the most useful software, etc; if we present them in a slipshod manner, they will be perceived as slipshod[.] If we present them in a creative, professional manner, we will impute the desired qualities."

supplied media equipment to schools, marketed a classroom version of the Apple II that was black with a tamper-resistant cover.

The influence of Markkula's attention to appearances, an attention shared by Jobs and Regis McKenna, was evident in abundance on December 12, 1980, when Apple Computer had its initial public offering. The board of directors was extravagantly well credentialed for such a young firm, including the CEO of Venrock, the Rockefeller family's venture capital firm; the CEO of retail giant Macy's; and the venture capitalist behind Fairchild Semiconductor, Scientific Data Systems, and Intel.* The offering's investment bankers—Hambrecht & Quist and Morgan Stanley— were among the most respected in the world, "the best names we could get," in the words of board member Arthur Rock, who noted that Morgan Stanley usually did not deign to take upstart tech firms public.[1] The size of the offering was enormous ($101.2 million raised versus the $35 million raised by Genentech two months earlier), and the list of underwriters included some of the top names in finance at the time, including Bear Stearns; Blyth Eastman Paine Webber; Alex. Brown & Sons; Drexel Burnham Lambert; Goldman Sachs; E. F. Hutton, Kidder Peabody, and Lehman Brothers Kuhn Loeb. Thirty years later, Apple would be the biggest company, by market capitalization, in the world, and several of its underwriters would be bankrupt or sold.

One name that did not appear on the investor list was Sequoia Capital, Don Valentine's venture capital partnership. The partnership had sold all of its 140,008 shares in Apple in the second round of private investment in 1979, the same round in which Xerox had bought 100,000 shares. "It's not

---

* Hank Smith, the original Venrock partner on the Apple board, was replaced by CEO Peter Crisp shortly before the IPO ("They wanted the managing partner on the board," Crisp says). In a 2008 interview with the National Venture Capital Association, Crisp also recalled that the night before Apple went public, David Rockefeller invited several people from Apple, including Jobs and Wozniak, for cocktails at Rockefeller's Manhattan town house. The next day, Crisp saw Rockefeller and thanked him for hosting the event. "It was really a pleasure. I enjoyed meeting them," Rockefeller said. "But next time, ask them not to leave decals on the mirrors." At some point during the party, someone had stuck a rainbow-striped Apple logo on a lavatory mirror.

a decision I'm happy with," Valentine said years later, explaining that he was in Africa when Sequoia had decided to sell.[2] It proved to be a costly vacation. The stock, which sold for $1.47 million in the summer of 1979, would have been worth nearly seventeen times that ($24.64 million) at the IPO sixteen months later.

The influence of semiconductor industry veterans such as Valentine on the young Apple is striking. In the three years that Apple Computer was privately held, the following key executive roles were filled by people who had worked in the semiconductor industry: board chair, president, CFO, and the vice presidents of manufacturing, sales, marketing, human resources, and communications. Of key executives, only Apple's chief counsel and two senior technical leads had not come from the semiconductor industry. Moreover, three of Apple's most important early investors—Valentine, Rock, and Venrock—came to Apple via Markkula's semiconductor industry ties. Regis McKenna, who played a vital role in positioning Apple for the public, was also a semiconductor veteran.

The pace of Apple's growth was stunning. The company had gone from $774,000 in sales in 1977, its first fiscal year, to ten times that in its second, to $40 million in 1979, to $117 million in 1980. Profits had zoomed from $42,000 to $11.7 million. From three employees at the end of 1976 (Jobs, Wozniak, and Markkula), Apple had grown to roughly 1,000. Through Gene Carter's redesigned distribution model, Apple had sold 131,000 Apple II systems via some 1,800 retail computer stores.[3] A month before the IPO, Apple had introduced the Apple III, a higher-end machine with an internal floppy drive, more sophisticated components, and more memory. Two new machines, the Lisa (which Tesler, formerly of PARC, was helping to develop) and the Macintosh, were also in development, though secret. Both computers would be released with a graphical user interface, a mouse, menus, icons, and other features inspired by the PARC Alto. Soon it would become apparent that Apple had too many different computers under development, but in 1980, the variety was cause for celebration.

Hundreds of software programs had been written for the Apple II. Although games comprised the most popular category, the single most important program was VisiCalc, the first easy-to-use spreadsheet program.[4] Available exclusively on the Apple II for a year, VisiCalc was of such value to Apple that the IPO prospectus devoted a full paragraph to the unnamed "financial modeling system." (One wag called the Apple II a

"VisiCalc accessory."[5]) With VisiCalc, a user could change a single number, and the spreadsheet would automatically rerun calculations that depended on it with the new value. Before, changing a number had meant recalculating every value. Gene Carter recalls that to have enough room for all of the cross-outs and recalculations, he would unroll butcher paper on his floor and hand-draw boxes large enough to hold all the changes.[6] Markkula, the planner, loved VisiCalc, which he called "the what-if program" for its ability to show him what would happen if he changed one variable in his model.[7]

VisiCalc's exclusive deal with Apple, like so many decisions that benefited the company in its early years, can be traced to the combined efforts of Markkula, Jobs, and Wozniak. The program's creators chose to write for the Apple II in part because Jobs had impressed the founder of VisiCalc's parent company (who had bought a discounted Apple II from him) and in part because the machine was likely to have a floppy drive—which existed thanks to Markkula's insistence that Wozniak build one.[8]

Apple had several competitors by the time of the public offering. The Tandy/Radio Shack TRS-80 had a larger market share than the Apple II, and Commodore's PET was a close third. In its prospectus, Apple anticipated the imminent arrival of other competitors, including IBM, with far greater financial resources. But Apple was the exciting, sexy new firm in an exploding market, and anticipation had run high in the weeks leading up to the public offering. Reporters praised "the finest management team of any company of its size" and gushed that "Apple has achieved more since 1977 than most companies have managed in the last century."[9] As investors' enthusiasm for Apple veered toward frenzy, the state of Massachusetts barred the sale of shares within its borders, deeming the offering too risky.[10] Would-be investors in Illinois were also out of luck, thanks to strict state laws on new issues.

The public offering on December 12 sold out within minutes, raising $90 million for the company and marking one of the most successful IPOs in US history up to that time.[11] The markets closed with Apple selling at $28.50. Markkula, who owned 14 percent of the company (second only to Jobs, who owned 15 percent) had stock worth more than $200 million ($540 million in 2016 dollars). In other words, every dollar that Markkula had invested four years earlier was now worth almost $2,200. Wozniak had about half the amount of his cofounders. His former wife owned more than 1 million shares originally granted to him. Moreover, he had given or sold stock to employees whom he thought deserved more.

By any measure, it was an astonishing rise from the garage. At the western-themed closing party a week after the IPO, Markkula wore a suit, a snakeskin-print tie, and a giant suede cowboy hat. Throughout dinner, one person after another stood to give a speech or receive a gag gift: Markkula and Jobs; the executive team; attorney Larry Sonsini, who had shepherded the offering. A group of bankers from Morgan Stanley challenged Apple to field a team against them in the next year's Bay to Breakers foot race across San Francisco.[12] Two employees of Hambrecht & Quist, Catherine de Cuir Schaefer and Garrett C. D'Aloia, donned T-shirts decorated with a giant rainbow-striped Apple logo and sang "You Picked a Fine Time to Bring a New Deal" to the tune of Kenny Rogers's hit song about Lucille. The pair also wrote new lyrics to "Somewhere Over the Rainbow" from *The Wizard of Oz*:

> Somewhere Over-the-Counter
> Not too high
> There's a stock that I heard of
> Waiting for me to buy
> Somewhere Over-the-Counter
> When I'm blue
> I'll have bought in at one
> And I'll sell at ten to you.

A few months earlier, Apple had hired financial planning and accounting experts to help employees plan for the wealth that would be theirs after the IPO.[13] But the party was pure celebration for the industry builders whom Jobs sometimes called "young maniacs."[14] The festival of jokes, silly hats, sillier gifts, and grinning attorneys, bankers, and Apple employees seemed as though it would never end. Then the troubles hit.

■ ■ ■ ■

In February, two and a half months after the IPO, Wozniak, a licensed pilot, crashed his Beechcraft Bonanza while taking off from a small airport.[15] He was badly injured. For about five weeks, he suffered from amnesia that prevented him from forming any new long-term memories. He had grown increasingly unhappy at Apple and suspected that his colleagues there would not miss him. With more than a hundred engineers at the company, Wozniak says, he did not feel needed.[16] After he recovered, he left Apple to enroll under the name Rocky Clark at the University of California, Berkeley, intending to finish his degree. Rocky was the first name of his dog, Rocky Raccoon, and Clark his wife, Candi's, last name.[17] Wozniak returned to Apple about a year later.

Meanwhile, the recently introduced Apple III was suffering from manufacturing and technical problems so severe that allegedly 20 percent of the computers did not work when they arrived at the dealerships. More failures hit when users took the machines home.[18] Apple pulled the Apple III from the shelves. The stumble irritated and embarrassed Markkula, who placed such high value on first impressions. "We learned how not to do it," he admitted at the time.[19]

Apple was hiring so quickly that almost half of its employees nine months after the IPO had not been there when the company went public. This trend alarmed Ann Bowers, the vice president of human relations, formerly of Intel, who circulated an urgent memo to the executive staff. She pointed out that although head count had doubled, productivity, as measured by sales per employee, had plummeted by 40 percent. "Sometimes we appear to be throwing bodies at the problem," she wrote. "Do people know what to do? Are they spending time on the right things?" She warned that Apple would suffer if it continued to grow this way.[20]

By 1981, Apple was manufacturing or developing four different computers: the Apple II, the Apple III, the Lisa, and the Macintosh. Each computer had its own staff and culture. Divisions of other sorts had arisen, as well. Employees who had had stock before the IPO were generally much wealthier than the newer hires, though recently recruited executives received large options packages—which proved another source of tension. Jobs, who had lobbied Scott for control of the Lisa project and been denied, was now pushing for control of the Macintosh, which was intended to be an inexpensive, user-friendly machine.

Employees had begun to complain about the new formal policies around engineering change orders, hiring, and decision making. Burnout was also a problem. "It was 12 hours a day and weekends," one person recalled. "I knew if I took a drink at a water fountain I would miss a beat and slip a schedule."[21]

Outside the company, the gee-whiz stories about Apple and its computers continued. A reporter wrote about how Industrial Light & Magic, a division of Lucasfilm, was using an Apple II to develop special effects. Steve Jobs, who would later buy the computer division of Industrial Light & Magic called Pixar, clipped the article and sent it to Apple's head of communications. Several papers carried the story of the minister who conducted marriage ceremonies via an Apple II. "Hello, my name is Reverend Apple. I am the world's first ordained computer," the screen read before displaying passages from Kahlil Gibran's *The Prophet* and asking bride and groom to input a *Y* as their "I do." Even the article about a pimp using an Apple II to store client names had a "What will they think of next?" tone.[22]

But now, for the first time in Apple's young life, the company received negative press, as well. Reporters probed the stumbles with the Apple III and raised questions about morale and the company's ability to develop new products. *Forbes* headlined one article "Apple Loses Its Polish."[23] The *Wall Street Journal* story "Apple Computer Takes a Bruising" quoted Markkula's insistence that "the polish isn't gone; it's getting brighter."[24]

According to the four-years-and-out plan that Markkula and his wife, Linda, had agreed to when he had joined, 1981 was to be the year Markkula retired from Apple. But troubles continued to mount, pulling him even further into the company.

On March 16, a month after Wozniak's plane crash, Markkula did something he later described as "one of the hardest things in my life." He asked his friend Mike Scott, Apple's president, to resign and take a nonoperating role as vice chairman of the board. It was a clear demotion. Three weeks before Markkula asked for the resignation, Scott had fired forty employees on a single day that even official Apple publications called "Black Wednesday."[25] Markkula says that the firings were justified and "not a reason" that he asked for Scott's resignation. Nor, he says, was the failed launch of the Apple III. To this day, Markkula declines to specify why he asked Scott to leave.[26]

Ever since the IPO, Scott had worried that Apple was coddling its employees. Anyone who had worked at the company for at least sixty days and demonstrated basic competence using a software program received a computer to take home.[27] On business trips, people flew first class or rented luxury cars.[28] The company had debuted the "Apple Collection, a catalog of clothing and lifestyle items bearing the Apple logo" for employee purchase. At one off-site, the executive staff had debated for a half hour whether the company should offer free decaf coffee to its employees, alongside regular. Markkula thought this type of conversation was important and showed that Apple cared about its employees. (There was concern at the time that caffeine might be harmful.) Scott thought the discussion was a waste of time.

Scott lasted a few months in his new role as vice chairman, but in July 1981, he left Apple for good. In his resignation letter, he deplored Apple's "excessive emphasis on titles and high salaries and not enough frugality." He wrote, "I quit, not resign to join a new company or retire for personal reasons." He added, "This is not done for those who fear my opinions and style, but for the loyal ones who may be given false hope."[29]

With that missive, Apple's first president, who still owned about 6 percent of the company, left. In a rare interview around the time of his resignation, Scott, photographed at home with two of his cats on the floor of his airy, high-ceilinged living room, told a reporter that he planned to "start two or three electronics companies," finance a few others, and buy real estate.[30] In recent years he has traveled the world, funded an effort to launch a rocket from water, and backed research to build what he describes as the "tricorder" from *Star Trek*: a device that can be pointed at a rock and instantly provide information about it.[31]

Scott was never the public face of Apple, even when he was president. But he was the mastermind behind Apple's flexible, scalable manufacturing operation that made it possible for the company to grow as quickly as it did. He did everything from writing manuals to creating the behind-the-scenes systems that supported Apple's growth. "Scotty was the daily rhythm of the company, the heart regularly pumping in the body," says Trip Hawkins.[32] Wozniak says, "The world never knew how important Mike Scott was."[33]

■ ■ ■ ■

Scott had not groomed a successor. Jobs, at twenty-six, was too inexperienced to take on the presidency of a company that had $80 million in sales in the previous quarter alone. Apple's frenetic growth and increasingly aggressive competitors, along with rumors that the computing giant IBM would soon introduce a personal computer, also meant this was not the time to bring in an outsider. "We need someone who knows the organization," human relations head Ann Bowers told Markkula.[34] She wanted him to replace Scott as Apple's president. She knew that he was loyal to Apple and its employees, and she also knew that Markkula had more than $100 million of his personal wealth in Apple stock.[35]

Markkula had never wanted to be Apple's president. He did not want extensive day-to-day management responsibilities, and this man, who so valued his privacy that he used a pseudonym to write software, did not relish the prospect of serving as Apple's public face with stock analysts and the press. But there was no one else to do the job.

In the middle of the board meeting to name Scott's successor, a secretary knocked and asked Markkula to step outside. His mother-in-law was on the phone. His father-in-law, Markkula's best friend, had died of a heart attack. Markkula's wife, Linda, did not yet know; his mother-in-law could not reach her.

Markkula hurried back into the conference room. "I'm sorry, guys, you're going to have to decide without me," he said. "If you want me to run the company, that's fine, but I've got a personal event that I just have to deal with." He drove from house to house, asking Linda's friends if they had seen her. Several hours later, he called his office and learned that the board had named him Apple's president.

Markkula calls the loss of his happy, athletic father-in-law and friend "a huge trauma for me in the middle of all this other stuff that was going on." He handled the funeral arrangements and comforted his wife while dealing with his own grief and assuming the presidency of one of the fastest-growing companies in American history. Alone, he says, "I would spend time thinking, 'How am I going to get through this?' . . . It was really hard. After it was all over, I looked back on it, and I said, 'I'm a lot tougher than I thought I was.'"[36]

Jobs, who was named Apple's chairman after Scott's departure, told a reporter that "Mike Markkula became president because his management style is more suited to the running of a multidimensional, multidivisional, multinational corporation."[37] Markkula told the *San Jose Mercury*,

"Within a couple of years, I would hope that I would not be in the same position."[38] He explained to the board that he would do the job "until we find somebody better." Whether that was six months, or a year, or several years away, he could not predict. But he hoped it would happen quickly. Being president "was not what I wanted to do with my life," he says.[39]

With Scott and Wozniak gone, Markkula and Jobs had each lost the man he had brought with him to launch Apple. But Markkula and Jobs worked well together. Trip Hawkins says that Markkula "absorbed stress like a sponge," a characteristic that made Apple's new president a stabilizing force for the chairman's more passionate temperament.[40] Markkula valued Jobs's ability to anticipate what a customer might want next, an ability that stood out even in a company filled with bright, creative people. Markkula also believed that with enough maturity and experience, Jobs could become a powerful leader. For his part, Jobs took Markkula as a model of business acumen, selling the same number of shares as Markkula every time the opportunity arose in the company's early rounds of private offerings.[41] He also credited Markkula as the source of his own belief that customers derive their sense of a company from every detail, no matter how small, that they associate with it.[42] He later said of Markkula, "His values were much aligned with mine. He emphasized that you should never start a company with the goal of getting rich. Your goal should be making something you believe in and making a company that will last."[43] The venture capitalist Don Valentine, one of Apple's first board members, calls Markkula "Steve's personal trainer."[44]

Apple's official publications began showcasing Markkula and Jobs as a team. Both men now signed the quarterly reports, though only Scott had done so before. The 1982 annual report includes a full-page shot of Markkula and Jobs, striding side by side into the light. Jobs, in a tie and dapper mustache, is gesturing and looking at Markkula, who, hands in his pockets, listens intently. Beneath their photo is a promise: "Bringing technology to individuals through personal computers is, we believe, the extraordinary business of this decade."[45]

Markkula and Jobs split responsibilities. Markkula rejiggered the company, bringing in four new vice presidents in three months.[46] Always the planner, he also introduced a companywide meeting during which he reviewed Apple's plans and priorities for the next year with employees.

"I wanted people to know how they fit in and where we were pointing the money guns," he says. "I wanted people to care about what their peers were doing, because the whole thing has to come together for it to succeed."[47]

Markkula, believing that "too much supervision is one of the most common management mistakes," took a hands-off approach. "He would camp in his office a lot, scheming away very thoughtfully," says Hawkins. "There was not a lot of management-by-walking-around."[48] Markkula liked to say that he wanted to hear people arguing in the halls. They argued, too, in his meetings. When two of his deputies could not agree on how many Apple IIs to build for the Christmas season, Markkula told them that if they couldn't choose a number, he would have to do it, "and you don't want me to decide [because] I don't know half as much as you know."[49] That was about as much of a threat as he ever made. He let people at Apple make their own decisions and mistakes.

Jobs, meanwhile, tried to see around corners. With Markkula's blessing—"My strategy from Day One with Steve was to keep him so busy that he would stay out of trouble," Markkula says—Jobs took over the Macintosh group, to the applause of some and consternation of others. When Jobs hung a pirate flag over the building where the Macintosh was being developed, many people at Apple interpreted it as a defiant move. Markkula felt differently: "I wanted [Jobs] to use as much of the Lisa technology as made sense. And for them to think they were pirating it away was just fine with me."[50] The Lisa team included the Macintosh group on its user interface memos.[51]

Regis McKenna, who continued to work closely with Apple on advertising and public relations, believed that the best way to make a complex technical company approachable was to associate it with an appealing person. "What really differentiates a business is people," he once said. "The idea of infusing personalities into this started back very, very, very early."[52] McKenna had already made Intel's Robert Noyce—father of the microchip, humble millionaire, midwestern preacher's boy made good—the public face of Intel and the microchip industry. At Apple, the obvious choice for the role was young, handsome, passionate Steve Jobs. McKenna dubbed Jobs Apple's "media man."

Though Markkula spoke occasionally to the media and financial analysts about day-to-day operations, Jobs became the company's—and the industry's—young philosopher-king. In private conversation, Mark-kula could spin inspiring dreams in a manner one admirer describes as "charming and intimate and warm . . . let me sit on the side of the bed and tell you how great it's going to be."[53] But Jobs possessed a spellbinding charisma whether he was in large groups or small, with friends, strangers, or reporters. He gave Apple's products an aura of inevitability and transformative power. To buy an Apple computer, he made clear, was to become part of something extraordinary, a company that, like its young chairman, exemplified the hope for tomorrow.

Jobs gave an extended interview that McKenna turned into a series of three full-page advertisements in the *Wall Street Journal*.[54] The advertisements featured four different shots of Jobs's face and quoted his big ideas, such as "A personal computer is more than just a small 'big' computer." In the ads, Jobs also recounted his favorite stories: the Apple II that had saved the business of a sewing machine repairman in England, the medical center that was using the machines to process ambulance reports, the teachers who had found that the computer's graphics "make it fun to learn." He further explained, "When we invented the personal computer, we created a man-machine partnership."[55] The presumably unintentional echo of Licklider's paper on "Man-Computer Symbiosis," which had so inspired Bob Taylor in 1963, staked a claim about the invention of the personal computer that would have infuriated Taylor and many others at Xerox PARC.

Jobs told a meeting of Apple users in Boston, "Civilization seems to have taken a notch forward every time some source of free energy became available. Petrochemicals were a source; the plow was a source.* People either leveraged mechanical energy, or they found a free source of intellectual energy, like language or mathematics. But there's never been much of an artifact that allowed the free use of intellectual energy. The Apple II saves two hours a day for me. That's free intellectual energy."[56] Jobs told the *Los Angeles Times* that "Personal computers will promote

---

* More than a year earlier, the analyst Ben Rosen, then with Morgan Stanley, had made similar comments about electronics providing free energy and free intelligence.

much more of a sense of individualism, which is not the same as isolation. It will help someone who is torn between loving his or her work and
loving the family."[57]

At times, the chairman of Apple's board sounded like the twenty-six-
year-old he was. "When we first started Apple, I was very apprehensive
about getting into business," he admitted.[58] But after five years, he had
changed his mind: "I think business is probably the best kept secret in
the world. It's viewed on the outside mostly negatively, but it really is a
wonderful thing." What he liked most, it seemed, was the excitement of
defining a new market: "It's like the razor's edge."[59]

In 1981, Apple spent $10 million on advertising.[60] In addition to print
advertisements, the company ran a series of prime-time television spots.
The commercials starred popular talk show host Dick Cavett, who in chatting about the easy-to-use Apple II became the tech industry's "first celeb-
rity spokesman," according to McKenna, who had convinced the star to do
the job.* By the middle of the year, Apple had sold 150,000 computers, and
80 percent of people surveyed could identify Apple's brand and business.[61]

■ ■ ■ ■

But there was no respite for Markkula. On August 12, 1981, five months
after he became Apple's president, IBM introduced its personal computer
at a press conference in the ballroom of New York's Waldorf Astoria hotel.
The IBM machines, with 16-bit Intel microprocessors, were faster, had
more memory, and were touted as capable of handling more complex
tasks than Apple's computers. IBM had developed its personal computer
in a single year at a secret skunk works in Boca Raton, Florida.

Apple responded to the IBM PC's introduction with a full-page ad in
the *Wall Street Journal* that declared in huge, bold-face type, "Welcome,
IBM. Seriously." The audacity of the headline was echoed in smaller print

---

* The commercials had a deliciously self-mocking tone, and one had a decidedly
  feminist feel: when Cavett explains the Apple II's versatility and memory capacity
  to a woman in terms of recipe storage, she smiles and nods before mentioning that
  she also uses the machine at the steel mill she runs and to track her investments in
  gold futures.

below: "We look forward to responsible competition in the massive effort to distribute this American technology to the world. And we appreciate the magnitude of your commitment." A year later, Alan Kay, the mustachioed organ player at PARC who would soon join Apple as an Apple Fellow, said, "Apple welcoming IBM to the market is like a caveman welcoming a saber-toothed tiger into his cave."[62]

Markkula, who had been anticipating IBM's entry, could only hope that Apple had staked out enough territory to stave off the attack.[63] Today he recalls IBM's introduction of the PC as "bittersweet"—sweet because it validated the market and bitter because Apple now had a formidable competitor. At the time, IBM's entry appeared not to faze him. "Apple Computer president Armas Clifford Markkula Jr. took a drag on his Winston cigarette and made a boast that grossly belied his soft voice and easy manner," a reporter noted five days after IBM's announcement. "Outside of a third world war, I can't see anything knocking us out of the box as one of the premiere makers of personal computers," Markkula said. "We not only can withstand IBM's new personal computer, we'll come out on top."[64]

Today, Apple is known for its self-representation as the iconoclastic outsider company. This reputation exists in large measure thanks to Apple's iconic 1984 advertisement for the Macintosh that compared IBM to George Orwell's Big Brother and Apple to a courageous female rebel (as well as later advertising campaigns, such as 1997's "Think Different"). But in 1981, Apple, led by Markkula, was the *establishment* company in the personal computer industry. Markkula's plan from the start had been to be first to market so that Apple could set the standard. In 1981, at least when it came to the personal computer, IBM was the disruptor. "IBM knows as well as anyone the power of an established base," Markkula told the *Wall Street Journal* in 1981. "Now the shoe's on the other foot."[65] He compared IBM to a new hamburger stand and Apple to McDonald's.[66] IBM ran advertisements featuring women, warm-fuzzy promises ("One nice thing about having your own IBM Personal Computer is that it's *yours*"), and Charlie Chaplin's Little Tramp from *Modern Times*, the movie about one small man's fight against big business and bureaucratic technological efficiency.[67]

The IBM PC proved so popular that within days of its launch, the company quadrupled production.[68] By the end of 1982, one IBM PC sold every minute of the workday, and the 150,000 machines that IBM manufactured that year came a close second to Apple's production of 225,000.[69] One year later, IBM's share of the personal computer market surpassed Apple's, and by 1985, the personal computer division of IBM was so successful that had it been a stand-alone company, it would have been the world's third largest computer company, behind Digital Equipment Corporation and IBM itself.[70]

Within a year of the PC's introduction, software developers around the world had written 753 programs for the machine.[71] VisiCalc's parent company wrote a version of the program for the IBM PC. Microsoft can trace much of its success to IBM's decision to adopt a Microsoft operating system for the PC. The flood of new software titles initiated a virtuous cycle for IBM: more software meant more sales, which in turn meant that more people wanted to write software for the fast-selling machine.

Since Microsoft licensed the PC's operating system to other computer manufacturers (who could also buy the same Intel microprocessors and other chips used in the IBM machines), Apple faced even more competition from IBM-compatible computers made by different companies. As Markkula later put it, IBM's entry "was good in that it legitimized the market for personal computers, but it was not so good in that we didn't have just one competitor, we had twenty."[72] At the end of 1980, roughly two dozen firms sold personal computers. By the end of 1981, the year of the IBM PC's introduction, forty-four did.[73] One month after IBM's PC launch, analysts lowered their earnings estimates for Apple's fourth quarter, and Apple's share price fell by 13 percent.[74]

The market that IBM and Apple were seeking to dominate was growing quickly, but it was still small. Roughly 100,000 homes in the United States had a computer in 1981, and offices and businesses housed another 275,000 personal computers.[75] That same year, there were 81 million televisions in the country.[76] For most people, computers were still foreign and vaguely menacing. "Today's computers are easygoing companions that want only to serve, educate and entertain us," a writer for the luxury magazine Town & Country promised at the end of 1981, but, he admitted, "most of us are still frightened of computers or react to them with outright hostility—they're the totally unreasonable monstrosities that continue to send us the same incorrect bill."[77]

If 1980, the year of Apple's IPO, had been a time of unprecedented growth and celebration, 1981 tested both Markkula and Apple. "Our progress did not eclipse problems, internal or external," Markkula wrote in the 1981 annual report. With a local reporter, he was more direct: "From the outside it looks like we walk on water, but on the inside we know we don't."[78]

What in the Hell Are You Trying to Say?

FAWN ALVAREZ

Around the same time that Mike Markkula began planning for Apple's IPO, Fawn Alvarez was promoted to the job of manufacturing supervisor at ROLM, the pioneering telecommunications and military computer company. Her new position was still considered a production job, but it entitled her to a private cubicle near the manufacturing line for ROLM's CBX telecommunications system. Alvarez loved having her own space, even if she did not have her own phone line like the engineers she worked with.

As supervisor, Alvarez was responsible for hiring, firing, and training the roughly one hundred women working on the assembly line. In addition, one of her most important responsibilities was implementing the changes on the line that the production engineers requested. There were around twenty production engineers in the telecommunications division where Alvarez worked. At least one change a day was requested, sometimes more. Perhaps a part needed to be substituted for a different one. Or the line needed to be reconfigured to build the version of the CBX system most in demand. Or maybe a brand-new assembly—a different power supply, for example—needed to be folded into production. Every day, Alvarez would receive copies of the change orders in the interoffice mail. For time-critical changes, an engineer would hand-deliver the order.

The change orders were typically a dozen pages long. Alvarez would scan through the boilerplate at the top, skimming over the signatures from various supervisors and department heads, to get to the change and what it meant for her. How many people would she need to assign to a new job? What did the new parts look like? Did the new build plan and drawings make it clear where to put the parts? Would she need to move tables? Could the changes be worked in without stopping the line, or did production need to stop temporarily?

Often Alvarez received a change order that she could not understand. "Impede the resistance of U4," the change order might instruct. Or "Speed up the channel driver." Alvarez would groan when she saw such language. It described the problem to be fixed, not the fix itself. She would have to get clarification—again. She would carry the sheaf of papers to the appropriate engineer's office, push the pages across his desk, and ask, "What in the hell are you trying to say? I don't know what you're saying. These are not words any of us know. What do you want us to do?" The engineers, who were used to Alvarez's direct approach, could always explain what needed to happen: swap out a chip, or sever a printed connection between components and wire in a new one. Meanwhile, Alvarez had to stand by and wait as the engineer rewrote the change order to explain the necessary steps. She found the experience, which happened at least a few times each week, very frustrating.

Looking back, Jeff Smith, a former ROLM production engineer, understands why. After forty years and many senior management jobs, Smith sees that back then, engineers wrote change orders "without regard for what the manufacturing process is. [It was] one department issuing edicts to another department without really understanding the effects of those edicts."[1]

Alvarez complained enough that her boss, Bill Rea, finally told her that if she thought she could do a better job writing change orders than the engineers, she should do it. Get a draft order, ask her questions, and then rewrite the change order in a way that made sense to her, rather than waiting for the engineers to do it. If the engineers approved her version, Rea was fine with it. Alvarez soon found that Rea's suggestion made work easier not just for her but also for the engineers, who could spend less time writing and rewriting change orders.

After a few weeks Alvarez told Rea that she no longer wanted to work as a manufacturing supervisor. She asked him to give her a different job, one that had not existed before. She wanted to write change orders full-time. Moreover, since this would be professional work, she wanted to move out of manufacturing and join the engineers in the production engineering group.

Rea, a fellow risk taker unafraid of upending convention—"At ROLM, the meek didn't inherit the stock options," Jeff Smith explains—was willing to try it. Rea pointed Alvarez to an office. She noted that it had a door, unlike her cubicle, and a phone with a dedicated number.

Twenty-two-year-old Fawn Alvarez had finally gotten her longed-for job behind a desk.

Some of the engineers who had thought Alvarez was cute and spunky when she questioned them from the other side of the production/professional divide now saw her as a potential threat. A woman who had never attended college was in the office next to theirs, doing a job—writing change orders—that they also did. Alvarez sensed that her very presence lowered their prestige.

Soon the engineers were giving Alvarez so many change orders to rewrite that she began to suspect a deliberate attempt to overwhelm her. She went back to Rea and proposed that she hire two women to work for her. She pointed out that the three women would be paid less than the engineers, and now the engineers would be freed up for other work.

Within weeks of the third woman hiring on, the engineers had stopped writing even the first drafts of their change orders. They instead gave the women only cursory notes. "When they saw that someone who didn't go to college and was a girl could do it, none of the engineers wanted to do it anymore," Alvarez recalls. From then on, the engineers approved the change orders but did not write them. She adds, "That was women's work now."[2]

■ ■ ■ ■

As Alvarez began her new job behind a desk, ROLM was marking its tenth anniversary with the highest profits in its history ($11.3 million) and sales of $114 million, up 128 percent over the previous year. More than two thousand people worked for the company, five times the number three years earlier. The stock split twice in as many years.

The company's phenomenal growth meant great wealth for the founders and for the executives and engineers with generous stock option packages. Moreover, the benefits filtered throughout the company. In January

1980, Alvarez's mother, Vineta, who had been buying ROLM stock at the discounted rate for years—and who had been granted options on 2,000 shares—retired. She had not been allowed to buy stock at the 50-cents-per-share price that Ron Diehl had promised when he had recruited her away from Sylvania; shortly after hiring on, she had been told that state law forbade women to buy shares in companies not listed on a stock exchange.[3] Nonetheless, without a high school degree, she had pulled herself and her daughters out of a financial situation so precarious that the girls had used to worry that their feet were growing, which meant having to buy new shoes they could not afford. Now, at age forty-three, Vineta Alvarez Holdridge felt she never had to work again.

■ ■ ■ ■

Among the most memorable days of Fawn Alvarez's two years writing change orders alongside the production engineers was the day she decided to leave. In her new job, she had begun paying attention to the other women at ROLM who worked in offices. She would see them walking along the campus's manicured paths or eating in the cafeteria. One woman was particularly striking, always dressed in elegant suits and often wearing a hat. Alvarez noticed the hats first; they were felt or feathered, often brightly colored. She had never seen anything like them. They looked expensive, as did the suits the woman was wearing. Deciding that the beautiful woman must make real money at ROLM, Alvarez wanted to know how she did it.

"I just walked up to her one day, out of the blue, introduced myself, and asked her what she did," Alvarez recalls. When the woman said that she worked in marketing, Alvarez thanked her and left. She was too embarrassed to ask what "marketing" was.

She went to the library and left with a book on business. After reading about the different organizations inside most companies (engineering, finance, manufacturing, sales, marketing, and so on), she decided that the only place where a woman could be guaranteed a good income was sales, where compensation was almost entirely a function of how much product was sold. In other areas, women could be paid less simply because they were women.

Now determined to jump into sales or marketing in as few steps as possible, Alvarez mapped out how ROLM was organized. Only one group interacted with production engineering, where she worked, and also with marketing and sales. A forecasting group translated sales and marketing forecasts ("We will sell X number of CBX phone systems with Y features") into manufacturing forecasts ("We will need to buy Z number of components of this type or that"). If she could transfer into forecasting, it would be an easy move into sales and marketing as a next step.

In 1981, a year after her trip to the library, Alvarez joined the forecasting group. She enjoyed the work but soon discovered that she was suffering the very discrimination that she had feared would hold her down in any job other than sales. She overheard one of her coworkers bragging about his salary—and realized it was twice her own. She was enraged. Her work, which involved forecasting for very large CBX systems, was more complicated than his, and a mistake on her part would be far more costly.[*]

She complained to Bill Rea, who had earlier created her job writing engineering change orders. Rea, who had received a number of promotions and become a good friend of Alvarez's, was now her boss's boss. "I want to make what that guy does, plus one buck," she told Rea. When he justified the discrepancy by saying that the coworker was married with children to support, Alvarez was furious. "Why should that matter?" she asked. "What matters is the work!" She got her raise.

■ ■ ■ ■

By 1981, ROLM had added more than 400,000 square feet to the flagship Silicon Valley campus and purchased seventy acres of land nearby. The company sold two categories of products: CBX phone systems for businesses and ruggedized computers for the military. Most of the new space housed administrative and engineering employees. ROLM had shifted

---

[*] ROLM made CBX systems of various sizes. Alvarez was responsible for systems for up to four thousand phone extensions. The man she overheard forecast for systems up to 144 extensions, systems that contained many fewer parts whose costs the group needed to forecast. Alvarez says, "I was pissed, because I had to forecast 90 percent more crap than he did, and if I made a mistake of 5 percent, it would cost the company a lot of money, versus 5 percent on what he did. I was being taken advantage of."

many assembly jobs like those Alvarez and her mother had once held to Colorado Springs, Colorado. Land for factories was cheaper there, and wages were lower. A strong military presence in the area also meant that trained assemblers were relatively easy to find.

Companies throughout the Valley were moving their manufacturing operations out of the area by the early 1980s. Intel had built plants in Oregon, Arizona, and New Mexico. National Semiconductor, where Apple's first president, Mike Scott, had once been a manufacturing manager, was building in Washington and Arizona. Other companies assembled products in Asia, often in the Philippines, where wages could be as low as 13 cents per hour. Apple was building a plant in Cork, Ireland.

Jobs outside of manufacturing were leaving, as well. ROLM moved a team focused on product development and product marketing to Austin, Texas. "We thought moving was pretty much an imperative," explains cofounder Bob Maxfield. "Like so many other companies, we were struggling with being able to bring in enough engineers and have them afford the cost of living."[4] Hewlett-Packard adopted a formal policy that all future growth would take place outside the Valley. A survey of sixty companies in Santa Clara County found that the firms expected to add 41 percent of their new jobs outside California.[5]

"There will be a change in the character of this area, from predominantly manufacturing-oriented to increasingly research-oriented, run by highly skilled professionals who can afford to live here," Hewlett-Packard president John Young predicted in 1980. He worried about what he called "the consequent pressure on blue-collar workers, the poor, aged, and disadvantaged."[6] His concerns were prescient.

We Don't Need Any Money

SANDRA KURTZIG

About the time Fawn Alvarez got her job writing change orders, Sandy Kurtzig was sitting down to lunch with a man she hoped would join ASK's board. It was hard for her to believe that only a year before, the company's future had seemed doubtful enough that she had considered selling ASK and finding a better-paying job. Now ASK was on track to triple its 1979 revenues of $2.8 million and quintuple its profits to $1 million.

The exploding market for minicomputers, like the Hewlett-Packard machines for which ASK's MANMAN software was written, was fueling ASK's growth. (In 1975, HP had sold about eighty of its top-of-the-line minicomputer systems. Five years later, five thousand systems had been installed worldwide.)[1] As hardware prices dropped and customers began to buy, ASK rode the surge. MANMAN was intended for managers and workers with no programming knowledge, but customers who coded could also easily add custom features, because MANMAN was written in a well-known language (FORTRAN) and shipped to customers with its source code.[2]

Though Kurtzig had found it exhilarating to lead ASK through its ramp-up, she also knew that as an engineer who had so far managed the company on what she called "gut feel," she could use some guidance.[3] Already she had enrolled in the Harvard Business School's Smaller Company Management Program (SCMP), a three-week intensive session for owners and CEOs of small companies that met every summer for three consecutive years. She had also hired a second-in-command, Tom Lavey, a thirty-five-year-old former army lieutenant. An experienced salesman, he had met Kurtzig when he had come to California to see if she wanted to sell ASK to the giant computer-leasing conglomerate Itel Corporation, his then employer. After spending the day with Lavey, Kurtzig declared, "I don't want to be bought by Itel, but I want to buy you. How much do you cost?"

Kurtzig was on the cusp of recognizing that ASK, a small business started at her kitchen table, might be a high-tech Silicon Valley startup. Certainly Paul Ely, a computing executive at Hewlett-Packard, saw ASK that way. When Kurtzig asked him to suggest potential board members—she, her father, and Arie were the ASK board at that point—he recommended men enmeshed in Silicon Valley's financial and entrepreneurial networks.*

One of those men was Burt McMurtry, now sitting across the restaurant table from Kurtzig. McMurtry headed a venture capital firm, Institutional Venture Associates, whose lead fund would skyrocket in value, from $19 million to roughly $200 million, in a space of six years. McMurtry was a forty-five-year-old native Texan who had moved to the Valley almost a quarter century earlier. He had an engineering PhD from Stanford, and before he had become a venture capitalist, he had run an important lab at Sylvania, where Fawn Alvarez's mother had also worked. McMurtry had recruited the four founders of ROLM to come to the Valley to work for Sylvania; shortly after they had launched ROLM, McMurtry had joined the startup's board. A humble man in a valley where humility was rare, McMurtry once attributed his success to his extensive technical training, which had convinced him that "you don't know very much, really, about anything... so the only way you make any progress is to say, 'I don't get it.'"[4]

Kurtzig knew little of McMurtry's background. Ely had recommended him for ASK's board because McMurtry was a director of a company, Triad Systems Corporation, with customers similar to ASK's. Triad, which

* Kurtzig had initially asked Ely to join the board. When he declined—he did join later—he gave her other names.

was backed by Hambrecht & Quist, as well as McMurtry's firm, sold an inventory management system to automotive parts distributors.

Since venture capitalists rarely join corporate boards without a financial stake in the business, McMurtry assumed that Kurtzig's invitation to join the board was also an invitation to invest in ASK. He was thus surprised when, after telling Kurtzig that he would be happy to invest and serve as a director, she seemed confused.

"What do you mean, you'd like to invest in the company?" she asked after a pause. "We don't need any money."

Kurtzig was now in her third year of the Harvard Business School management program. She had been running a software company in Silicon Valley for nearly eight years. It may thus seem improbable that she had set up a lunch with Burt McMurtry scarcely aware of the venture capital industry in which he worked.

But Kurtzig had not thought of herself, or ASK, as part of the world of high-tech entrepreneurs or venture capital. She was not pursuing management education at Stanford's business school but at Harvard, three thousand miles from Silicon Valley, where most of her classmates were running family businesses, not tech startups. Nearly all of the press she had received had appeared in manufacturing industry magazines or newsletters, with a nod here or there from a Hewlett-Packard publication.

She was not part of the crowd from Intel, National, or AMD who were drinking together at the Wagon Wheel bar. She did not join the scraggly-bearded gang with roots in the Homebrew Computer meetings. She had no connection to PARC or the Computer Science Department at Stanford or any other of the origin points of various Silicon Valley networks. Instead, she belonged to groups such as the American Production and Inventory Control Society. She had no mentor like Mike Markkula, no one to push her like Tom Perkins or Don Valentine. She had been going it alone, getting advice from her father-and-husband board of directors or her advisory board composed of customers. Every year she combed through the Hewlett-Packard annual report on her own, calculating HP's gross margins and after-tax profit percentages, to come up with reasonable targets for ASK.

Kurtzig's status as an outsider was due, in some part, to the industry in which she worked. Most tech companies in Silicon Valley in the 1970s built tangible products: instruments, computers, microwave devices, chips, disk drives. Venture capital deals around software were unusual before 1980. The first software products company IPO was not until 1978, when Cullinane Data Systems went public, underwritten by Hambrecht & Quist. No software company was listed on the New York Stock Exchange until Cullinane was listed in 1983.[5]

Software had even less meaning to the general public. When *BusinessWeek* ran a major story on the industry in 1980, the magazine had to define the term ("the long lists of commands or instructions that tell the computer what to do").[6] Entrepreneurs complained that even as the personal computer was gaining attention, the software that controlled the machines was little appreciated. "You can't kick it," one explained. "People just don't understand it. They don't like it."[7]

At the beginning of 1980, when Kurtzig invited McMurtry to lunch, venture capital, IPOs, and stock options were esoteric concepts outside of a small group of financiers and executives.[8] (That would change within months, after the Genentech and Apple IPOs.) McMurtry found it plausible that Kurtzig was not familiar with venture capital, and he was impressed that she had wanted to meet with him because she thought he had relevant board experience. "That she was not focused on financing and liquidity issues made me more interested in investing," he says. "She was focused on building a great company."[9] At the end of the lunch, after declining to join the board (since he could not invest), he recommended another person she might ask.

In the course of their conversation, Kurtzig asked McMurtry: If he were to invest in the company, how much would he want to put in, and what would he expect in return? He replied that in exchange for a million dollars, he would expect to receive about twice as much stock as he would be able to buy at the same price "if you went public."[10]

Kurtzig says that that was the first time she had heard the phrase "go public."[11] She asked McMurtry to explain public offerings. Looking back nearly four decades, it is natural to wonder if at times Kurtzig manipulated similar situations because she knew that as a young woman, feigned cluelessness could help garner assistance and support from powerful men. Perhaps she did exaggerate her lack of understanding with McMurtry or at other times. But context matters. Women were negotiating a professional world still dominated by men. Kurtzig's lunch with McMurtry happened around the same time that Fawn Alvarez had a supervisor stick a Post-it note on her chest and say, "I was looking for a flat surface to put this on." The bestselling advice book for career women in the late 1970s included a chapter on "What If They Call You a Castrating Bitch or a Lez" and suggested that the best way to avoid a pass was to say, "I'm flattered by your invitation, but..."[12] A woman CEO of a technology company was nearly unheard of, so Kurtzig was inventing almost every step she took.

She listened carefully to McMurtry's explanation of an IPO and quickly grew convinced that going public would be "the next logical step to making ASK a better company, the next challenge for me personally." As a publicly traded company, ASK would have a certain gravitas, a seriousness of image and reputation that would reassure customers and attract employees. A public offering would also enable the company, with $8.3 million in revenues, to expand operations, and Kurtzig and her employees to sell stock.

Even before returning to ASK after lunch, Kurtzig began researching the next steps to an IPO. She took the same approach that Nolan Bushnell, Chuck Geschke, John Warnock, and Fawn Alvarez had taken when they had questions about business: she read a book. In this case, she drove to the public library.

Back in her office, reading and making phone calls, she learned that ASK was smaller than most companies then going public.[*] She hoped, however, that its rate of growth, its lack of debt, and the building excitement around high-technology ventures would make up for the company's size, especially since fast-growing technology companies were beginning to be valued as a hedge against inflation.

She learned something else as well: no woman had ever taken a high-tech company public. "All I heard was about men taking companies public. It was the macho thing to do, men measuring themselves against other men," she recalls. The discovery fired her. "Why not women?" she asked herself.

■ ■ ■ ■

Several months later, in June 1980, Kurtzig was hosting a dinner after the first formal meeting of the new board she had assembled in the wake of

---

[*] Much of ASK's revenue went to pay HP for the computers that ASK sold loaded with MANMAN software. Computers were still expensive: ASK's revenues of $8.3 million represented sales of only about forty systems, at around $200,000 each.

her conversation with McMurtry. As the waiters stood discreetly by, she surveyed the directors at her table. She had added several marquee names, including the investment banker Tommy Unterberg, Tymshare executive Ron Braniff, and Ken Oshman, the CEO of ROLM.

Kurtzig had planned the dinner to celebrate what she anticipated would be a unanimous vote to take ASK public eight months later, in March 1981.[13] The naïf who had never heard of an IPO now planned to lead one. She had fielded underwriting offers from several investment banks, including Hambrecht & Quist, which was already co-managing what would be the top three high-tech IPOs of the year.[*] But Kurtzig thought that Hambrecht & Quist was "greedy" for investing in a startup before taking it public; the bank thus made money as both an investor and an underwriter. She instead chose the New York–based L. F. Rothschild, Unterberg, Towbin, the firm that had taken Intel public.[14]

Kurtzig was proud of ASK. She expected that within a few weeks, the company would, for the first time, sell more than $1 million in products in a single day, a milestone achievement. Once dependent on a single computer supplier (Hewlett-Packard), ASK was now negotiating with a second company, the Massachusetts-based minicomputer manufacturer DEC. The move to acquire a second source was inspired, in part, by HP's decision to release a competing manufacturing software package, with a name—MM/3000—suspiciously similar to MANMAN. "We had this definite feeling that HP's hardware was all responsible" for ASK's success, recalls one HP executive who also described the company's relationship with ASK as "a little bit of a love-hate."[15]

The dinner began as the wineglass-clinking, multicourse event that Kurtzig had expected. But when the dessert dishes had been cleared, Ken Oshman, the ROLM CEO, laid his napkin on the table and said something she did not want to hear.

"You can go public if you want to go public. Everyone else is going public," Oshman said. "But I simply don't think ASK is ready." As Kurtzig and the rest of the table went silent, Oshman enumerated the reasons

---

[*] The top three high-technology issues of 1980 were Apple, Genentech, and Monolithic Memories.

against an IPO. ASK didn't need the money, and if Kurtzig wanted outside investors, they would be easy to find privately. A public offering was risky. At $8.3 million, ASK's revenues were well below most companies' at the IPO stage. (Apple, admittedly an outlier, had $117 million in revenues at its IPO.) With the exception of Tom Lavey, the sales executive Kurtzig had "bought" in 1978, ASK's management team lacked experience outside the company. Oshman said that ASK should not go public until it had grown and hired more senior managers.

"I know you want a unanimous vote, and I'd be willing to resign from the board," Oshman said, "but my recommendation is to wait."[16]

Kurtzig was stunned. She heard herself asking the other board members what they thought of Oshman's points. She heard them saying, after a bit of hemming, that they guessed they agreed with him. And when Kurtzig thought about it, she did, too.

She excused herself. In the ladies room, where she knew no other board member could follow her, she let herself cry.[17] She had been so close, and public markets were fickle. Right now the market was churning out one successful IPO after another, at a pace more than triple the previous year's. Every month brought a new record for the number of IPOs.[18] What if the window of opportunity slammed shut while she followed Oshman's advice? She would not be able to take ASK to the next level. She would not be able to cash out.

After a few minutes, she wiped her face and took a deep breath. She went back to the table. She paid the bill. She went home. The next morning, she postponed ASK's public offering by six months, to October 1981.

■ ■ ■ ■

The white-hot market for public offerings in the months after Kurtzig delayed ASK's IPO felt like a tease. Genentech's record-setting IPO three months after Kurtzig's fateful board meeting kicked off even more excitement as some fifty companies—Apple among them—went public in the final quarter of 1980. The graph of NASDAQ performance from February 1980 through the end of the year is almost a vertical line.[19]

In the midst of the frenzy, a week before Apple's December 12 IPO, Kurtzig drove to Santa Clara to attend an American Electronics Association seminar on "going public." Sitting in the audience with some two hundred other aspiring entrepreneurs, she listened to an investment banker caution that the thirst for new technology companies had not been so great since the bubble in 1969, when fully half of the companies that went public had had the words "data," "computer," or some derivative of "electron" in their names. "You better get out there fast," he said, mostly joking. Already *BusinessWeek* was worrying that venture capitalists were pushing out companies with "inadequate knowhow, personnel, and track records" and that Wall Street did not have the scientific acuity to analyze these new issues properly. The banker assured Kurtzig and the other entrepreneurial hopefuls, each of whom had paid $155 to attend the conference, that the excitement ought to last a few more months.[20]

Kurtzig needed it to last nearly a year while she followed Oshman's advice to bring in experienced management and increase revenues and profits. She hired a seasoned head of R&D from Hewlett-Packard and a CFO from a company called Advanced Electronic Design. With Tom Lavey, she spent one night brainstorming straight through to morning, trying to imagine a way to bring in customers that could not afford a full $200,000 ASK computer system. The solution she and Lavey devised, called ASKNET, allowed customers that installed terminals (not full computers) at their companies to use MANMAN software, paying a monthly fee for processing (which would be run on computers at ASK), rather than making a large up-front purchase. In some sense, ASKNET was a throwback to the old time-sharing model. It also anticipated what today is called "software as a service," in which companies—Salesforce.com is among the best known—offer their services as software that a user accesses through a browser over the web, rather than installing a program on a computer's hard drive.

ASKNET was intended as a stepping-stone for companies too small, new, or timid to invest in their own computers, but businesses ended up staying with ASKNET long past the time it made economic sense. An executive at one computer manufacturer told Lavey that he was willing to pay a premium to stay on ASKNET, since the alternative, buying a standalone system, would mean buying HP or DEC minicomputers. "I sell computers," he said. "You think I'll have an HP computer in my building? We compete with them."[21] By 1984, ASKNET would account for 20 percent of ASK's customer base and 15 percent of revenues.[22]

By July 1981, with the IPO scheduled for October, Kurtzig had molded ASK into a more typical-looking company on the brink of an IPO. ASK had a seasoned executive team; relationships with two major computer suppliers; multiple revenue streams; an increasingly complex and integrated version of the MANMAN software package that now included graphing capabilities; nearly a hundred employees spread among company headquarters in Los Altos and sales offices in New York, Boston, Chicago, and Southern California; and revenues of $13 million, well above the $10 million target Oshman had recommended she hit. ASK's performance was even more impressive in light of the significant competition that had arisen. More than seventy suppliers, from other software businesses to mainframe manufacturers, sold materials resource planning software that competed with ASK's flagship MANMAN product.[23]

Meanwhile, the mainstream press had picked up the ASK story. Kurtzig was the subject of admiring articles in *BusinessWeek*, *Informatics*, *San Francisco Business Journal*, *Executive SF*, the *San Jose Mercury*, and *Computer Systems News*. The single-page typed newsletter *Executive Woman* praised her as having "absolutely no hang-ups about anything . . . self-assured, easily capable of handling a family and a business, and not the slightest bit concerned about being a woman in a man's world."[24] Kurtzig confided to one reporter that she had a private bet riding on her plan to make ASK a $100 million company in five years: "I tried to get a little handicap because I was a woman but he wouldn't give it to me. We'll do it because I just have to win that bet."[25]

She asked Regis McKenna, the public relations, marketing, and strategy expert whose best-known clients included Genentech, Apple, and Intel, to help develop the slides and story for ASK's IPO road show. "He was famous, and of course, we couldn't afford him," Kurtzig says.[26] She paid him in reduced-price stock: 1,000 shares at $6 each, about half the price at which ASK would go public.[27]

ASK had begun giving its entire customer list to any prospective client who promised to require other companies bidding on a project to do the same. The message—ASK is so confident of its ability to satisfy customers that we encourage anyone to call any of them—not only was good for ASK but also helped to build community among customers who came to know one another from the reference calls. In 1980, ASK brought its customers together for an over-the-top user-group conference. Hollywood-themed, this amped-up version of the company's Friday beer bust featured a Burt

Reynolds look-alike and a near river of alcohol. Customers presented to one another on how they were using MANMAN or had modified the source code that ASK gave to every customer. "It was like a reunion," recalls Liz Seckler, one of ASK's first employees.

The spirit remained, but much had changed. The company was no longer a ragtag team of Deadheads and new college graduates dragging their sleeping bags into Hewlett-Packard. Lavey, though loved and respected, was nicknamed "the suit" for his natty attire. Ken Fox, the newly hired head of R&D, drove a Mercedes, and Kurtzig had a Jaguar. ASK's headquarters, which had tripled in size for two consecutive years and featured a showpiece glass-walled computer room, now occupied two one-story buildings off El Camino Real.[28] A sign in front, emblazoned "ASK Computer Systems," featured a loop-de-loop logo Kurtzig intended to represent diffuse "information coming in, getting churned up by ASK software, and coming out as manageable information."[29]

At the company Christmas party in 1980, Howard Klein, who had joined ASK when there were only four other employees, sang Simon and Garfunkel's "I Am a Rock" with revealing new lyrics that he had written: "I need a loan/Gazing from my window, I wonder how it's done./There's a new Mercedes, musta cost a ton" and "Sales that are soaring is money we don't see/We spend it all to buy some new V.P.s."

That Kurtzig kept a copy of those lyrics in her company scrapbook is a measure of both her sense of humor and the easygoing relationship she maintained with her employees. She knew that some on her staff resented the changes the company was undergoing—a few members of Marty Browne's team, though not Browne himself, wore a T-shirt emblazoned "ASK Me If I Care" on the first day the new R&D head came to work[30]—but Kurtzig saw the objections as inevitable growing pains.

■ ■ ■ ■

In August, ASK submitted its preliminary prospectus, the so-called red herring, to the SEC.* One month later, Kurtzig headed to Europe for a road

---

* The document is called a "red herring" because it features a prominent statement in red on the front stating that the information contained within is not complete and may change.

show similar to the back-to-back-different-luxury-meal-in-a-different-hotel-every-night effort Genentech and Apple had undertaken.

ASK's road show, however, had a unique feature, notes Marty Browne, who joined the trip. Investors "were there to see Sandy, the woman CEO."[31] The marketing vice president at ASK's lead underwriter had warned Kurtzig that many in attendance would be there "because they've never seen a woman do it before. And you just have to be prepared for that." Both he and she felt that her status as a curiosity was a fact to be dealt with, not an insult. So when the vice president recommended that she trim her long red fingernails and paint them a pale, nearly invisible shade to appear more professional, she did.[32]

The weeks between the end of the European road show and ASK's IPO, scheduled for October 8, were fraught. The S&P index had dropped 17 percent in the previous nine months, seeming to fulfill Kurtzig's fear that she might miss an opportune market window.[33] Analysts were already calling it a crash.

As the day of the offering neared, no one at ASK or L. F. Rothschild, Unterberg, Towbin spoke of canceling. Kurtzig and the bankers were concerned enough, however, that they priced the stock at $11, the lowest amount proposed in the preliminary prospectus (which had given a range of $11 to $13). The highest price would have meant an additional $2.7 million, but the uncertain market meant that Kurtzig and the bankers needed to play it safe.

■ ■ ■ ■

Kurtzig was asleep at home when Unterberg called to tell her that ASK's offering, under the symbol ASKI, had been a success, and the company was valued at nearly $50 million. Kurtzig sold 546,550 shares of stock, worth roughly $5.6 million, at the IPO and still retained 3 million shares, worth more than $33 million ($92 million in 2016 dollars). She owned 87 percent of ASK before the IPO and 66 percent after—remarkably high percentages that can be traced, in part, to her decision not to bring in private investors along the way to her public offering. Most of the other companies that had gone public in the previous years had done so with shareholder equity three to twelve times that of ASK's.[34]

Kurtzig had shown that women could, indeed "compete in this league."

Two weeks after ASK's offering, another woman-owned tech company, the Los Angeles–based computer manufacturer Vector Graphic, also had its IPO.[35]

■ ■ ■ ■

At ASK headquarters, reaction to the IPO was muted. Kurtzig admits that part of her wanted to cash in her millions for a pile of single dollar bills and "frolic naked through it, like Scrooge McDuck in the Disney comics of my childhood." She restrained herself. She did spend $40,000 at a jewelry shop at the Stanford Shopping Center, but it was to buy watches for her management team.[36] Beyond that and a luncheon for a few managers, the celebration consisted of a low-key company party in the new on-site cafeteria, with a cake iced to read "Congratulations, ASKI."

The quiet nature of the celebration might have reflected the distribution of stock at ASK. The company had instituted a stock option plan in 1974, and Kurtzig says that "virtually everyone" in the company was a shareholder. Liz Seckler remembers that when ASK was still small, Kurtzig had offered her the choice of a $50 monthly raise or fifty more shares. Seckler had chosen the $50. Kurtzig's father had had a similar reaction to an offer of stock. When it came time for Kurtzig to repay his loan in 1979, she offered him shares in lieu of cash. "What am I going to do with it? Wallpaper the bathroom?" he joked, referring to the paper stock certificates. "That bathroom would have been worth $12,500,000," Kurtzig later pointed out.[37]

Though many ASK employees may have had stock options, few, including even the earliest hires, owned significant amounts of stock. When ASK went public, Kurtzig's shares were worth $40 million. She owned 3.6 million shares, seven times the combined stock of all other employees.[*38]

By the time ASK went public, it had many trappings of a classic

---

* Kurtzig says, "It would have been easier for me to give stock than cash, but at that time employees didn't want stock. They wanted the cash. There was no expectation that the stock was going to be worth anything. The thinking was, you can't eat stock—especially stock in a women's lingerie shop."

Silicon Valley company. Its improvisational work culture featured jokes, first names, and dispersed authority. ASK's board included the CEO of ROLM, one of the Valley's most successful firms; the managing senior partner of a major investment bank; and an attorney from Wilson Sonsini Goodrich & Rosati, the law firm that Larry Sonsini had by now successfully built into a one-stop shop for entrepreneurs.[39] Young programmers in jeans worked in ASK's low-slung headquarters building.

When it came to stock distribution, however, ASK was structured like an old-line company or a typical small business at the time, with the founder/CEO owning the vast majority of the company, while even the earliest employees received little. As Klein's fondly joshing Christmas song put it, "I'll buy up all the stock options I worked for/And I will be worth hundreds, maybe more."

Sharing equity was common in Silicon Valley at the time but not yet as pervasive as today. It certainly was rare in the manufacturing industries most familiar to Kurtzig. Moreover, though Kurtzig by the time of the IPO had well-connected CEOs, bankers, and attorneys on her board, in the beginning of the company's life she worked outside the networks of financiers and specialist suppliers that might have pushed her toward a different equity structure.[40] The one Silicon Valley company that Kurtzig did choose as a model, Hewlett-Packard, already mature and publicly held for fifteen years when Kurtzig began analyzing it, had a generous stock purchase plan but limited its options grants to its highest-level employees.* Ken Fox, the R&D manager Kurtzig hired (and to whom she granted options on 45,000 shares of restricted ASK stock at $1.67 per share), had never had stock options at HP, even after twelve years.[41] Chuck House, a computer executive at HP, says that he received as much stock in his first year with a new employer as he had in twenty-nine years at HP combined.[42]

---

* HP's stock option plans, adopted in 1964, 1966, and 1969, limited grants to "unusually well qualified personnel" and "key personnel whose long-term employment is considered essential to the company's continued progress." HP did have a profit-sharing bonus, and employees could buy stock at a generous 25 percent discount, a practice ASK mirrored, though the discount was 15 percent.

Stock distribution at ASK remains a sensitive subject for early employees. Several of them spoke of their great love for the company, their great pride in its success—and their great bitterness and confusion at having shared in relatively little of the upside.

■ ■ ■ ■

Kurtzig was undoubtedly the only CEO in history who, during the road show for her IPO, was handed a bright yellow T-shirt, which she held up for photos in front of her hired limousine, reading "ASK me if I go down."[43]

But $40 million makes up for a lot of sexist baloney. Of the thousands of entrepreneurs and employees who made money from the 448 companies that went public in 1981, only eight emerged from the public offerings with stock worth more than Kurtzig's nearly $40 million stake in ASK. (Nolan Bushnell's holdings in Pizza Time Theatre, which went public six months before ASK, were worth roughly $28.5 million.)[44]

Kurtzig never did roll around in a pile of money, but she did do the Silicon Valley equivalent. She bought a red Ferrari 308 GTSI, calling from one dealer to another to negotiate the best price. She divorced her husband within a few years of the IPO. The settlement—reportedly some $23 million to her ex—was, she says, "one of the largest, if not the largest, cash and stock settlement in the United States ever given by a wife to her husband."[45]

On the first day Kurtzig brought the Ferrari to work, Marty Browne, her first programmer, asked to drive it. Kurtzig tossed him the keys.

TRANSITION
1983–1984

Even in the face of bad news—pollution, overcrowding, skyrocketing divorce rates and housing costs—Silicon Valley continued to capture the public's imagination in 1983, three years after Genentech's and Apple's record-setting IPOs. In a regional equivalent of Hollywood's "map of the stars" bus tours, enterprising guides offered driving tours past the original garage offices of Hewlett-Packard and Apple. A company built rows of garages and offered to lease them to would-be entrepreneurs eager to begin their ventures in the Silicon Valley tradition.[1] US companies on average spent 38 percent of their capital equipment budgets on information technology, up from 22 percent fifteen years earlier. Magazines and newspapers previously uninterested in startup technology operations launched "Small Business" columns and "Up-and-Comers" sections.[2] In his 1983 State of the Union address, President Reagan praised Silicon Valley's "Pioneers of Tomorrow," saying "Surely as America's pioneer spirit made us the industrial giant of the 20th century, the same pioneer spirit today is opening up on another vast front of opportunity, the frontier of high-technology."[3]

For many in Silicon Valley, the middle years of the 1980s were a time of transition. Some founders left their companies, while the profiles of others skyrocketed. One major technology industry collapsed. Another matured. The modern Internet came a step closer to reality, *thanks to the widespread adoption of a common network operating protocol.**

Meanwhile, an ecosystem comprising more than three thousand Silicon Valley firms had evolved to provide new businesses with startup expertise.[4] *These venture capitalists, chip designers, glassblowers, fabrication houses, die cutters, equipment suppliers, and specialized law, recruiting, and public relations firms were themselves entrepreneurial ventures. They helped launch a new generation of entrepreneurs, many of whom built on the breakthroughs and precedents set by the troublemakers who had come before.*†

---

* The protocol was TCP/IP, which helped make it possible for different networks (the Arpanet, satellite networks, and other data networks, for example) to communicate with one another.

† Adobe, Cisco Systems, Cypress Semiconductor, Electronic Arts, Silicon Graphics, Sun Microsystems, and Sybase were among the firms that launched in the mid-1980s.

The Rabbits Hopped Away

■■■■■■■■■ BOB TAYLOR

In May 1983, Bob Taylor received an email from a panicked employee working on the Xerox STAR, the pricey computer system that Xerox had brought to market instead of the Alto.*

"About a dozen of us went to a lecture/demo of the Apple Lisa at UCLA this afternoon, and WE'RE SCARED TO DEATH!" the employee, Bruce Hamilton, wrote. The Lisa, introduced four months earlier, had a few problems—applications started slowly, and email and networking were allegedly "a few months away"—but these, in Hamilton's opinion, were of little consequence when stacked against Lisa's advantages. The Apple machine incorporated the mouse, a graphical user interface, and the interactivity of the Alto and Star. Moreover, Apple was allowing outside programmers to write software for the Lisa. As a result, "There is NO SIGNIFICANT STAR FEATURE WHICH WILL NOT HAVE A LISA COUNTERPART WITHIN 12–18 MONTHS," Hamilton predicted. "It is a MYTH to think Xerox can identify some 'safe market sector' where we can maintain our expertise." He wrote with alarm about any "dirt ball . . .

---

* The Star was unveiled on April 27, 1981. It sold for almost $17,000 but had to be installed as a system with laser printer and Ethernet, bringing the price to roughly $30,000. Apple's Lisa sold for $9,995 when it debuted on January 19, 1983.

company of two guys in a garage" that will be able to create and release new products in less time than Xerox needed to write specifications for a new feature.[1]

Hamilton's fears, which he broadcast to two Xerox email lists, were exaggerated but not unfounded. It had taken Xerox more than seven years to move from the first working networked Alto system to a product. Apple had made essentially the same move in half the time. The scenario was what Larry Tesler, the PARC researcher who had left for Apple in 1980, had told Taylor would unfold. Others were refining, improvising on, and popularizing PARC's innovations . . . and then moving beyond them.

One of Taylor's employees had forwarded Hamilton's email with the subject line "For your amusement," but Taylor was not amused. His team had proved what was possible years earlier. Someone else at Xerox—not Apple—should have determined, in Taylor's words, "what will work in a constrained consumer environment." In general, Taylor liked a clean division between research and production. "You do not want a researcher building your roads, your subways, your airlines, your houses," he wrote.[2] The Lisa rankled nonetheless. Xerox should have owned personal computing by now, not ceded it to unpedigreed corporate upstarts. Moreover, the success of the IBM PC and its clones had stunned Taylor and many PARC researchers. "We never thought people would buy crap," Alan Kay explained.[3]

The past few years—and the last months in particular—had been tough for Taylor. His marriage was falling apart. He would soon leave the Craftsman house in Palo Alto with the neighborhood tennis court and breezes blowing through, to move to an isolated house in Woodside at the end of a winding road overlooking the valley. At work, problems that had been brewing for a long time were now boiling over. He continued to battle with other PARC divisions and PARC administrators. Key employees were frustrated and leaving. Budget freezes had become cuts. At his last performance review, he learned that his annual raise would be less than the cost-of-living increase generally given as a matter of course at Xerox. At $118,440, his annual salary was at the top of his pay grade.[4] He could not expect more.

Through all the troubles, though, he had found an unlikely ally: his immediate boss for the past five years, an engineer named Bob Spinrad.

Spinrad was named director of PARC after George Pake had been promoted to head of corporate research. Spinrad had worked in Xerox corporate, beginning his career at the company with Scientific Data Systems, the computer firm that Xerox's headquarters had acquired in 1970 and shut down $84 million later.

Spinrad understood Taylor. He wrote of Taylor's "sharply bimodal" performance: the outstanding leader within his own lab, who was at the same time a source of much of the interlab tension at PARC. One day, Taylor, feeling particularly unappreciated, used a ruler and graph paper to chart his lab's resources. According to his calculations, the amount of money his lab received per employee had plummeted from $37,000 to $2,000 during his tenure. He sent the graphs to Spinrad with a cover note that asked, "Is this message one of appreciation?" The tone was not challenging but plaintive, a request for help in prying more money out of Pake.[5] Spinrad appreciated the importance of the computer science lab and its work. "It is a source of distress to me that we cannot adequately support all the excellent and varied work that the CSL staff is capable of," he wrote in early 1981. Around the same time, he sent a five-year plan to Pake at headquarters proposing that the PARC budget be reallocated to send more money to the computing labs.[6]

Shortly after Spinrad submitted his five-year plan, Pake, in effect, demoted him, splitting PARC in two and making Spinrad head of only the computing side. Rumors in Taylor's lab held that Pake had removed Spinrad from the directorship out of fear that Spinrad was falling under Taylor's sway. Spinrad left PARC for a job on the corporate staff the next year.

With Spinrad gone, the buffer between Taylor and Pake had disappeared. A new one appeared in the form of the physicist Bill Spencer. A tall, handsome Kansan, Spencer had come to Xerox in 1981 to run a new integrated circuits lab at PARC. He and Pake got along well. After several months, Pake asked Spencer to serve as PARC's director.

Spencer had a willful streak that rivaled Taylor's own. At Sandia Laboratories, where he had worked before Xerox, he had enjoyed provoking a boss he detested. Forty years later, he chuckled as he recalled a time when he had made the man "so angry that we had to cancel a meeting."[7] At another job, he received a gift of a giant hypodermic needle—because he was always "needling" others.[8]

Spencer was so similar to Taylor that Taylor would either enjoy or despise him. He did both. For the first months after Spencer arrived, when he was running the integrated circuits lab, the men played tennis on Saturday mornings. Their wives were friendly, and the Taylors attended the fiftieth birthday celebration for Spencer's wife. All in all, the two families shared "a lot of beer and Dr Pepper nights," according to Spencer.[9]

But after Pake promoted Spencer to head PARC, making him Taylor's boss, the friendship faded. Taylor was soon pointing out that Spencer's undergraduate degree was in physical education. It did not help that Spencer was a physicist with expertise in semiconductor electronics. Taylor's disdain for physicists was well known at PARC, as were his feelings about semiconductors, which he thought made for boring research, since Moore's Law had already mapped the expected rate at which the density of components on a chip would increase. He also thought microchips received too much credit for the digital revolution. For Taylor, semiconductors were a necessary but pedestrian base upon which the elegant and complex dance of software could perform. "Without software, hardware is just a piece of hot iron," he liked to say.[10]

Spencer's first move as head of PARC was to schedule a meeting of the lab leaders. Taylor offered his beanbag room for the meeting. Spencer was shocked when a representative from another lab said that none of the researchers from her lab would attend a meeting held in Taylor's territory. "They were afraid of being physically attacked," Spencer says, though he acknowledges that he knew of no physical violence at PARC. "It was worse than I had ever seen in my twenty-five years of being in places like Bell Labs and Xerox."[11] The meeting never happened.

Spencer soon decided that Taylor was standing in the way of inter-lab cooperation at PARC. Moreover, thanks to the clean division between research and development that Taylor championed, he was impeding Xerox's ability to commercialize PARC technology. "Getting technology picked up by the rest of Xerox was my problem," Spencer recalls. "That requires that people who are experts in [the technology] have to spend time with people in other parts of Xerox. [Taylor] was totally unwilling to participate in that at all."[12]

As Taylor saw it, a number of his researchers were already moving technology into other parts of Xerox. People from his lab had worked on the Star, and he had encouraged one key researcher (Geschke) to launch a new imaging lab within Xerox. But Spencer had a different model in mind: he envisioned Xerox researchers helping an outside company develop products based on PARC technology in exchange for an ownership stake for Xerox. Spencer would later successfully promote this model with Spectra Diode Laboratories and Synoptics.[13]

Taylor thought that Spencer's real objective was neither interlab cooperation nor product development. His true goal, in Taylor's opinion, was to seize control of Taylor's lab.

Looking back four decades later, Spencer says with a sigh, "Had I not been so anxious to leave [Sandia], I probably would have done a better job looking into PARC—and gone to Hewlett-Packard."[14]

Taylor's defensiveness and Spencer's frustration peaked at the end of August 1983, when Spencer handed Taylor a typed memo after a heated meeting in Spencer's office. "Your management and your personal attitude are counterproductive," Spencer wrote. If things did not change, he said, Taylor would face "disciplinary action that may include your termination."[15]

Spencer gave Taylor three weeks to reorganize the computer science lab to look like the other labs at PARC. No longer would everyone report to Taylor. Instead, Taylor should divide the lab into several groups, each with a dedicated manager. Spencer felt that the move would "help develop future Xerox managers and to improve our contacts with other Xerox organizations." He directed Taylor "to make no derogatory remarks about Xerox programs or organizations" and "not to discuss the formation of an outside organization with Xerox employees . . . [or] induce them to leave Xerox." Finally, Taylor needed to meet with Spencer every Monday morning at 9:00 "to go over your compliance and progress on these actions."*

---

* Taylor also needed to alert Spencer of all staff and planning meetings, so Spencer could participate. Moreover, Taylor was required to meet with members of the recently dissolved imaging lab to assure them they could join his lab if they wished (whether Taylor wished it was irrelevant).

No one, including the conflict-averse Pake, had ever treated Taylor so imperiously. Not even Jerry Elkind, the nominal manager of the computer science lab in the early years, had ever told Taylor how to run his lab.[16] Spencer had designed his letter to put Taylor in his place.

■ ■ ■ ■

"As I read and read again your memo," Taylor wrote to Spencer six days later, "I was increasingly appalled and shocked by its approach, its tone, and its content. . . . When I began to recover from [the memo's] effect, my reaction was simply to resign immediately. However, I have worked for a long time, and over one of the most productive periods of my life, to bring the very best people and technology to Xerox. I owe to this investment much more than just walking away from it without responding to your dismaying memo in more detail."[17] In the following five single-spaced pages, Taylor defended his lab and his management.

He dismissed the allegation that he had encouraged people to leave Xerox or to join him in a new venture. ("It is clear from the tone of your memo that I should have been interviewing for employment elsewhere, but I have not," he said.) He wrote that Spencer had always been welcome at lab meetings and disagreed that reorganizing the lab would foster new managers. PARC researchers were hired for their research, not their management skills and interests, Taylor noted. Taylor also offered a list of ten technologies that the lab had made "work on a daily basis for ourselves and for thousands of others inside Xerox." His list included local area networks, an internetwork serving more than 4,400 electronic mailboxes, integrated circuit design aids, and the laser printer.* Any responsibility for

---

* Taylor's list further included a systems implementation language with a matching optimized machine architecture; programming environments; personal workstations; interactive text formatters; server stations for printing, filing, and mail; bitmap displays; electronic mail with transport and authentication facilities; and "a complete, distributed systems architecture accommodating all the above."

Taylor also noted that "the real challenge has been the transfer of an entirely new and quite different framework for thinking about, designing, and using information systems. This is immensely more difficult than transferring technology. Opportunities for pioneering completely new ways of thinking about large

the failure of these and others of his lab's accomplishments to reach the general public lay beyond Taylor's responsibilities.

As for derogatory comments: "I have a fundamentally positive, optimistic spirit; I am not interested in being derogatory," Taylor wrote. He admitted that he could be critical, but "truth and open expression should be the coin of the realm" in a research environment. PARC, by contrast, had become "more politicized than I would ever imagined was possible."

In closing, he requested that he be allowed to share Spencer's memo and his own response with his senior researchers (denied), as well as with Xerox CEO David Kearns and other top executives (permitted). In a note to Kearns and two senior executives, Taylor wrote, "It is absolutely clear that I can no longer report to Spencer and, through him, to Pake." Taylor did not know it, but Kearns was well aware of the content of Spencer's original memo. Before confronting Taylor, Spencer had discussed its major points with Kearns, warning him that Taylor might quit in response.[18]

After receiving Taylor's note, Kearns deputized his second-in-command to meet with Taylor and Spencer at Xerox's East Coast headquarters. After four hours, Taylor understood that Xerox senior management was backing Spencer. The antagonists shared a silent ride back to the airport in a Xerox corporate car.[19]

When he returned to California, Taylor knew what he had to do. He briefed a few members of his inner circle and then called a lab meeting for the next morning.

Standing in front of his team, with Bill Spencer tall at the back of the room, Taylor read a short message that he had handwritten in tiny looping letters on a page torn from his Xerox notepad:

Thirteen years and 19 days ago, I formed a laboratory with the hope of creating an entirely new form of Information processing that

---

collections of ideas are rare. Over the past twenty years, I have been fortunate to have been a leader in three: timesharing; long distance, interactive networking; and personal, distributed computing. Each of these required large upheavals in the way people think about information systems."

would someday serve millions of people. You have created that new form, and I will always be indebted to you. . . .

By the mid 70s, you had created an excellent base upon which I expected corporate research to build. That expectation has not been fulfilled. PARC continued to invest less than 20 percent of its resources in computer systems research . . .

So I will leave and have no immediate plans. [I will be on] vacation until 30 Sept. I do not know what I will be doing next.

It has been an incredibly great experience to work for you.[20]

Stunned silence followed, as the researchers watched Taylor lower his notes and walk away. Bill Spencer stepped into the space where Taylor had stood, ready for questions. What he got instead was fury. With the exception of the few lead researchers whom Taylor had briefed the night before, Taylor's firing—anyone who knew him knew he would never voluntarily resign this way—came out of nowhere. The researchers began shouting at Spencer. The passion of the best Dealer meetings turned black. What had just happened? What had Spencer done to Taylor? Was the lab being dissolved?

Chuck Thacker stood up. He said he was quitting and left the room. Later that day, he submitted his formal resignation, telling Spencer that he doubted "your motives as well as your wisdom" and explaining that he did not wish to "witness or participate in" the dissolution of the lab that would inevitably follow Taylor's departure.[21]

Spencer, meanwhile, was dealing with his own shock at how Taylor had handled the situation. Though he says "my intent was never to force Bob to leave PARC," deep down, he did not think it would be the worst thing for Xerox if the computer science lab got a new manager and even a new core research staff. "I think they had made almost all the contributions they were going to make," he later reflected about the computer science lab under Taylor. "I think their accomplishments probably came between 1971/72 and 1980, and then they were kind of fine-tuning the things they had done."[22]

But Spencer had not expected Taylor to play his hand the way he had. "I had not thought he would leave abruptly. I thought we'd sit down and talk, and if we couldn't reach a compromise, we'd have to say, 'Okay. Now what?'"[23] That misapprehension, more than anything, shows how little Spencer understood Taylor. Compromise was not Bob Taylor's style.

■ ■ ■ ■

A rumor began to circulate through the computer science lab's email lists that Spencer had been ordered to shut down the lab. Within a few days, copies of Spencer's memo and Taylor's response had been posted electronically (by whom remains a mystery). Soon both posts were quietly taken down.

Several of the most senior computer scientists flew to Xerox headquarters to urge CEO David Kearns to keep Taylor at Xerox.[24] They wanted Kearns to understand that if Taylor left, others would leave, too. Moreover, even if the lab somehow remained intact—and it would not, they assured Kearns—Spencer would never be able to run it. It was vital to understand that Taylor "works for the laboratory, rather than vice versa," they said. Moreover, by "deflecting external distractions," Taylor freed the researchers to focus on research at the highest levels.

Spencer, by contrast, possessed a different—and unsuitable— management style, the researchers told Kearns. "He has preconceived ideas about technology transfer, does not understand systems research, does not take account of CSL's demonstrated success, [and] commits the lab to research without consultation."

Taylor, they promised, "is manageable," and they, as a team, had found ways to "compensate" for his "imperfections." The group's suggestion for salvaging the situation was to induce Taylor to return, have him report to someone other than Spencer, and increase the funding of the computer science lab.[25]

It is difficult to imagine that the researchers thought their appeal would have any real effect. To bring Taylor back to PARC with more resources and more power would have exacerbated the problems perceived by Xerox management.

While the group was making its case to Kearns, other members of the lab contacted the world's leading computer scientists to ask them to write to the CEO and pressure him not to let Taylor leave. Within days, the letters were on Kearns's desk.

"A mind-boggling event," Donald E. Knuth, the distinguished computer scientist at Stanford, called it. "Bob has accomplished such miraculous things. I had always assumed he was considered one of Xerox's most important people." With Taylor gone, Knuth wrote, the world's finest assembly of computer scientists would likely dissolve, which would "spell the end of Xerox's leadership in computer systems."[26]

Similar letters arrived from Dana S. Scott at Carnegie Mellon ("The matter is absolutely crucial to the future of Computer Science, and I could not remain silent"[27]), Richard Karp at the University of California, Berkeley ("I know that a great many people throughout the computer science community share my concerns"[28]), and Brian K. Reid at Stanford ("From my perspective, the very survival of Xerox as a company depends heavily on Mr. Taylor, and you obviously do not realize this or you would have fought harder to keep him"[29]). Taylor's mentor, J.C.R. Licklider, wrote a long letter from his laboratory at MIT, worrying that "if the resignation is allowed to become effective, it will surely disrupt and could even destroy what many computer scientists (including me) believe is the best computer science laboratory in the world." The lab, Licklider wrote, was "much more than the sum of its parts," thanks to the "atmosphere, esprit, critical mass, and favorable interpersonal interactions" Taylor had fostered. "If Taylor does leave and you are nevertheless able, somehow, to keep together the group Taylor assembled, you are going to need another Taylor. And I am afraid that there may not be another Taylor."[30]

At best, progress in computer science would be delayed as the PARC researchers either dispersed or regrouped elsewhere. At worst, the momentum within the lab would be lost for good. Computing progress in the United States, at a time when the Japanese were feared as posing a threat to the nation's technical primacy, would suffer.

The onslaught of letters stunned Kearns—and enraged him. "If I had known this would happen, I wouldn't have agreed to the changes you proposed," he told Spencer. Spencer reminded him that he *had* known it might happen; Spencer had told him so.[31]

But Kearns had said he would support Spencer, and he did. Bob Taylor was gone.

Less than a month later, Taylor founded the Systems Research Center for Digital Equipment Corporation in Palo Alto. "I said I would consult for them, and the next thing I know, some guy shows up asking where I want to build the lab," Taylor says. His pay was $145,080, a 22 percent increase over his PARC salary. He received 2,500 stock options, and he reported to an accomplished computer scientist he liked and respected: Sam Fuller, Digital's vice president of research and architecture.[32] Today Taylor calls Fuller "the best boss I ever had."

Back at the PARC computer science lab, an anonymous author circulated a gleeful "Parable of Rabbits" beautifully printed on a PARC printer. In the story, a group of rabbits (some experts in software, others in hardware) rejoice to hear that the "one amber rabbit, their leader, who knew all about how to help rabbits work together in harmony, so that they cooperated with one another and built wonderful things," had gone to an "Other Company"—and now every rabbit could "hold the warm thought in his heart that he would go to the amber rabbit and say, 'May I work for you in the Other Company?'"[33]

Once year-end bonuses had been issued at Xerox, the rabbits hopped away. Within eight months of Taylor's departure, twenty-eight people, more than half the technical staff of the computer science lab, had resigned.[34] Of those, fifteen joined Taylor at Digital Equipment Corporation. Once again, researchers under Taylor's direction helped to lay the foundations for a new wave of technical innovations based on easy-to-use distributed computing systems. Among other projects, his team at DEC helped to develop reliable computer networks, multiprocessor machines, and an electronic book. One researcher, Mike Burrows, was also an essential contributor to AltaVista, the world's first blazingly fast search engine. AltaVista came to market three years before Google was founded.[35]

Video Nation
AL ALCORN

By 1983, video games, and Atari in particular, had captured the nation. Fifty new video game arcades opened in the United States every week.[1] One in five American convenience stores had at least one arcade game on site.[2] A case arguing that children under seventeen should be restricted from playing video games reached the Supreme Court.[3] *The New England Journal of Medicine* reported a new type of ligament strain: "*Space Invaders* wrist," named for the popular arcade game manufactured by Taito Corporation and turned into the bestselling home video cartridge by Atari's Rick Maurer. Steven Spielberg told *Newsweek* that he had eight arcade games in his house and wondered aloud if he might be in the wrong business. Atari ran a series of ads around the concept of a nation addicted to video games, even setting up an "Atari Anonymous" toll-free hotline.[4]

Thanks almost entirely to the Atari division, Warner stock soared 3,000 percent in the six years following the acquisition. The division was so profitable that the other parts of Warner, while also making money, diluted Atari's performance.

Then it all crashed. In June, Warner announced losses of $310 million in three months, more than $3 million per day. Even Warner CEO Steve Ross seemed stunned, admitting that while he had expected a loss, one "of this magnitude clearly was not anticipated."[5] Atari had begun missing its earnings estimates at the end of 1982, but the second quarter results for 1983 were nonetheless mind-boggling. By the end of the year, the company had lost roughly $500 million. The previous year, it had recorded profits of $323 million.

Steve Ross and Manny Gerard blamed competition for game cartridge sales, by far the most profitable part of Atari's business, for the devastating losses. (It did not help that companies such as Mattel were selling more advanced game systems with better graphics.) Atari had introduced several new games—including *PAC-MAN* and *E.T. the Extra-Terrestrial*, a tie-in with the Universal film—with graphics and design that many players considered primitive. Of *E.T.* one critic wrote, "It wasn't a game. It was a thing waddling around on a screen."[6] *E.T.* had been rushed to market after Steve Ross had made a deal with Steven Spielberg, reportedly guaranteeing the filmmaker $23 million in royalties, without consulting anyone at Atari.*

By the end of 1982, some twenty competitors were selling cartridges for Atari's VCS game system—but Atari made only one-third of the three hundred cartridges available for the console.[7] There were games from serious companies such as Activision, but also offerings such as *Chase the Chuck Wagon* (from Purina, the pet food company) and *Custer's Revenge*, a pornographic game in which a sexually aroused George Custer dodged arrows to advance toward a naked Indian maiden tied to a post. The game manufacturer's appalling slogan: "When you score, you score!" Atari sued the manufacturer of *Custer's Revenge*. The game also generated protests by women's and Native American groups who said it promoted rape.[8] Atari had lost control of the market it had created.

Only hours before the first of Warner's losses was announced in December 1982, Atari president Ray Kassar had sold 5,000 shares of stock.

---

* Many people point to the size of the *E.T.* deal with Spielberg—$23 million was roughly $22 million more than anyone had paid for video game rights to a movie—as the beginning of the end for Atari. But if the point of reference is not what the deal did to Atari but what it did for Warner, Ross seems to have gotten a bargain, since the agreement helped bring Spielberg, who had been making films for Universal, to Warner.

Investigated by the SEC for insider trading, he had to return the profits. Then he was fired. Warner CEO Steve Ross, who had sold stock worth $21 million in the months leading up to the announcements of Atari's slipping earnings, faced no investigations.[9] Atari laid off three thousand workers in 1983, after which the company faced a unionizing attempt that quickly failed. "It's an agonizing time emotionally here," one executive said. "This was the best place to be in Silicon Valley, and people don't understand how all this happened."[10]

"It was as if the entire industry disappeared," explains Warner's Manny Gerard. Bally's profits fell 85 percent. Mattel, the maker of Intellivision, laid of 37 percent of its workers.[11] Some two thousand arcades closed in the first nine months of 1983. It seemed as though video games might be only a passing fad.

The crash nearly destroyed the rising generation of video game startups. Activision, the company founded by four disgruntled Atari game designers, suffered through a lackluster public offering in 1983 before losing millions of dollars. Imagic, another Atari spinout, folded. "Atari's meltdown created a tsunami that wiped out public interest in games . . . and gave gaming a stigma that lasted a decade," says Trip Hawkins.[12] Hawkins, one of Mike Markkula's "diamonds in the rough" at Apple, had left the company in 1982 to found Electronic Arts to build video games for personal computers. The first outside backer of Electronic Arts was Don Valentine, the venture capitalist who had funded Atari and introduced Mike Markkula to Steve Jobs and Steve Wozniak. Ben Rosen, the optimistic technology analyst, was another early backer.*

Electronic Arts had its first big hit in October 1983 with *Doctor J and Larry Bird Go One-on-One*, a basketball game featuring two of the sport's biggest stars. The title was one of the most popular games for the Apple II and an early indicator of the enormous market for licensed sports video

---

* Valentine gave Hawkins an office at Sequoia Capital and helped him secure a $2 million round of funding from, among others, Kleiner Perkins Caufield & Byers and Sevin Rosen, a fund started by Ben Rosen. The original name of Electronic Arts was Amazin' Software.

games. (Electronic Arts would exploit this market with great success five years later with the release of *John Madden Football.**) But in the wake of the 1983 crash, even Electronic Arts, which made games for computers, not dedicated consoles or arcades, had to retrench. "We had to operate like the Fremen of Dune, recycling our own saliva to survive," Hawkins says. "We had to rebuild the industry brick by brick over a period of years."[13]

Atari never recovered from the crash. Just as Alcorn and others had feared, when sales of Atari games began to decline, the company had nothing new to offer. A higher-end game system released at the end of 1982 played only a limited number of games (earlier VCS cartridges, for the 2600 system, would not work with it) and cost as much as some personal computers that could do more.† The other two divisions of Atari—coin-op arcade games and the personal computer division—also lost money.

Too late, Atari tried to pay serious attention to research and development. After Alcorn left, Kassar hired PARC's Alan Kay to oversee "long-term research" and gave him what Kay called "essentially an infinite" budget.[14] In early 1983, Ted Hoff, a coinventor of the microprocessor and among the most senior engineers at Intel, joined Atari to focus on "short- and medium-term research."[15] Both men quit within a year, far too early for their efforts to have had an effect on Atari's bottom line or technical direction.

In 1984, Warner split Atari, keeping the arcade division but selling the home computing and game console divisions to Jack Tramiel, the former CEO of Commodore Computer. The next year, Warner sold the arcade division to Namco. "It's clear that the old Atari was dead," one analyst told the *New York Times*. "It just wasn't worth anything."[16]

■ ■ ■ ■

---

* As of 2013, Electronic Arts had sold more than 100 million copies of the Madden football game (today called *Madden NFL*), generating $4 billion in revenue.

† Some low-end home computers sold for $200; the new Atari 5200 was priced at $269.

The video game industry would begin to resuscitate in 1985, when Nintendo introduced its NES game console. The NES, which in 2011 was named the best video game console of all time, was a closed system; it played only cartridges made or approved by Nintendo.[17] The Japanese company had learned from Atari's stumble.*

---

* Nintendo's rise marked an important shift in the video game industry, from open to closed systems. Apple made a similar shift around the same time, moving from an open operating system in the Apple II and its descendants to a closed system in the Macintosh.

Knew It Before They Did

NIELS REIMERS

In 1983, Stanford earned more than any other university from technology licensing: $2.5 million ($6.1 million in 2016 dollars).[1] The recombinant DNA patent generated almost 60 percent of this amount. Niels Reimers, with his inspirational messages to himself and his dogged persistence, was among the first people to recognize how to make a fortune from academic innovation. He was also the very first to see the financial potential of recombinant DNA. He figured it out before the inventors. He knew it before the entrepreneurs and venture capitalists. The fortune did not go into his own pockets; he says that he did not receive a bonus, a promotion, or even a particularly large raise for his daring decision to patent the recombinant DNA process. But he nonetheless knew that he had helped launch the $270 billion biotech industry and redefine the nature of invention in universities around the world.[2]

After the excitement of the recombinant DNA license, Reimers returned to his day-to-day work at the Office of Technology Licensing. From this perch, he observed the rise of a new generation of technology companies with roots in Stanford innovations, including the workstation giant Sun Microsystems and Cisco Systems, a pioneering telecommunications firm. The birth of fast-growing companies like these, with innovations developed at Stanford at their core, motivated Reimers to engage in yet another battle with the top leadership of the university. Again millions of dollars hung in the balance.

Reimers wanted Stanford to be able to accept stock in lieu of cash payments for university innovations patented by the Office of Technology Licensing. Accepting equity would enable young startup companies, so often short on cash, to acquire licenses. The move would also benefit Stanford. Because the pace of change in the young tech firms was so rapid, Reimers feared that licensed technology might be designed out of a product even before it reached the market. (Stanford's licensing income, often a percentage of sales of products using the technology, would thus decline.) It would be better, he reasoned, for Stanford to hold an ongoing equity stake in startup companies. By the mid-1980s, he was estimating that by not taking equity, the Office of Technology Licensing had "left over $50 million on the table."[3]

Stanford president Don Kennedy had a different perspective. He worried that by accepting equity, and thus ownership, in private companies, Stanford could inadvertently become a competitor against corporations whose gifts accounted for 18 percent of the funds the university raised each year. In defense of his concerns, he quoted a high-tech CEO who said that he would welcome universities into his trade association, "but I'm not likely to contribute to their efforts."[4]

Companies founded by faculty, staff, and students who remained affiliated with Stanford presented an even thornier problem. "It is difficult for me to see how a university can be an equity participant in the work of one faculty member or a few, and yet be seen at the same time by all of its faculty as an even-handed supporter of scholarship and allocator of institutional resources," Kennedy said.[5]

Stanford had supported the surrounding entrepreneurial community for decades through open lectures, distance-learning initiatives, and a generous consulting policy for faculty. Moreover, the university had been an early limited partner in several venture capital funds, and in 1983 alone, invested $64 million with venture capitalists.[6] But Reimers was proposing something new. As a limited partner in a venture capital fund, Stanford neither selected the individual companies for investment nor held a significant portion of their stock. Reimers wanted Stanford to do both for companies built around Stanford innovations.

Reimers claimed not to be worried about the problems Kennedy foresaw. Asked if researchers might be inclined—or pushed by the university—to focus on work that could yield marketable products (and by extension, money for Stanford), Reimers called the situation unlikely. Scientists, he said, were motivated not by money but by "being first scientifically."[7] Kennedy, a biologist, was less convinced.

After years of debate within the university, Reimers's perspective prevailed. The Office of Technology Licensing began accepting equity in young companies in 1989.[8] Over the next two decades, the university took equity in 136 companies in exchange for licenses.[9] Among the companies were VMware, which had among the most successful public offerings of 2007, and Google, which granted Stanford 1.8 million shares of stock in exchange for a license to the search algorithm developed by Sergey Brin and Larry Page while they were graduate students working under National Science Foundation funding. Google equity alone brought $336 million to Stanford.[10] As of 2015, Stanford held equity in 121 companies as a result of licensing agreements.[11]

Reimers played an important role in helping Stanford define its relationship to the entrepreneurial landscape around it. Today Stanford sponsors so many efforts to support entrepreneurship that *The New Yorker* nicknamed it, rather unfairly, "Get Rich U."[12]

Reimers remained at the helm of the Office of Technology Licensing until 1991, when he resigned. The battle over equity ownership, plus an organizational change that meant Reimers no longer reported directly to a dean of research or the president, led him to take an early retirement offer. Although Reimers "strongly recommended" a specific person to replace him, the university made a different choice.[13]

Even as he was leaving, the Office of Technology Licensing was proving the success of Reimers's leadership and vision. Three inventions—recombinant DNA, FM sound synthesis, and a patent fundamental to developing the magnetic resonance imaging (MRI) machine—were bringing in more than $1 million each in royalties to Stanford every year.[14] By the time of his resignation, the little office that Reimers had launched as a pilot project twenty-one years earlier had brought in more than $87 million ($177 million in 2016 dollars).[15] The value of the licensing office as a tool for attracting and retaining faculty, students, and staff with entrepreneurial tendencies has been incalculably higher.

No One Thought They Would Sell

FAWN ALVAREZ

In 1982, ROLM introduced a breakthrough innovation: voice mail. For the first time, callers could leave extended recorded messages. Voice mail helped send the sales of ROLM's CBX phone system skyrocketing. Fawn Alvarez and other employees at companies that installed voice mail now needed to learn how to use it. The earliest messages were variations of "It's me. Call me back." It took a while for people to understand that they could leave the important part of the message in the recording, with no return call necessary.

Alvarez's work in the forecasting group had grown even more interesting over the past two years. Her boss, Dave Ring, who would later serve as the vice president of manufacturing at Cisco Systems, was overhauling the forecasting process. He wanted the cooperation of people from finance and product management, as well as sales, marketing, and manufacturing. He relied on Alvarez, with her long history at ROLM and contacts in multiple divisions, to help identify and interest the people with leverage or influence throughout the company.[1]

Ring had graduated from Harvard and Stanford. Soon a manufacturing expert would join the forecasting group; her graduate degree was from Yale. Everyone had at least a bachelor's degree, except Alvarez. "I think Fawn felt some risk around her lack of degree," says Ann Magney Kieffaber, the Yale School of Management graduate who joined the group and the only other woman. "She shouldn't have. Her confidence and her contacts at the firm were truly hard to beat. She was very sharp, and she knew all the players and what their objectives were."[2]

Alvarez says, "Even though I maybe exuded confidence, it was always in my mind that maybe I was—I don't want to say a fraud, but maybe I wasn't as smart as I thought I was. Maybe I was just a butthead with a big mouth. I'd see these people with all their degrees, and I'd think, 'If I'm going to bring something up, I'd better know what I'm talking about.' I was afraid I would say or do something stupid."[3]

■ ■ ■ ■

In 1983, around the same time that the video game industry crashed, IBM bought a 23 percent stake in ROLM as a defensive move. The previous year, the Department of Justice had forced AT&T to divest itself of the Regional Bell Operating Companies that provided local and long-distance telephone service. Many people believed that with the loss of the operating companies, AT&T would enter the commercial computer market.[*] IBM reasoned that if the country's largest telecommunications provider was going to start selling computers, the largest computer manufacturer needed to be in the telecommunications business. ROLM provided a way in.

In October 1984, IBM announced that it was fully acquiring ROLM for $1.8 billion ($4.3 billion in 2016 dollars). It was Big Blue's first acquisition in twenty-two years and the largest in its history.[4] "Every one of us remembers the day we heard about the sale," recalls Ron Raffensperger, a high-level engineer at ROLM who joined in 1976. "It's, like, 'Where were you the day Kennedy died?'"[5]

For weeks, Alvarez had heard rumors that "something big" was going

---

[*] AT&T Information Systems was founded in 1984. Bell Labs, which had developed the UNIX operating system and had been designing computers for decades, remained part of AT&T, as did Western Electric, which had long built computers for internal use deep within the telephone network. Another effect of the 1982 consent decree was increased competition in long distance as companies like MCI and Sprint gained market share.

to happen, but the acquisition nonetheless shocked her. ROLM had recently joined the Fortune 500. The company had $37 million in profits on nearly $660 million in sales.[6] "No one thought the founders would sell," Alvarez recalls. She had not known that although ROLM's sales had continued to climb, its profit margins were shrinking. The founders, somewhat to their own surprise, had decided that a sale was inevitable, and the sooner the better for ROLM shareholders.

On paper, ROLM and IBM seemed a fine match. Nearly all of the top one hundred companies on the Fortune index owned computers (either large machines or the year-old IBM PC) from IBM and phone systems from ROLM, which had 15 percent of the switchboard market, second only to AT&T.[7] For years, analysts and forecasters had debated whether computer systems or telephone systems—data or voice—would be at the heart of the "office of the future."[*] A ROLM-IBM pairing would obviate that question with systems that seamlessly integrated data and voice. On news of the acquisition, both companies' stock prices rose.

"All over ROLM there was euphoria, followed by 'Oh, my God. What next?'" recalls Ron Raffensperger.[8] The euphoria came from the financial bonanza the acquisition represented for ROLM shareholders. Anyone who had bought stock at ROLM's IPO in 1976 saw a fortyfold return on that investment in only eight years. ("They are going to pay *how* much?" Raffensperger remembers asking.) One person recalls that even the women on the assembly line talked about the value of ROLM's stock and where it might go.[9] Fawn Alvarez's mother, Vineta, who had not sold a single share of the stock for which she had paid $7,000 during her years at ROLM, discovered that her equity was worth $360,000 (nearly $830,000 in 2016 dollars).[10]

Fear quickly followed the elation. How would ROLM, with its 9,000 employees, retain its identity and "great place to work" culture after IBM, with 350,000 employees, absorbed it? At ROLM, many people wore jeans to work, and Friday-afternoon beer bashes were sacrosanct. On Halloween, ROLM's vice president of marketing dressed as Supermanager, complete

---

[*] The voice-data distinction is a bit oversimplified, since the ROLM CBX stored voice messages as data.

with red tights and a red cape. By contrast, IBM had an unwritten dress code (white oxford-cloth shirt, tie) and a corporate policy that forbade drinking on the job.[11] IBM's success with its personal computer—developed in secret by what was essentially an internal startup operation—offered some hope that the giant company would appreciate ROLM's innovative and entrepreneurial spirit. A visit by three top IBM executives, who dressed down and hoisted beers at one of the Friday bashes, was also reassuring. Nonetheless, it was impossible not to wonder, as Raffensperger put it, "What will IBM do to us?"

Alvarez, too, experienced the seesaw of euphoria and anxiety. She had never received stock options. ("I'm sure if I had asked for them, I would have gotten them," she says, "but I didn't know to ask for them.") She had, however, bought as much ROLM stock as she could over the previous five years, once she was out of her "young and dumb" stage. Her mother's example had inspired her, and she knew that most of the engineers and executives bought as much stock as they were allowed. Alvarez had cashed in some of her shares so she could travel—she estimates she visited twenty-five countries in the early 1980s—but she retained enough that after the IBM acquisition she could sell what she needed to buy a house. "I was not wealthy," she says. "But still: I was a woman, on my own, buying a home in Silicon Valley."[12]

Then came the anxiety. "Okay, this is going to be really hard," she recalls thinking. "We've done something amazing here. Now we are going to be bought by the most conservative company in the world. This will be hell." She asked herself how IBM would be able to manage the acquisition, and then she decided, "I will help them. I can tell these 'beamers a thing or two about how not to wreck the place."[13] A member of the forecasting group, Alvarez did not occupy a position that traditionally had much influence—but that had never stopped her before.

The Entire World Will Never Be the Same

MIKE MARKKULA

On a sunny afternoon in February 1983, more than three thousand Apple employees spilled out of an enormous tent in the parking lot at company headquarters. A race car painted in Apple's distinctive colorful stripes was parked nearby. Inside the tent, giant bushels of apples and red glasses filled with snacks rested on long tables alongside napkins in Apple's rainbow colors. A large ice sculpture of the company logo sat in a prominent position. The Mountain View High School jazz band played, after which the Stanford band marched through the crowd to a podium where Mike Markkula stood, wineglass in hand, grinning.

One month earlier, *Time* magazine, for the first time in its sixty-year history, had given its "Man of the Year" award to an inanimate object, the computer. The accompanying article singled out the many novel uses to which the Apple II had been put: the Grateful Dead had used an Apple machine for accounting and scheduling; an exhausted father had programmed his computer to rock the baby's cradle. *Time* averred that, thanks to the computer, "the entire world will never be the same." Although only 30 percent of Americans had access to a computer at home or work, 80 percent expected that in the near future, computers would be as common as televisions.[1]

But Apple was celebrating something different: the prospect of $1 billion in sales over the next year. "We sold $88.7 million in December," Markkula said after the band stopped playing and crowd members shushed one another to hear him. "That means a billion-dollar run rate. We should celebrate!"* Markkula introduced board chair and Macintosh group leader Steve Jobs, who, dressed in a white linen shirt, quoted Robert Kennedy: "There are those that look at things the way they are, and ask why? I dream of things that never were, and ask why not?" Jobs then singled out two engineers, Bill Atkinson and Rich Page, for their contributions to the new Lisa computer. (The tent also contained an ice sculpture of that machine.) Jobs handed each inaugural "Apple Fellow" a Superman cape and a large medal. Later, they would receive bonus checks, as well.

Markkula stepped back to the podium. Every Apple employee, he said, would receive a pair of crystal goblets like the one he held, to commemorate the company's "billion-dollar achievement." He thanked the employees for their work and left the stage to shouts and applause.[2]

Markkula was celebrating privately as well. Soon, he hoped, he would be able to end his two-year tenure as Apple's president. "I didn't dislike being president," he recalled years later, "and I did, I think, a damn good job." But he had always been ready to step down as soon as Apple found "somebody better." At the end of 1982, he thought the company had found somebody better in Don Estridge, the mastermind behind the successful launch of the IBM personal computer. But Estridge did not want to leave IBM. Apple had also considered candidates from minicomputer manufacturers such as DEC, HP, and Data General, but the fit had never been right. The company broadened its search beyond electronics and computing, and now it seemed that Apple had found its new president: John Sculley, the head of PepsiCo.

■ ■ ■ ■

With high hopes and no small measure of relief, Markkula left Apple's presidency in April. His replacement, John Sculley, had been lured to the position, in part, after Jobs asked him, "Do you want to sell sugar water for the rest of your life, or do you want to come with me and change the

---

* Apple would not have its billion-dollar year. Revenues for 1983 were $989 million, and by the third quarter of that year, profits had begun to slide in the face of competition from the IBM PC and the worse-than-expected reception of the Lisa machine.

world?" Sculley knew little about computers. But in the mid-1980s, Apple
was one of several technology firms that thought it wise to hire leaders
with general, but not technical, business experience. After Kassar was
fired, Atari named as CEO a former Philip Morris executive. Osborne
Computer gave the top job to the president of Consolidated Foods Corporation.[3]

The hiring of Sculley, a world-class consumer marketer, made further
sense, given how desperately Apple needed to appeal to people unfamiliar with computers. Today nearly half of Americans say that they cannot
live without the handheld computer that we call a smartphone, but in the
mid-1980s, when consumers were presented with a list and asked which
products they "don't know how they could get along without," only 2 percent of respondents named a home computer. Twice that number named
"do-at-home hair color" essential, while 38 percent said they could not
imagine life without aluminum foil.[4]

In 1983, computers were not easy to use. When *Newsweek* assigned a
reporter without computer experience to learn to use a personal computer,
she spent a total of seventy hours with three different machines (including
an Apple II and an Apple III), only to conclude, "I remained wholly inca-
pable of utilizing them to suit the needs I would have bought them for."[5]
Apple needed to figure out how to sell to the general public—by 1983,
IBM had 26 percent of the personal computing market, while Apple had
only 24 percent[6]—and Sculley seemed to be someone who might know
how.

Markkula remained a director of Apple after Sculley's hiring. Meanwhile, Sculley and Jobs entered a reporting relationship nearly identical to
the one that Markkula and Mike Scott had concocted at Apple's launch. As
president, Sculley reported to Jobs, the board chair. But Jobs, as head of
the Macintosh group, also reported to Sculley. The unconventional structure had worked well enough for Markkula and Scott, but it would prove
problematic for Jobs and Sculley.

One month after Markkula stepped down, in May 1983, six-year-old
Apple, trading at a record high, became the then youngest company ever
to enter the Fortune 500, just as Markkula had predicted.[7]

■ ■ ■ ■

The magnitude of Markkula's influence on Apple is not widely recognized today, and he has generally liked it that way. After Jobs, Markkula was the first person to perceive the business opportunity latent in Wozniak's Apple II. He used his connections to staff and finance the company. He wrote Apple's business plans and pushed for key technical developments. He set the company's marketing focus, as well as its attention to detail and first impressions. He led Apple to one of the most successful public debuts in US history, mentored Steve Jobs (who once called Markkula "like a father"), stepped in as president despite not wanting to do it, and escorted Apple into the Fortune 500.[8] To be sure, Apple's profits began slipping within six months of his resignation, evidence that the products and plans put into place under his watch were not adequately meeting the challenge from IBM and other competitors. Moreover, he served on, and briefly chaired, the Apple board during the years 1983–1997, when the company struggled almost to the point of collapse. Yet his early, vital importance cannot be diminished. Without Mike Markkula, there would be no Apple.

She Works Hard for the Money

SANDRA KURTZIG

By 1983, ASK was the fastest-growing software company and the eleventh-fastest-growing public company in the country.* From just under 100 employees and 150 customers at its 1981 IPO, ASK had shot to 350 employees and 700 customers. During the same period, its revenues zoomed from $13 million to $65 million and its profits from $1.6 million to $6.1 million. MANMAN blossomed from one stand-alone product with six modules to ten integrated products comprising thirty-three modules, compatible with VisiCalc and the ever-more-popular Lotus 1-2-3 spreadsheet program.[1]

Kurtzig now moved among Silicon Valley's elite. A poll of analysts named her one of the top three CEOs in the software industry. *Working Woman* magazine called her the "Queen of Silicon Valley."[2] At her beautiful, sprawling new home in the rolling hills near Sand Hill Road, she hosted parties attended by Steve Jobs and Don Valentine. She dated California governor Jerry Brown.

She and Tom Lavey folded ASK into the heart of the Silicon Valley

---

* Altos Computer Systems (no relation to the PARC Alto) was the second fastest growing; Bushnell's Pizza Time Theatre (filled with video games and singing electronic characters) ranked third; and Apple ranked fifth.

ecosystem she had spent so much of her early career outside of. They targeted the rising generation of entrepreneurs as customers. Lavey assigned someone full-time to work with venture capitalists and bankers, so the financiers would be more likely to recommend ASK's services to promising young companies. ASK landed one of its biggest clients thanks to this outreach, when a banker phoned Lavey to say that he had just left a meeting with a startup that planned to build computers. "I said that before they bought anything, they needed to call ASK," the banker told Lavey.[3]

Lavey had a sales person at the company, Compaq Computer, the next morning.[*] Compaq became a customer on ASKNET, the remote computing service that Lavey and Kurtzig had designed to allow customers access to MANMAN software without the up-front cost of a hardware purchase. Lavey estimates that Compaq accounted for $100 million in revenues to ASK over the next years.[4]

Kurtzig, whose first purchase for her company had been a filing cabinet, wanted an ASK system to be "the first thing people bought after they got funded." She mimed plucking an investor's check from an entrepreneur's hand: "Don't let the ink get too dry on that!"[5]

ASK's customer list included some of the most exciting names in high technology in the 1980s, including Cisco, Seagate, and Sun Microsystems, as well as Compaq. At the important American Electronics Association conference at the Del Monte Hyatt in Monterey, Kurtzig sent every one of ASK's customers a dozen long-stemmed roses to display in their rooms while they met with investors. "I could do that because I was a woman," Kurtzig says, adding that every bouquet served as an advertisement. Investors would ask companies without the flowers, "Why aren't you a customer of ASK?"[6]

Kurtzig kept her team together for years, even though many of ASK's employees lacked the large stock option grants with delayed vesting that serve as "golden handcuffs" to keep even wealthy employees tethered to companies after a successful IPO. The ribbing about Kurtzig's wealth

---

[*] The analyst-turned-venture-capitalist Ben Rosen had funded Compaq and would chair its board for eighteen years.

persisted. A joke video shot in Kurtzig's office in 1983 shows a woman's hands dumping a purse on her desk. As hundreds of dollar bills spill onto the wood, the hands dig through the pile to retrieve a bottle of nail polish. One hand begins painting the nails on the other, and when a bit of polish bleeds onto a cuticle, the hand grabs a bill as if it were a tissue and uses the money to wipe away the smear. When the nails look good, the hands begin shoving the bills off Kurtzig's desk, into her trash can. Through it all, the chorus of Donna Summer's "She Works Hard for the Money" blares in the background.[7]

The hands were not Kurtzig's, but she played along with the joke. The sequence ends with Kurtzig, her fingernails perfectly polished, faking a phone call for the camera and laughing.

Conclusion

The heart of Facebook's Silicon Valley headquarters, nine buildings containing more than 1 million square feet of office space, resembles a small town—except that, for employees, nearly everything is free. There are restaurants, a dry cleaner's, a bike shop, an arcade, and a barber. Most weekends, the company hosts a farmer's market, where employees and people from surrounding communities can buy fresh produce, play games on the grounds ("Human Foosball" is popular), watch cooking demonstrations, and listen to local musicians just steps away from San Francisco Bay.[1]

The company's unmistakable thumbs-up icon dominates the sign at the main entrance off Hacker Way. But the bright Facebook logo is a facade, a cheerfully printed swath of heavy-duty vinyl stretched taut over the front of a metal sign. On the reverse side of the sign, like a record track played backward, are the name and logo of another company: Sun Microsystems.

The Facebook campus once belonged to Sun Microsystems, a computer hardware and software company founded in 1982 that grew to be one of Silicon Valley's most successful and best-known firms.* Sun reached

---

* Sun built powerful, sophisticated, networked computers called workstations that were about three times the cost of personal computers. Workstation users were not

$1 billion in sales in six years—as rapidly as Apple, a company Sun once considered buying. Between 1985 and 1989, Sun was the fastest-growing company in the United States.[2] Cofounders Andy Bechtolsheim and Vinod Khosla were celebrated in their home countries of Germany and India. (There were two American-born founders as well, Bill Joy and Scott McNealy.) The company was so successful and iconic that a US senator visiting Silicon Valley in the mid-1980s wonderingly asked CEO McNealy, "Who told you that you could do this?"[3]

The innovations and human networks developed in the previous decade underpinned Sun's success. The computer workstation's technical design was based on technology from Xerox PARC, where Bechtolsheim had worked while in graduate school.[4] Kleiner Perkins Caufield & Byers, the venture capital firm that had launched Genentech and whose cofounders had worked at Fairchild and Hewlett-Packard, was an early and significant investor in Sun. Bechtolsheim had designed the workstation while a student at Stanford, so Niels Reimers's Office of Technology Licensing received an invention disclosure.[5] For years, Sun used MANMAN software, running on ASKNET, to track its materials resource planning.[6]

Sun, which at the turn of the millennium marketed itself as "the dot in dot-com," crashed after failing dot-com companies canceled their orders for computers and servers. Sun, also facing competition from less expensive servers running Linux and Berkeley Unix operating systems, never recovered. In 2010, Oracle Corporation acquired Sun for $7.4 billion. Oracle's cofounder and CEO was Larry Ellison, a former Ampex employee who had worked on the same Videofile system as Atari's Al Alcorn, Nolan Bushnell, and Ted Dabney.

When Facebook CEO Mark Zuckerberg left the Sun sign angled so that employees pulling out of the main campus parking lot could see it, he was sending them a message: take none of Facebook's success—the wealth, the accolades, the free lunches, the pride you feel in your company's name—for granted. Even a once iconic company can disappear

the personal computer users that Markkula called "Joe Averages" but scientists and engineers in computation-heavy fields such as computer-aided design, where the work demanded speed and significant storage capacity.

in the wake of relaxed vigilance, a few bad quarters, a series of poor decisions, or the rise of a new competitor.

■ ■ ■ ■

If Sun drew power from the waves of innovation that preceded it, the company also created momentum for those that came after. Sun's legacy is more than a fading logo on the back of a sign. Sun's Java programming language runs beneath millions of websites and applications, in web-based applications such as Google Docs, in financial trading systems, and in video games such as *Minecraft*. Among the more than 250,000 people who worked at Sun are the former CEOs of Google, Yahoo!, and Motorola.[7]

Sun's cofounders went on to fund new generations of technology companies. Vinod Khosla has backed dozens of startups as a venture capitalist, first at Kleiner Perkins Caufield & Byers and later at Khosla Ventures. In 1998, Andy Bechtolsheim met two young Stanford computer science graduate students who were thinking of starting a company. They had not incorporated and so did not have a name for the company, they told him, but Bechtolsheim was so excited that he wrote a $100,000 check to the name they thought they were most likely to choose.[8] "Pay to the order of," Bechtolsheim wrote, "Google, Inc."*

This chain of cross-generational support—in which technology, alumni, and resources from one innovative generation support the next— is one of Silicon Valley's great contributions to the world economy. It is the "baton pass" Steve Jobs referenced in his 2005 commencement address at Stanford.

It is easy to play the "begat" game in the Valley. Stanford and Xerox PARC begat Sun, which begat countless other companies and technologies. Fairchild begat Intel and National Semiconductor, which, through Markkula, Scott, and Carter, begat Apple, which through Trip Hawkins, who worked for Markkula, begat Electronic Arts. Stanford begat Cisco, Google, Hewlett-Packard, IDEO, Instagram, MIPS Computer Systems, Netscape, NVIDIA, Silicon Graphics, Snapchat, Sun Microsystems, Varian

---

* In 2012, Bechtolsheim's stake in Google had an estimated value of more than $1.5 billion.

Associates, VMware, and Yahoo!—and those companies, many funded by Don Valentine's Sequoia Capital or Kleiner Perkins Caufield & Byers, have begotten others. (Stanford has been a limited partner in both venture firms' funds, as well.) Ampex begat Memorex and Oracle, as well as Atari, which begat many other video game companies including Activision, the company behind *Call of Duty*. Research staff from the Computer Science Laboratory at Xerox PARC begat Microsoft Word, as well as AltaVista, Ethernet, the pioneering networking company 3Com, and Adobe. As recently as 2016, several of the most senior members of the research staff at Google were former PARC employees. Even companies that cannot trace their lineage to a Valley predecessor—Facebook, for example—have moved to the region to take advantage of its unique network and resources.

It takes a certain kind of audacity to think that you can launch a company, much less invent an industry—and audacity often veers into arrogance. The waves of innovation that have sustained Silicon Valley for the past sixty years have not been an unmitigated good. Waves crash. Undertows are strong. Silicon Valley is more crowded and more expensive than ever, and people who have lived in the Valley for decades can no longer afford to stay. The manufacturing and assembly jobs that supported Vineta and Fawn Alvarez as well as tens of thousands of others have largely disappeared from Silicon Valley, shipped overseas to cheaper markets or replaced by automation. The digital world has exposed, and some would say contributed to, the widening gaps between rich and poor. In Silicon Valley itself, men as a group earn more than women at the same level of educational attainment, and the lowest-earning racial/ethnic group earns 70 percent less than the highest-earning group.[9] Waste from electronics is a major cause of pollution around the world. High-tech companies spend tens of millions of dollars lobbying federal, state, and local governments, and the tech industry has what some see as disproportionate political clout as a result.[10]

Silicon Valley's record is not unblemished, but the impact of the innovations from the region is unmatched. The years covered in this book saw the launch of five major industries: video games, personal computing, biotechnology, modern venture capital, and advanced semiconductor logic. In the decades since, Silicon Valley has reinvented itself again and again. The consumer electronics wave of video games and personal computers became the wave of powerful software and networking companies, which was followed by web and search businesses, and now, cloud, mobile, and social networking industries.

The most recent waves of innovation have been sustained by the constant refresh of immigrants to the valley. This mixing of new ideas and new perspectives has been essential to Silicon Valley's continuing vitality throughout its history. In 1969, the immigrants came mostly from other parts of the United Sates. By 1980, the percent of the Silicon Valley population born outside the country was roughly double that of the American population as a whole. Today, one person born outside the country moves into Silicon Valley every half hour. Thirty-seven percent of the population is foreign born, nearly three times the US population overall.[11] More than half of the people in Silicon Valley over the age of five speak another language than exclusively English at home. Two-thirds of the people with bachelor's degrees working in science and engineering in the Valley were born in another country.[12] Immigrants are key members of management or product development teams in more than 70 percent of American startup companies valued at $1 billion or more—and founders at more than half of those companies.[13] The next great "Silicon Valley idea" can pull its energy from, or export its energy to, anywhere in the world.

Bob Taylor (holding football) plays touch football with fellow Xerox PARC employees in the early 1970s.

Venture capitalist Don Valentine in his Sequoia Capital office in the mid-1980s. Valentine was an early backer of Atari and Apple, as well as Oracle, Electronic Arts, LSI Logic, and Cisco.

Sandy Kurtzig at her desk at ASK Computer Systems, around 1979.

ASK employees at the company's offices in 1979. Back row (*left to right*): Howard Klein (*standing*), Marty Browne, Stewart Florsheim, Ann Rehling, Liz Seckler, Jeremy Lutzker, Sandy Kurtzig, Susan Glass. Front row: Tom Lavey, David Yee, Trish Clinton, Mark Ripma, Arlene Hill.

Niels Reimers (right) and a staff member of Stanford's Office of Technology Licensing assess a student's idea for a ski binding.

Mike Markkula and Steve Jobs in Europe, three months before Apple's IPO.

Regis McKenna, around 1985. McKenna helped introduce the world to the microprocessor, personal computer, and recombinant DNA. He was also instrumental in launching Silicon Valley's presence in Washington, DC.

Courtesy: Mike Markkula

The Apple executive team celebrates Mike Markkula's fortiethth birthday in 1982. Back row (*left to right*): Gene Carter, Del Yocam, Fred Hoar, Carl Carlsen, John Couch, Mike Markkula, Joe Graziano, John Vanard, Rod Holt. Front row: Ann Bowers, Al Eisenstat, Ken Zerbe, Steve Jobs. Most of these people had worked in the semiconductor industry before coming to Apple.

Courtesy: Fawn Alvarez Talbott

Fawn Alvarez began a rapid rise at ROLM after only a few months on the assembly line. By 1982, when this picture was taken in Amsterdam, she was traveling the world.

Mike Markkula, with an
Apple II, in his backyard in
the late 1970s or early 1980s.

Sandy Kurtzig at her ASK desk in the mid-1980s.

The author in front of the Facebook sign, May 2016.

The back of the Facebook sign reveals the original Sun Microsystems sign, May 2016.

Al Alcorn (*left*) and Nolan Bushnell (*right*) discuss Atari's history, 2016.

One of the last orchards in Silicon Valley, 2017.

Postscript

Al Alcorn

After leaving Atari in 1981, Al Alcorn worked on a project, funded in part by Nolan Bushnell's new technology incubator, that enabled buyers to load games onto blank cartridges dispensed through a vending-type machine. It was a very rudimentary version of downloading software in real time.

In 1986, Alcorn joined Apple as an Apple Fellow. It was a coveted appointment with a job description that thrilled him: change the industry.[1] He worked on video compression technology and a project to put the Macintosh operating system on an IBM PC. His boss was Larry Tesler, the Xerox PARC employee who had given Jobs the demonstration of the Alto and then moved to Apple.

After five years at Apple and another four as vice president of engineering at a company that built multimedia slot machines, Alcorn joined Interval Research Corporation, a Palo Alto–based lab and incubator launched by Microsoft cofounder Paul Allen and David Liddle, formerly of Xerox PARC. Alcorn ran engineering at an Interval-backed company, later sold to Lego, that built "smart" play sets with electronics embedded in figurines and blocks.[2] The physical movements of the toys could be mirrored on the screen of an attached personal computer.

Alcorn has since helped to found Hack the Future, a hackfest for middle school and high school students. Soldering iron in hand, he often appears at the local events in a T-shirt with the word MENTOR printed in capital letters across his back.

Fawn Alvarez Talbott

Fawn Alvarez continued to advance at ROLM after its acquisition by IBM. In 1987, she accepted a job as the assistant to the president of ROLM-IBM, Ray AbuZayyad. When ROLM was sold to Siemens a year later, she did the same job for Peter Pribilla, the CEO of ROLM Systems. These were not secretarial jobs; she functioned more as a chief of staff, calling on her long tenure at ROLM and her experience in production, engineering, finance, and marketing. She had a secretary of her own. She worked for Siemens in various capacities until 1997, after which she became a marketing consultant, finally enjoying a long tenure in a role she had wanted ever since seeing the ROLM employee who had worn lovely suits and fancy hats. In 2005, Alvarez married and changed her name to Fawn Alvarez Talbott. Ten years later, she retired.

The house in Cupertino that Alvarez Talbott's mother bought for $22,000 in 1970 would sell for more than $1 million today. Stevens Creek Boulevard, the two-lane street that neighborhood kids raced down on bikes, is now six lanes wide, with high-tech companies lining both sides. The orchards are gone.

Alvarez Talbott has left the Valley, as has her mother, Vineta. Both women now live in the Sierra foothills, about an hour and a half from each other. Vineta Alvarez Eubank, vigorous at eighty-one, began investing after retiring and follows the market closely. Fawn Alvarez Talbott is an active volunteer and gardener. She grows much of her own food, including plums not too different from those she picked as a child.

Sandra Kurtzig

Sandy Kurtzig moved to chair and CEO of ASK in 1984, when board member Ron Braniff became president. Before the move, she had acquired Software Dimensions for ASK for $7.5 million in 1983 and sold its constituent parts for less than $1 million a year later.[3] Despite that misstep, by 1991, ASK, with sales of $315 million, was the tenth largest independent software vendor in the world.[4] In the decade after Kurtzig left ASK's presidency, she moved into and out of leadership at the company, at times stepping in as CEO.

The rising wave of personal computing ultimately swamped ASK. Businesses moved away from the minicomputers so central to ASK's model, but the company never successfully developed a version of MANMAN for personal computers.[5] In May 1994, in a series of mergers and acquisitions, ASK, along with five others of the top fifteen independent software companies, was acquired by Computer Associates International. ASK, by then described in the *New York Times* as "a struggling software supplier," had two thousand employees and was valued at $310 million.[6]

Some one hundred MANMAN systems remain in use around the world. ASK's corporate DNA lives on in today's enterprise software industry, as former employees have taken on key roles at Oracle, Workday, and Rootstock Software. Kurtzig had several career turns—memoirist, business correspondent for *Good Morning America*—before returning to entrepreneurship in 2010, when she launched a cloud-based enterprise software company called Kenandy with backing from Salesforce.com, Kleiner Perkins Caufield & Byers, and the venture arm of Larry Sonsini's firm Wilson Sonsini Goodrich & Rosati. She currently serves as executive chairman of Kenandy.

Mike Markkula

In 1984, one year after leaving Apple's presidency, Mike Markkula started a company, ACM Research, to investigate distributed intelligence control systems.[*] Today the publicly held company is called Echelon. Its first CEO was Ken Oshman, the ROLM cofounder who told Sandy Kurtzig that ASK was not ready for an IPO in 1980. Early board members included Larry

---

[*] Those systems were precursors to today's Internet of Things. In the Internet of Things, items ranging from household thermostats and lights to industrial robots are connected to both a network and computing power to be more useful or valuable.

Sonsini, as well as former Apple directors Arthur Rock and Venrock's Peter O. Crisp.[7]

Markkula launched other successful businesses—a 14,000-acre cattle ranch and native grass farm on the Northern California coast, a jet center for private aircraft—but most important to him is the Markkula Center for Applied Ethics at Santa Clara University. He and his wife, Linda, seeded the center in 1986 and several years later donated $5 million to establish an endowment.[8] The center has developed a much-cited framework for ethical decision-making and runs programs for businesspeople, educators, students, government officials, and medical providers. When the ethics center launched, Markkula feared that the business world was populated by "two generations of ethical agnostics." He believed that they were not necessarily ill intentioned but made decisions that prioritized "dollars and cents, or their own advancement" without considering ethics. He calls the ethics center "the most bang for my philanthropic buck that I've ever gotten."[9]

Markkula remained on Apple's board until 1997, when Steve Jobs, who had left the company a dozen years earlier to launch NeXT, returned and asked him, along with every other Apple director save two, to resign. Markkula and Jobs had been somewhat estranged for years. Jobs felt betrayed by Markkula's backing John Sculley over him in the power struggle that had pushed him out of Apple in 1985. Markkula thought that Jobs had left Apple in a way that was "unethical," recruiting employees for NeXT while still chairing the Apple board.[10]

But the conversation upon Jobs's return after Apple purchased NeXT for $429 million was long and cordial. The two men walked and talked under under the redwood trees on Markkula's Woodside estate. Jobs was pensive and genial, seemingly no longer the "John McEnroe of American capitalism," as *Newsweek* had once dubbed him.[11] Markkula agreed to resign from the Apple board. Inside, he says, he rejoiced. He had wanted to leave for years but feared that the company had been in straits too dire to survive the resignation of the last person linked to its founding.[12]

As their conversation wound down, Markkula says, he offered Jobs some parting advice: Apple needed to be "opportunistic," to reinvent itself in much the same way that Hewlett-Packard had transformed itself from an instrument business to a calculator firm to a computer and printer company. He told Jobs that he did not know how Jobs could pull off a similar metamorphosis at Apple—but he needed to, or the company would die. Markkula, the great planner, had no specific plan for such a move; he just wanted his "diamond in the rough," one of the handful he had hoped would "learn everything I knew and add their own stuff to it and become world leaders," to make something happen.

Of course, Jobs did. He transformed the struggling computer business into the most valuable company in the world, one that made coveted music players and mobile phones, as well as computers. In the process, he rewrote the rules for three major industries: computing, music, and telephony.

Regis McKenna

Regis McKenna, the advertising and public relations guru who demystified high technology for the world, sold the advertising side of his business to Chiat\Day in 1981 so he could focus his own firm on strategy consulting and public relations. Throughout the 1980s and well beyond, he continued to work with some of the most important companies in Silicon Valley, including Apple, Intel, Tandem, and ROLM. He also played an integral role in helping the Semiconductor Industry Association make its successful case in Congress for the importance of high-technology industries.

Apple remained particularly important to McKenna. In the mid-1980s, he was the only non-Apple employee to attend executive staff meetings. In that role, he served as a confidant to both John Sculley and Steve Jobs as they battled for control of the company in 1985. Tensions rising, Sculley told the executive staff, "I never had a problem with leadership; I was president of everything since my first grade class, yet I know you question my leadership."[13] He called McKenna into his office and asked, "What will Steve do? Will he leave? Will he blow his brains out? Will he go to India and become a monk? Will he try to tear me down?"[14]

Jobs, meanwhile, bared his soul: "I need help in growing up that I never had," he said. "If I were at Kodak, for example, and I acted like an asshole, they would tell me to change or throw me out. But I spent my life at Apple with no one challenging me."[15]

Throughout Jobs's life, McKenna remained among his most trusted counselors and friends. The day after Jobs was pushed out of Apple, he presciently told McKenna, "Maybe we can develop a new, successful product line that would enhance the Apple product line, and they will buy us."[16] When Jobs particularly loved an Apple product, he sent one to McKenna. When a product had problems—when the antenna in the iPhone 4 had reception difficulties, for example—or when Jobs was concerned about Apple's image, he called McKenna for advice.

In 1995, McKenna sold the public relations side of his company to its employees. He retired from his eponymous firm in 2000 but continued to consult and served for several years as a consulting partner at Kleiner Perkins Caufield & Byers. An accomplished author (five books and dozens of articles), McKenna today serves as a director for several not-for-profit organizations and as an informal adviser to startup companies. He also travels widely, speaking on marketing, strategy, and the history of Silicon Valley.

Niels Reimers

After leaving Stanford, Reimers founded the licensing program at the University of California, San Francisco, and managed it from 1996 to 1998. He later served as a consultant to universities around the world that were planning to open or operate their own technology transfer offices. In 2008, he was inducted into the IP [Intellectual Property] Hall of Fame in recognition of his role in changing how universities use patents and transfer research to the private sector.

In 2016, cumulative income from the Office of Technology Licensing was nearly $2 billion.[17] But Reimers's influence has extended far beyond the Stanford campus. Before he established the Office of Technology Licensing, most universities used patents simply as a tool to help transfer technology beyond the campus. After, universities used patents to make money, and companies licensed academic intellectual property to build products. When Reimers launched the Office of Technology Licensing, there were nine other such offices in the country. Today, university licensing offices, nearly all established in the wake of the breakthrough recombinant DNA patent, are the norm.[18] In 2014 alone, academic research in the United States led to some $28 billion worth of product sales. In the same year, companies created 965 new products based on licensed university patents.[19]

Reimers has retired to a tidy cottage steps from the beach on the

California coast. He spent much of his childhood in this same town, and every day, at the post office or walking down the street, he sees dear friends who have known him for more than seven decades and yet have no idea of the impact he has had on the world around them. He likes it that way.

Bob Swanson

Bob Swanson stepped down as CEO of Genentech in 1990, after Roche bought 60 percent of the fourteen-year-old company for $2.1 billion. In 1999, three years after retiring from Genentech's board, he succumbed to brain cancer at the age of fifty-two. The man who at twenty-eight had wanted to change the world and had given himself the courage to do so by imagining his eighty-five-year-old self looking back at his life never made it to eighty-five. "One of the ironies here is he devoted so much of his life to applying this technology to generate medical breakthroughs, drugs that saved countless lives," says Arthur D. Levinson, a former CEO of Genentech and director of Google and Apple, where he also chairs the board. "But unfortunately, his wasn't one of them."[20]

Genentech was the first company to produce a human protein by splicing a gene into bacteria, the first to produce a drug by genetic engineering, and the first biotechnology company to go public.[21] It also set a new standard for the pharmaceutical industry when it allowed its scientists to publish papers on their research, rather than keeping it secret. Today many seminal publications in biology are written by scientists employed by industry, an unimaginable prospect when Genentech launched.[22]

Over the past four decades, Genentech has produced not only human insulin but human growth hormone, the cancer drugs Avastin and Herceptin, the antianxiety drug Klonopin, the anti-inflammatory Naprosyn, and the anti-influenza drug Tamiflu.[23] In 2009, Roche fully acquired Genentech for $46.8 billion. At the acquisition, every share of stock bought at the IPO for $35 in 1980 was worth $4,560.[24]

Bob Taylor

In 1996, Bob Taylor retired from Digital Equipment Corporation, where he had started the company's Systems Research Center. By then, there were around 725 million personal computers in the world, many of them with large screens, graphical user interfaces, a mouse, a word processor, email, and networked printing. The Arpanet had spawned the Internet. The first website was already six years old.

Taylor died on April 13, 2017, at the age of 85. For decades, he had lived in a secluded house overlooking Silicon Valley. Every summer, he hosted a tomato and oyster fest that brought dozens of the world's leading computer scientists—all of whom had worked for him—to his modest yard on a steep hill.

Taylor's feelings about technology were complicated. He thought the Internet was generally good for democracy and called the easy access to information that Google provides his "favorite part of the Internet." He was a witty and regular email correspondent who kept his Kindle full of books. He believed Facebook and Twitter were "wastes of time," though in many ways the applications do what Taylor dreamed a network of human-centered computing machines could do: create communities that transcend geographic boundaries.

He refused to try to predict the next big trend in computing.

Money never motivated Taylor, who saw lawyers and marketing people as distractions from the main event: research. When people he respected from his lab started or joined companies, he might have bought a bit of stock—that was how he ended up with early stakes in Adobe, Google, and 3Com—but the investments were so small and so not a focus of his attention that when asked about his investing, he talked about the stocks in his 401(k) plan.

Gaining recognition for the right people was the dream of Taylor's retirement. He had received a number of awards, including the National Medal of Technology in 1999 and the National Academy of Engineering's Draper Prize five years later, but he did not attend either ceremony. When he was inducted into both the Computer History Museum's Hall of Fellows and the Internet Hall of Fame, he skipped those ceremonies, as well. He did, however, send along comments to be read. "I have . . . some trouble with the way in which we, as a society, implement the idea of awards," he wrote. "Awards are usually given to individuals. But in computer research, especially in computer systems research, significant achievements are accomplished by teams of people, not just one or two." To the extent people know his name, he said, it is because the teams he led at Xerox PARC and Digital Equipment Corporation "made me look good."

Taylor had long pushed for an award to honor group creativity. He liked to quote a Japanese proverb: "None of us is as smart as all of us."[25] It is a fitting epitaph for Taylor and a message for innovators everywhere.

Acknowledgments

I hope I've made it clear in these pages that innovation is a team sport. The same could be said about the effort to get *Troublemakers* out of my head and onto the page. While I wrote every word and bear responsibility for every error, the team that I was able to call on for assistance was the best anyone could ever hope for.

I owe a huge debt to the North 24th Writers. Allison Hoover Bartlett, Leslie Crawford, Frances Dinkelspiel, Kathy Ellison, Sharon Epel, Susan Freinkel, Katherine Neilan, Lisa Wallgren Okuhn, Gabrielle Selz, Julia Flynn Siler, and Jill Storey: thank you for your wisdom, your humor, and your edits!

The people listed on pages 391 to 392 have been essential sources of information and inspiration as they sat through interviews and pointed me to other sources. To all of you: thank you for sharing your time, your stories, your papers, your memories, and your ideas.

What do you call someone who advises your dissertation, goes on to become your most important mentor, and somehow manages simultaneously to devastate and encourage you every time he reviews an early draft of your work? David M. Kennedy. Thank you, David, for reading and dissecting a draft of the full manuscript. Julia Flynn Siler did the same, even coming to my house to sit on my sofa and talk through her excellent recommendations. And after Randy Stross spent hours reviewing the draft, he for some reason felt obligated to take *me* to lunch. Thank you all for your help.

Thank you, too, to the five hearty and gracious souls who conducted technical reviews of select chapters and saved me from much embarrassment: Raiford Guins, Sally Smith Hughes, Mark Seiden, Doogab Yi, and Henry Lowood. Henry, my boss at Stanford for nearly fifteen years, has been endlessly supportive of this book and much more. *Danke*, Henry.

A deep bow to these people who shared their expertise and assistance: the Center for Advanced Study in the Behavioral Sciences 2012–2013 fellows, Janet Abbate, Bob Andreatta, David Brock, Carolyn Caddes, Martin Campbell-Kelly, Catherine de Cuir, Beth Ebben, Benj Edwards, Bret Field, Terry Floyd, Daniel Hartwig, the HP Alumni Association, Paula Jabloner, Kathy Jarvis, Laurene Powell Jobs, Kris Kasianovitz, Mike Keller, Chigusa Kita, Greg Kovacs, Steven Levy, Sara Lott, Anna Mancini, Natalie-Jean Marine Street, John Markoff, Pam Moreland, Mary Munill, Tim Noakes, Bill O'Hanlan, Margaret O'Mara, Sue Pelosi, Nadine Pinell, Sarah Reis, Paul Reist, Nora Richardson, James Sabry, Larry Scott, Lenny Siegel, Lisa Slater, Kurt Taylor, Bill Terry, and Fred Turner.

Two men who have no idea I exist have been an important part of writing this book. John August and Craig Mazin host a podcast called *Scriptnotes*. It's about "screenwriting and things that are interesting to screenwriters." I am not a screenwriter, but I have learned a great deal about story telling (and movies!) from John, Craig, and their guests.

At Simon & Schuster, I've been lucky enough to work with a wonderful editor, Ben Loehnen, along with Jon Karp, Cat Boyd, and Amar Deol. Many thanks to my agents Christy Fletcher, Don Lamm, and Sylvie Greenberg; and to Mark Fortier and Pamela Peterson.

My family is the axis on which my life turns, and I am so lucky to be able to call you all mine. Rick, Corbin, and Lily Dodd: you are my stars. Steve, Vera, Jessica, and Loren Berlin; Jim, Liz, Ryan, and Rob Dodd; Debbie, Brian, John, Olga, Katie, Lukas, Trevor, Sadie, Fiona, and James: you are the best.

Selected Bibliography

This book relies heavily on unpublished primary sources loaned to the author, materials and interviews housed in archival collections, and interviews that the author conducted with more than seventy very generous people over a period of six years. These and all other sources are listed in the Notes. The following books and articles were particularly helpful sources of information or inspiration.

Abbate, Janet. *Inventing the Internet*. Cambridge, MA: MIT Press, 1999.

Berg, Paul and Janet E. Mertz. "Personal Reflections on the Origins and Emergence of Recombinant DNA Technology." *Genetics* 184 (2010): 9–17.

Berlin, Leslie. *The Man Behind the Microchip: Robert Noyce and the Invention of Silicon Valley*. New York: Oxford University Press, 2005.

Biskind, Peter. *Easy Riders, Raging Bulls: How the Sex-Drugs-and-Rock 'n' Roll Generation Saved Hollywood*. New York: Simon & Schuster, 1998.

Brand, Stewart. *II Cybernetic Frontiers*. New York: Random House, 1974.

Brand, Stewart. "*Spacewar!*: Fanatic Life and Symbolic Death Among the Computer Bums." *Rolling Stone*, Dec. 7, 1972.

Bruck, Connie. *Master of the Game: Steve Ross and the Creation of Time Warner*. New York: Simon & Schuster, 1994.

Caddes, Carolyn. *Portraits of Success: Impressions of Silicon Valley Pioneers*. Palo Alto, CA: Tioga Publishing Company, 1986.

Campbell-Kelly, Martin. *From Airline Reservations to Sonic the Hedgehog: A History of the Software Industry*. Cambridge, MA: MIT Press, 2003.

Campbell-Kelly, Martin and William Aspray. *Computer: A History of the Information Machine*. New York: Basic Books, 1996.

Carter, Gene. *Wow! What a Ride: A Quick Trip Through Early Semiconductor and Personal Computer Development*. Raleigh, NC: Lulu Press, 2016.

Cohen, S. N., A.C.Y. Chang, H. W. Boyer, and R. B. Helling, "Construction of Biologically Functional Bacterial Plasmids *in Vitro*," *Proceedings of the National Academy of Sciences*, 70 (November 1973): 3240–4.

Cohen, Stanley N. "DNA Cloning: A Personal View After 40 Years." *Proceedings of the National Academy of Sciences*, 110, no. 39 (Sept. 24, 2013): 15521–9.

Copeland, Alan. *People's Park*. Phoenix: Ballantine, 1969.

Donovan, Tristan. *Replay: The History of Video Games*. Sussex, UK: Yellow Ant, 2010.

Draper, William H. III. *The Startup Game: Inside the Partnership Between Venture Capitalists and Entrepreneurs*. New York: St. Martin's Press, 2011.

Eymann, Marcia, and Charles M. Wollenberg. *What's Going On: California and the Vietnam Era*. Oakland, CA: University of California Press, 2004.

Freiberger, Paul, and Michael Swaine, *Fire in the Valley: The Making of the Personal Computer*. Berkeley, CA: Osborne/McGraw-Hill, 1984.

Gertner, Jon. *The Idea Factory: Bell Labs and the Great Age of American Innovation*. New York: Penguin Books, 2012.

Goldberg, Adele, ed. *A History of Personal Workstations*. New York: Addison-Wesley, 1988.

Goldberg, Martin, and Curt Vendel. *Atari Inc.—Business Is Fun*. Carmel, NY: Syzyg Press, 2012.

Guins, Raiford. "Beyond the Bezel: Coin-Op Arcade Video Game Cabinets as Design History," *Journal of Design History*, Oct. 7, 2015.

Hafner, Katie, and Matthew Lyon. *Where Wizards Stay Up Late: The Origins of the Internet*. New York: Simon & Schuster, 1996.

Harragan, Betty Lehan. *Games Mother Never Taught You: Corporate Gamesmanship for Women*. New York: Warner Books, 1978.

Herman, Leonard. "The Untold Atari Story." *Edge*, April 2009.

Hiltzik, Michael. *Dealers of Lightning: Xerox PARC and the Dawn of the Computer Age*. New York: HarperCollins, 2000.

House, Charles H., and Raymond L. Price. *The HP Phenomenon Innovation and Business Transformation*. Stanford, CA: Stanford Business Books, 2009.

Hughes, Sally Smith. *Genentech: The Beginnings of Biotech*. Chicago: University of Chicago Press, 2011.

———. "Making Dollars Out of DNA." *Isis* 92 (2001): 541–75.

Isaacson, Walter. *Steve Jobs*. New York: Simon & Schuster, 2011.

Jacobson, Yvonne. *Passing Farms, Enduring Values: California's Santa Clara Valley*. Cupertino, CA: William Kaufmann, 1984.

Katz, Barry M. *Make It New: A History of Silicon Valley Design*. Cambridge, MA: MIT Press, 2015.

Kearns, David T., and David A. Nadler. *Prophets in the Dark: How Xerox Reinvented Itself and Beat Back the Japanese*. New York: HarperBusiness, 1992.

Kent, Steven L. *The Ultimate History of Video Games: From Pong to Pokémon and Beyond—The Story Behind the Craze That Touched Our Lives and Changed the World*. New York: Three Rivers Press, 2010.

Kurtzig, Sandra. *CEO: Building a Four Hundred Million Dollar Company from the Ground Up*. New York: W. W. Norton & Co., 1991.

Lane, Frederick S. *American Privacy: The 400-Year History of Our Most Contested Right*. Boston: Beacon Press, 2011.

Levering, Robert, Michael Katz, and Milton Moskowitz. "Bill Hambrecht," in *The Computer Entrepreneurs: Who's Making It Big and How in America's Upstart Industry*. New York: New American Library, 1984.

Licklider, J.C.R. "Man-Computer Symbiosis," *IRE Transactions on Human Factors in Electronics*, March 1960: 4–11.

Licklider, J.C.R., and Robert W. Taylor. "The Computer as a Communication Device." *Science and Technology*, April 1968: 21–31.

Lowood, Henry. "Video Games in Computer Space: The Complex History of Pong." *IEEE Annals in the History of Computing* 31 (July–September 2009): 5–19.

Markoff, John. *What the Dormouse Said: How the Sixties Counterculture Shaped the Personal Computer Industry*. New York: Penguin Books, 2005.

Maxfield, Katherine. *Starting Up Silicon Valley: How ROLM Became a Cultural Icon and Fortune 500 Company*. Austin, TX: Emerald Book Company, 2014.

McElheny, Victor K. "Animal Gene Shifted to Bacteria; Aid Seen to Medicine and Farm." *New York Times*, May 20, 1974. As of March 2015, the article appears online under the title "Gene Transplants Seen Helping Farmers and Doctors."

McElheny, Victor. "Revolution in Silicon Valley." *New York Times*, June 20, 1976.

Moritz, Michael. *Return to the Little Kingdom: How Apple and Steve Jobs Changed the World*. London, UK: Overlook Press, 2010.

Mukherjee, Siddhartha. *The Gene: An Intimate History*. New York: Scribner, 2016.

Norberg, Arthur L., and Judy E. O'Neill. *A History of the Information Processing Techniques Office of the Defense Advanced Research Projects Agency*. Minneapolis: Charles Babbage Institute, 1992.

Noyce, Robert N., and Marcian E. Hoff, Jr. "A History of Microprocessor Development at Intel." *IEEE Micro*, February 1981: 8–21.

O'Mara, Margaret Pugh. *Cities of Knowledge: Cold War Science and the Search for the Next Silicon Valley*. Princeton, NJ: Princeton University Press, 2004.

Packer, George. *The Unwinding: An Inner History of the New America*. New York: Farrar, Straus and Giroux, 2013.

Perkins, Tom. *Valley Boy*. New York: Nicholas Brealey Publishing, 2007.

Perry, Tekla, and Paul Wallich. "Design Case History: The Atari Video Computer System," *IEEE Spectrum*, March 1983: 45–51.

Pollock, Christopher. *Reel San Francisco Stories: An Annotated Filmography of the Bay Area*. San Francisco: Castor & Pollux P., 2013.

Price, Rob. *So Far: The First Ten Years of a Vision*. Cupertino, CA: Apple Computer, 1987.

Rogers, Michael. "The Pandora's Box Congress." *Rolling Stone*, June 19, 1975.

Rosen, Ben. *Morgan Stanley Electronics Letter/Rosen Electronics Letter*, 1979–1982.

Rosenfeld, Seth. *Subversives: The FBI's War on Student Radicals, and Reagan's Rise to Power*. New York: Farrar, Straus and Giroux, 2012.

Salus, Peter, ed. *The ARPANET Sourcebook: The Unpublished Foundations of the Internet*. Charlottesville, VA: Peer-to-Peer Communications, 2008.

Schlender, Brent, and Rick Tetzeli. *Becoming Steve Jobs: The Evolution of a Reckless Upstart into a Visionary Leader*. New York: Crown Business, 2015.

Schulman, Bruce J. *The Seventies: The Great Shift in American Culture, Society, and Politics*. New York: Da Capo Press, 2002.

Sellers, John. *Arcade Fever: The Fan's Guide to the Golden Age of Video Games*. Philadelphia: Running Press, 2001.

Smith, Douglas K. *Fumbling the Future: How Xerox Invented, Then Ignored, the First Personal Computer*. New York: iUniverse, 1999.

Sporck, Charles E., and Richard Molay. *Spinoff: A Personal History of the Industry That Changed the World*. Saranac Lake, NY: Saranac Lake Publishing, 2001.

Strassmann, Paul A. *The Computers Nobody Wanted: My Years with Xerox*. New Canaan, CT: Information Economics Press, 2008.

Sullivan, Andrew. "I Used to Be a Human Being." *New York*, Sept. 18, 2016.

Surveillance Technology. Joint Hearing Before the Subcommittee on Constitutional Rights of the Committee on the Judiciary and the Special Subcommittee on Science, Technology, and Commerce of the Committee on Commerce, United States Senate, 94th Congress, First Session on Surveillance Technology, June 23, Sept. 9–10, 1975.

Taylor, Robert W. "Man-Computer Input-Output Techniques." *IEEE Transactions on Human Factors in Electronics*, March 1967.

The "People's Park": A Report on a Confrontation at Berkeley, California, Submitted to Governor Ronald Reagan. Sacramento, CA: Office of the Governor, July 1, 1969.

Turner, Fred. *From Counterculture to Cyberculture: Stewart Brand, the Whole Earth Network, and the Rise of Digital Utopianism.* Chicago: University of Chicago Press, 2006.

Vettel, Eric. *Biotech: The Countercultural Origins of an Industry.* Philadelphia: University of Pennsylvania, 2006.

Waldrop, M. Mitchell. *The Dream Machine: J.C.R. Licklider and the Revolution That Made Computing Personal.* New York: Viking, 2001.

Watson, James D., and John Tooze. *The DNA Story: A Documentary History of Gene Cloning.* New York: H. W. Freeman and Co., 1981.

Watson, Lucinda. "Ray Kassar: Former CEO, Atari," in *How They Achieved: Stories of Personal Achievement and Business Success.* New York: John Wiley & Sons, 2001.

Wayne, Ronald G. *Adventures of an Apple Founder.* Valencia, CA: 512k Entertainment, 2010.

Wozniak, Steve, and Gina Smith. *iWoz: Computer Geek to Cult Icon: How I Invented the Personal Computer, Co-Founded Apple, and Had Fun Doing It.* New York: W. W. Norton & Co., 2006.

Yi, Doogab. *The Recombinant University: Genetic Engineering and the Emergence of Biotechnology at Stanford, 1959–1980* (doctoral dissertation, Princeton University, 2008). In 2015, the University of Chicago Press published a version as *The Recombinant University: Genetic Engineering and the Emergence of Stanford Biotechnology.*

Abbreviations Used in Notes Section

AA   Al Alcorn
ACM   Courtesy Mike Markkula
AR   Courtesy Arthur Rock
CBI   Charles Babbage Institute, Oral History Collection
CHM   Computer History Museum, Oral History Collection
EKF   Courtesy Eugene Kleiner Family
GC   Courtesy Gene Carter
HK   Courtesy Howard Klein
KK   Courtesy Kathy Ku
MB   Courtesy Marty Browne
MG   Courtesy Manny Gerard
NVCA National Venture Capital Association, Oral History Project collection
PSC   Pacific Studies Center Archives (now at SUSC)
RM   Courtesy Regis McKenna
ROHO Bancroft Library, Oral History Collection (formerly Regional Oral History Office)
RWT   Courtesy Bob Taylor
SB   Courtesy Steve Bristow (now at SUSC)
SK   Courtesy Sandra Kurtzig
SSH   Courtesy Sally Smith Hughes

SUOTL Stanford University, Office of Technology Licensing (now at SUSC)

SUSC Stanford University, Special Collections

SUSG Stanford University, Silicon Genesis interview collection

SV Interview for *Something Ventured* video production

XPA Xerox PARC Archives

Interviewees

Al Alcorn
Bobby Alvarez
Jack Balletto
Andy Barnes
Ann Bowers
Steve Bristow
Marty Browne
Nolan Bushnell
Brook Byers
Gene Carter
Lisa Kleiner Chanoff
David Cochran
Stan Cohen
Joe Decuir
Jerry Elkind
Vineta [Alvarez] Eubank
Doug Fairbairn
Terry Floyd
Manny Gerard
Chuck Geschke
Laura Gould
Andy Grove

John Hall
Trip Hawkins
Sally Hines
Steve Jobs
Pitch Johnson
Alan Kay
Joe Keenan
Steve Kent
Howard Klein
Robert Kleiner
Kathy Ku
Sandy Kurtzig
Butler Lampson
Tom Lavey
Ann Kieffaber
Marty Manley
Natalie Jean Marine-Street
Mike Markkula
Bob Maxfield
Kathie Maxfield
Steve Mayer
Regis McKenna

Burt McMurtry

Carla Meninsky

Bob Metcalfe

Rich Moran

George Murphy

Severo Ornstein

Bill Osborn

Chuck Peddle

Bill Pitts

Jim Queen

Ron Raffensperger

Niels Reimers

Arthur Rock

Eric Schmidt

Lizbeth Seckler

John Shoch

Dick Shoup

Lenny Siegel

Jeff Smith

Phil Smith

Bill Spencer

Fawn Alvarez Talbott

Bob Taylor

Larry Tesler

Chuck Thacker

Don Valentine

Kurt Wallace

Barry Wessler

Bryant York

Notes

Introduction: A Bit Like Love

1. Steve Jobs, Stanford Commencement Address, June 12, 2005, http://news.stanford.edu/2005/06/14/jobs-061505/. It was not inevitable that one successful generation of innovators would support its successors. In Hollywood, for example, Steven Spielberg, whose breakthrough film *Jaws* was released the year before Apple launched, recalled a very different attitude, though he used the same metaphor Jobs had. "It's not like the older generation volunteered the baton," he said. "The younger generation had to wrest it away from them. There was a great deal of prejudice if you were a kid and ambitious. . . . I got the sense that I represented this threat to everyone's job." Spielberg, quoted in Peter Biskind, *Easy Riders, Raging Bulls: How the Sex-Drugs-and-Rock 'n' Roll Generation Saved Hollywood* (Simon & Schuster, 1998): 20.

2. Steve Jobs, interview by author, May 24, 2003.

3. See Zuckerberg's comments on Jobs on *Charlie Rose*, Nov. 7, 2011. Zuckerberg has also said that he visited a temple in India on Jobs's recommendation. See, e.g., http://www.businessinsider.com/mark-zuckerberg-visited-india-thanks-to-steve-jobs-2015-9.

4. "OTL Financial Data, 1970–2016." KK; Martin Campbell-Kelly, *From Airline Reservations to Sonic the Hedgehog: A History of the Software Industry* (Cambridge, MA: MIT Press, 2003): 115.

5. Gabriel Metcalf, "Beyond Boom and Bust: Where Is Silicon Valley Taking Us?," *The Urbanist*, April 2016, Figure 3.

6. In 2015, the median home price in Palo Alto was $2.5 million. Xin Jiang, "A Perspective on Chinese Home Buyers," *Palo Alto Weekly*, Oct. 27, 2016.

7. Lenny Siegel, *Testimony Prepared for the Subcommittee on Science, Research, and Technology of the House Committee on Science and Technology and the Task Force on Education and Employment of the House Budget Committee, June 16, 1983*, 1100–1, PSC.

8. Burt McMurtry, interview by author, Nov. 26, 2012.

9. This was the landmark *Diamond v. Chakrabarty* ruling of 1980.

10. As of February 10, 2017, the largest companies by market capitalization were Apple, Alphabet (Google), Microsoft, Berkshire Hathaway, Amazon, and Facebook.

11. Mark Muro, Jonathan Rothwell, Scott Andes, Kenan Fikri, and Siddharth Kulkarni, "Executive Summary," in *America's Advanced Industries: What They Are, Where They Are, and Why They Matter* (Washington, DC: Brookings Institution, 2015): 2.

12. Andrew Sullivan, "I Used to Be a Human Being," *New York*, Sept. 18, 2016.

13. Radicati Group, Email Statistics Report, 2015–2019 (Executive Summary), http://www.radicati.com/wp/wp-content/uploads/2015/02/Email-Statistics-Report-2015-2019-Executive-Summary.pdf.

14. "Total domestic US revenues generated by biotech in 2012 reached at least $324 billion." Robert Carlson, "Estimating the Biotech Sector's Contribution to the US Economy," *Nature Biotechnology* 34 (2016): 247–355.

15. Manufacturing employment as a share of the total US economy has undergone a steady decline in the past fifty years, dropping from 25 percent in 1960 to under 10 percent in 2010. Martin Neil Baily and Barry P. Bosworth, "US Manufacturing: Understanding Its Past and Its Potential Future," *Journal of Economic Perspectives* 28, no. 1 (Winter 2014): 3–26, Fig. 1.

16. Center for Responsive Politics, https://www.opensecrets.org/lobby/top.php?indexType=i&showYear=2016.

17. Jane E. Brody, "Hooked on Our Smartphones," *New York Times*, Jan. 9, 2017.

18. Alan Kay, interview by Michael Schwarz, May 20, 2014. Thanks to Kay and Schwarz for allowing the author to sit in on the interview.

Arrival: 1969–1971

1. "Wherever We Look, Something's Wrong," *Life*, Feb. 23, 1968.
2. "Electronics Industry Failures Fall to Lowest Level Ever," *Electronic News*, June 10, 1968.
3. Nilo Lindgren, "The Splintering of the Solid-State Electronics Industry," *Innovation* 1, no. 8 (1969): 2–16.
4. Population figures are for the period 1950–1970.
5. Wallace Stegner, introduction to Yvonne Jacobson, *Passing Farms, Enduring Values: California's Santa Clara Valley* (Los Altos, CA: William Kaufmann, 1984).
6. For more on the birth and rise of the microchip industry in Silicon Valley, see Leslie Berlin, *The Man Behind the Microchip: Robert Noyce and the Invention of Silicon Valley* (New York: Oxford University Press, 2005).
7. "What Made a High Flier Take Off at Top Speed," *BusinessWeek*, Oct. 30, 1965: 118–22; "Exchange Calls FC&I Pacer," *Electronic News*, Feb. 7, 1966.

Prometheus in the Pentagon — Bob Taylor

1. A great blow-by-blow account of this transmission is Katie Hafner and Matthew Lyon, *Where Wizards Stay Up Late: The Origins of the Internet* (New York: Simon & Schuster, 1996): 152–4.
2. Leonard Kleinrock, "Memoirs of the Sixties," in *The ARPANET Sourcebook: The Unpublished Foundations of the Internet*, ed. Peter Salus (Charlottesville, VA: Peer-to-Peer Communications, 2008): 96. See also "The First Internet Connection with UCLA's Leonard Kleinrock" at http://www.youtube.com/watch?v=vuiBTJZfeo8, in which Kleinrock says the "Lo" marks "the day the infant Internet uttered its first word."
3. M. Mitchell Waldrop, *The Dream Machine: J.C.R. Licklider and the Revolution That Made Computing Personal* (New York: Viking, 2001): 266. The IPTO budget rose from $15 million to $19.6 million during the years Taylor served as director; Arthur L. Norberg and Judy E. O'Neill, *A History of the Information Processing Techniques Office of the Defense Advanced Research Projects Agency* (Minneapolis, MN: Charles Babbage Institute, October 1992): 119.
4. Bob Taylor to Eugene G. Fubini, March 31, 1967, RWT.

5. This story is from Wessler's comments at Taylor's retirement party.

6. Waldrop, *The Dream Machine*: 265.

7. Taylor's adviser was Lloyd A. Jeffress.

8. For a hilarious depiction of the waiting game that was batch processing, see the two-minute "Ellis D. Kropotchev Silent Film," created by Arthur Eisenson and Gary Feldman, at http://www.computerhistory.org/revolu tion/punched-cards/2/211/2253.

9. Bob Taylor, CHM interview.

10. J.C.R. Licklider, "Man-Computer Symbiosis," *IRE Transactions on Human Factors in Electronics*, March 1960: 4–11.

11. Taylor, CHM interview.

12. Louise Licklider to Bob Taylor, January 1990, 4, RWT.

13. M. M. Davis [United Air Lines Sales Manager] to Bob Taylor, June 2, 1970, RWT.

14. Taylor, CBI interview.

15. John R. Rice and Saul Rosen, "History of the Department of Computer Sciences at Purdue University," cs.perdue.edu/history/history.html.

16. In July 1966, Taylor wrote a letter in which he said that he had agreed to serve as director since none of the "several individuals we recommended for the job accepted." Bob Taylor to Anthony G. Oetinger, July 29, 1966, RWT.

17. "San Antonio Boy, 5, Given Genius Rating," *San Antonio Express*, Aug. 2, 1937.

18. "Supercommunity": J.C.R. Licklider and Robert W. Taylor, "The Computer as a Communication Device," *Science and Technology*, April 1968: 21–31; "metacommunity": Taylor, CBI interview.

19. Robert Taylor, "Plans for an Experimental, Interactive Computer Network," paper to be presented at the 2nd Workshop on National Systems of the Task Group on National Systems for Scientific and Technical Information, Front Royal, VA, n.d., but probably 1968.

20. Banks and airline reservation systems also used a different type of remote computing.

21. Taylor, interview by John Markoff, Dec. 9, 2008, at https://www.youtube .com/watch?v=vqsTpNtziE8&list=PL653B57BD7DA5B890&index= 1&feature=plpp_video.

22. Taylor, interview by author, April 22, 2013. Whether Licklider envisioned the Intergalactic Network as a centralized system (imagine a giant

time-sharing system, with a single machine at the core and nodes taking the place of individual users) or a decentralized one (which is how the Arpanet worked and the Internet does now) is the subject of some debate. The historian Chigusa Kita believes that Licklider was envisioning a "centralized network to share resources, both informational and computational," whereas Licklider's biographer M. Mitchell Waldrop believes that Licklider likely envisioned a decentralized system similar to the Arpanet. A 1990 interview between Taylor and William Aspray contains the following exchange: TAYLOR: Recently, in fact I said [to Licklider], "Did you have a networking of the ARPANET sort in mind when you used that phrase [intergalactic network]?" He said, "No, I was thinking about a single timesharing system that was intergalactic. ASPRAY: Very large, like an octopus? TAYLOR: Right.

23. Robert Taylor, "Recollections and Reflections on the ARPANET," unpublished, n.d., but appears to be mid-1970s, RWT. This document is the earliest I've found in which either Taylor or Herzfeld describes the conversation they had.

24. For an example of an ex post facto justification for the network, see Stephen Lukasik, quoted in Waldrop, *Dream Machine*, 279.

25. Noah Schachtman, "How Pacific Island Missile Tests Helped Launch the Internet," Aug. 27, 2012, http://internethalloffame.org/blog/2012/08/27/how-pacific-island-missile-tests-helped-launch-internet.

26. Charles Herzfeld, quote from Robert Taylor, "Recollections and Reflections on the ARPANET." Norberg and O'Neill also use the $500,000 figure. Since those were reapportioned funds, exactly how much was promised is unclear. Most accounts say $1 million, but they are all based on the same source: Taylor's memory. The $500,000 figure seems more likely, particularly since the Program Plan written in 1968 calls for $560,000 for that fiscal year before jumping to $1 million in the following year.

27. Al Blue, CBI interview.

28. Stephen D. Crocker, "Foreword," in *ARPANET Sourcebook*, ed. Peter Salus: 98.

29. Both Taylor and Roberts have used the term "blackmail" to describe Taylor's hiring of Roberts. Taylor, CHM interview; Roberts, "The ARPANET and Computer Networks," in *A History of Personal Workstations*, ed. Adele Goldberg (New York: Addison-Wesley, 1988), editor's note, p. 145.

30. Roberts tells this story at http://www.ibiblio.org/pioneers/roberts.html.

31. In his CBI oral history, Roberts says he was hired as a special assistant to the director, with an understanding that he would go on to run the Information Processing Techniques Office. Taylor has said that Roberts was hired as a program manager, and it is under this title that Roberts is listed in the ARPANET Program Plan.

32. Roberts makes this point in his CBI interview and in Waldrop, *The Dream Machine*, 268–9.

33. Taylor, "Recollections and Reflections on the ARPANET."

34. Blue, CBI interview.

35. Taylor, "Recollections and Reflections on the ARPANET." Roberts's take on the meeting is laid out in his CBI oral history and the generally excellent account in Waldrop, *The Dream Machine*, 272–4. Doug Engelbart, a participant, describes a similar lack of interest among principal investigators in John Markoff, *What the Dormouse Said: How the Sixties Counterculture Shaped the Personal Computer Industry* (New York: Penguin Books, 2005): 166. Taylor has said that neither IBM nor AT&T was interested in the network; indeed, that they were hostile to it.

36. Robert Kahn, CBI interview. Kahn was part of the network buildout from very early on and went on to direct the Information Processing Techniques Office from 1979 to 1985.

37. Wes Clark, CBI interview. Leonard Kleinrock, another key participant (and the man arguing for the significance of "'Lo!' As in lo and behold!"), says, "Bob set the tone for Larry's *modus operandum*. Bob Taylor is a great administrator." Kleinrock, CBI interview.

38. Those routers were called IMPs. Both Waldrop and Hafner/Lyon go into this ride at some length, though both say it was a taxi ride (unlikely, given the number of people in the vehicle: Blue, Clark, Dave Evans, Roberts, and Taylor). Both Roberts and Taylor say that Clark had mentioned the idea to him before mentioning it to anyone else, both thus implying that he had sought his approval before offering the idea to others.

39. Paul Baran to Bob Taylor, Oct. 16, 2004, RWT.

40. Licklider and Taylor, "The Computer as a Communication Device": 21–31. See also Robert W. Taylor, "Man-Computer Input-Output Techniques," *IEEE Transactions on Human Factors in Electronics*, March 1967.

41. Norberg and O'Neill, *A History of the Information Processing Techniques Office*. All information about the bidding process is classified, according to this document.

42. Taylor, CHM interview.

43. Taylor, CHM interview and interview by author, April 24, 2013.

44. The account in Hafner and Lyon's *Where Wizards Stay Up Late*, though not footnoted, is consistent with Taylor's version of events: "In the middle of December, Roberts entered into final negotiations with Raytheon for the IMP contract. Raytheon officials answered ARPA's remaining technical questions and accepted the price. So it surprised everyone when, just a few days before Christmas, ARPA announced that the contract to build the Interface Message Processors that would reside at the core of its experimental network was being awarded to Bolt Bernanek and Newman, a small consulting firm in Cambridge, Massachusetts" (81).

45. More on Taylor's role in the birth of the graphics Center for Excellence at Utah is in Norberg and O'Neill, *A History of the Information Processing Techniques Office*: 281.

46. Markoff, *What the Dormouse Said*: 51. The NASA funding for Engelbart's NLS oNLine Systems was about $80,000. Taylor also told Licklider about Engelbart's work, which led to ARPA (where Licklider was then director) also funding the research.

47. Doug Engelbart, interview by Lowood, March 4, 1987, at http://stanford.edu/dept/SUL/sites/engelbart/engfmst1-ntb.html.

48. Taylor, interview by author, March 18, 2013. Markoff's account corroborates Taylor's description of his role.

49. Waldrop, *The Dream Machine*, 289.

50. Taylor, interview by author, March 18, 2013. In his CBI interview, Al Blue says, "Bob Taylor certainly was responsible for our continued support of Doug and his work out there."

51. The 1968 demo in its entirety is at http://sloan.stanford.edu/mousesite/1968Demo.html.

52. "In those days, the concept was that nobody stayed in ARPA very long. You brought a guy in and got the best of his brains and then he moved on, and the next prodigy came along, if you will. So I never viewed the Taylor departure as being under any kind of duress." Blue, CBI interview.

53. Taylor's ID identifying him as a general was issued on January 20, 1967.

54. Taylor, CHM interview. It seems likely that one person who joined Taylor on these trips was Colonel Clair L. Shirley, USAF, who worked in the office of the Joint Chiefs of Staff and whom Taylor listed as a reference in a job application.

55. Barry Wessler, interview by author, March 28, 2013. Wessler accompanied Taylor on later trips.

56. Blue, CBI interview.

57. Ibid.

58. ARPANET program plan.

59. Wayne Morse to Harold Howe, Commissioner of Education [re: Taylor], July 18, 1968, RWT. Morse, who set a filibuster record in the Senate, possessed a kindred spirit to Taylor's, according to an official Senate biography: "His admirers called him 'The Tiger of the Senate.' His many enemies, including five presidents, called him a lot worse. Today he is remembered as a gifted lawmaker and principled maverick who thrived on controversy." http://www.senate.gov/artandhistory/history/minute/Wayne_Morse_Sets_Filibuster_Record.htm.

60. Taylor, interview by author, March 18, 2013 and April 22, 2013;

Nerd Paradise — Al Alcorn

1. The *Berkeley Barb* article ("Hear Ye, Hear Ye," by "Robin Hood's Park Commissioner") is quoted in its entirety in *The "People's Park": A Report on a Confrontation at Berkeley, California, Submitted to Governor Ronald Reagan* (Office of the Governor, State of California, July 1, 1969): 2–3.

2. Reagan, quoted in "You Say You Want a Revolution? Records and Rebels 1966–1970" exhibit at the Victoria and Albert Museum, London.

3. The *"People's Park"*: 5, 16.

4. Eldridge Cleaver, in Marcia Eymann and Charles M. Wollenberg, *What's Going On?: California and the Vietnam Era* (Oakland, CA: University of California Press, 2004): 53.

5. Al Alcorn describes Rackarock as a potent combination of ammonium nitrate and nitrobenzene similar to the explosives used in the Oklahoma City bombing in 1995. Alcorn, interview by author, Oct. 25, 2011.

6. Alcorn, interview by author, Oct. 25, 2011.

7. Alcorn's size: résumé in tabbed business plan, tabbed business plan, Atari Business Plans Collection, SUSC.

8. Statistics for and accounts of People's Park have been checked across a number of sources. When in doubt, I relied on Seth Rosenfeld, *Subversives: The FBI's War on Student Radicals, and Reagan's Rise to Power* (New York: Farrar, Straus and Giroux, 2012). Rosenfeld's definitive account corroborates Alcorn's description of his experiences on May 15.

9. "Nation: Occupied Berkeley," *Time*, May 30, 1969.

10. Ibid. It says 482 marchers were arrested in a single day. Frederick Berry, Thomas Brooks, and Eugene Commons, "Terror in a Teapot," *The Nation*, June 23, 1969, claims that "between May 15 and May 24, almost 1,000 arrests were made on the streets of Berkeley."

11. The image appears in Alan Copeland, *People's Park* (New York: Phoenix/Ballantine, 1969). Alcorn would end up working with the ACLU's investigation into the People's Park violence, and he would testify before the Alameda County Grand Jury.

12. "Starting before World War I, the United States and German governments began conducting climate surveys and gathering meteorological data. Their findings revealed Redwood City to be at the center of one of the world's three best climates." http://redwoodcityhistoryroom.com/redwood-city-history.html.

13. For Ampex's role in the Apollo 8 transmissions, see http://www.hq.nasa.gov/alsj/ApolloTV-Acrobat5.pdf and http://blog.longnow.org/02009/05/03/digital-recovery-of-moon-images/.

14. Alcorn, interview by author, Oct. 25, 2011.

15. Guild analogy suggested in Steve Mayer, interview by author, Feb. 3, 2012.

16. Alcorn, interview by author, Oct. 25, 2011.

17. "Videofile Training Manual: Introductory Concepts," Ampex Corp. Records M1230, Series 2, Box 3, accn 2001-241, SUSC. General sources for Ampex: interviews with Alcorn, Bristow, Bushnell, Mayer, and Kurt Wallace, as well as the Ampex collection, M1230, SUSC.

18. Leonard Herman, "The Untold Atari Story," *Edge*, April 2009; Martin Goldberg and Curt Vendel, *Atari Inc.—Business Is Fun* (Carmel, NY: Syzygy Press, 2012).

19. Nolan Bushnell, interview by Steve L. Kent. Steve L. Kent collection relating to the video game industry, M1872, SUSC.

20. Bushnell, interview by author, Aug. 1, 2012.
21. Alcorn, CHM interview.
22. Bushnell at "Atari's Impact on Silicon Valley, 1972–1984," panel organized by the IEEE SV Tech History Committee, Sept. 8, 2016, Santa Clara, California.

Eight Quarters in Her Pocket — Fawn Alvarez

1. Vineta Alvarez Eubank, interview by author, Feb. 14, 2017.
2. Cal. Labor Code § 1350 (1955). The section limiting women to eight-hour workdays was repealed in 1984.
3. Eubank, interview by author, Feb. 14, 2017.
4. Fawn Alvarez Talbott, interview by author, July 24, 2013.
5. Katherine Maxfield, *Starting Up Silicon Valley: How ROLM Became a Cultural Icon and Fortune 500 Company* (Austin, TX: Emerald Book Company, 2014): 28.

The Fairchildren — Mike Markkula

1. Mike Markkula, CHM interview.
2. Markkula, interview by author, Dec. 2, 2015.
3. Gene Carter, interview by author, Jan. 7, 2016.
4. Carter, CHM interview; Carter, interview by author, Jan. 7, 2016.
5. "slugged a coworker": Jack Gifford, SUSG interview.
6. Carter, CHM interview; Mike Markkula, interview by author, Dec. 2, 2015.
7. Charles E. Sporck, *Spinoff: A Personal History of the Industry That Changed the World* (2001): 219, 222.
8. Gifford, SUSG interview.
9. Markkula, CHM interview; Markkula, interview by author, Dec. 2, 2015.
10. Markkula, interview by author, Dec. 4, 2015.
11. Markkula, interview by author, Dec. 2, 2015; this interview is also the source of the footnoted story.
12. Markkula, CHM interview; Gifford, SUSG interview; Markkula, interview by author, Feb. 24, 2016.
13. Markkula, SUSG interview.
14. Markkula, interview by author, Dec. 4, 2015.

15. Ibid.
16. In 1971, Intel amended its stock option plan, which had originally included 175,000 shares, to add another 100,000 shares (Intel S-1).

What Do We Do with These? — Niels Reimers

1. Ron Goben, "Police Break Up Palo Alto Jam-in by Stanford Mob," *Palo Alto Times*, May 16, 1969.
2. Stanford University News Service Chronology of Events, May 16, 1969, and "What Is Happening Here?" [flyer distributed May 18 but using the same language] at http://a3m2009.org/archive/1968-1969/68-69_may16 _sri/files_68-69_may_16/A3M-5-16_What_is_Happening.pdf.
3. Stanford University News Service Chronology of Events, May 16, 1969, at http://a3m2009.org/archive/1968-1969/68-69_may16_sri/files_68-69_ may_16/A3M-5-16_Chronology_p1-2.pdf; Notes from Bill and Margie at http://a3m2009.org/archive/1968-1969/68-69_may16_sri/files_68-69_may _16/A3M-5-16_Notes_2.pdf.
4. "What to Do If You Are Arrested," http://a3m2009.org/archive/1968-1969 /68-69_may16_sri/files_68-69_may_16/A3M-5-16_If_Arrested.pdf.
5. Goben, "Police Break Up Palo Alto Jam-in by Stanford Mob."
6. Marc and Carrie Sapir, "Stanford, May 1969: Students Shut Counterinsurgency Center," *Peninsula Observer*, through May 16, 1969: 1, 3. "Cease all classified" (in footnote): "Guidelines for Research at Stanford and SRI," Documents of the April 3rd Movement, SC 841, 1:1, SUSC.
7. The cache of documents at the April Third Movement's online archive is invaluable: http://a3m2009.org/archive/1968-1969/68-69_may16_sri/68-69_may16_sri.html.
8. Ron Goben, "Stanford Boycott Hits Humanities Classes," *Palo Alto Times*, May 12, 1969; Jenny Matthews, "Movement Awaits Trustees' Action," *Stanford Daily*, May 13, 1969; Marc and Carrie Sapir, "Stanford, May 1969: Students Shut Counterinsurgency Center," *Peninsula Observer*, through May 16, 1969: 1, 3.
9. "In general, Stanford students practiced a more voyeuristic form of activism than their Berkeley counterparts," writes Eric Vettel in *Biotech: The Countercultural Origins of an Industry* (Philadelphia: University of Pennsylvania, 2006): 116.

10. On interdisciplinary centers: Cyrus C. M. Mody and Andrew J. Nelson, "'A Towering Virtue of Necessity': Interdisciplinarity and the Rise of Computer Music at Vietnam-Era Stanford," *Osiris*, Volume 28, Number 1, Jan. 1, 2013. http://www.journals.uchicago.edu/doi/10.1086/671380. The Air Force ROTC program at Stanford was terminated in June 1971; the Army and Navy ROTC programs concluded in June 1973. "Towards an On-Campus ROTC Program at Stanford University: A Report and Recommendation by the Ad Hoc Committee [to the Faculty Senate]," Stanford University, April 2011.

11. Niels Reimers, interview by author, Oct. 27, 2014; "Encina Hall: Leland Stanford's Grand Hotel," *Sandstone & Tile*, Winter 2000.

12. The inventor was William S. Johnson; the name of the patent application filed for the invention was "Synthesis of Juvenile Hormones." Niels Reimers to William S. Johnson, "Royalty Income Distribution," May 21, 1969, SUOTL.

13. Agreement between the Board of Trustees of the Leland Stanford, Jr. University and Research Corporation, Jan. 1, 1956, SUOTL. The contract notes that for Research Corporation, "no part of the net earnings . . . inures to the benefit of any private shareholder or university." The inventor received $100 and 16.66 percent of the gross dollars received by Research Corporation; Stanford received 50 percent of the net. Today Research Corporation exists, in a different form, as Research Corporation for Science Advancement; see http://www.rescorp.org/about-rcsa/history. For an excellent history of Research Corporation, see David C. Mowery and Bhaven N. Sampat, "Patenting and Licensing University Inventions: Lessons from the History of the Research Corporation," *Industrial and Corporate Change*, June 1, 2001.

14. "Data for Patent Licensing Program Regarding Research Corporation," n.d., but must be 1968, SUOTL.

15. Niels Reimers, "Subj: Individuals performing patent admin. functions at Stanford (per conversation with KB)," June 27, 1968.

16. Yaell Ksander, "The Invention of Fluoride Toothpaste," http://indianapublic media.org/momentofindianahistory/the-invention-of-flouride-toothpaste/.

17. Darrell Rovell, "Royalties for Gatorade Trust Surpass $1 Billion," http:// www.espn.com/college-football/story/_/id/13789009/royalties-gatorade -inventors-surpass-1-billion.

18. Karen W. Arneson, "Frank Newman, 77, Dies; Shaped Education," *New York Times*, June 4, 2004; "In Memoriam: Frank Newman," http://www.uri .edu/library/special_collections/exhibits/newman/newman_memoriam .htm.

19. Sally Hines, interview by author, Nov. 11, 2014.

20. Niels Reimers, "Commercialization of Ideas in a Research Environment," Oct. 13, 1988, SUOTL.

21. Niels Reimers to File, Oct. 29, 1968, SUOTL; Reimers to File, Patent Licensing Program Notes, Nov. 14, 1968, SUOTL.

22. "Invention, Events, and Licensing at Stanford," February 1972, SUOTL.

23. Samuel Baron, University of Texas, quoted in Hal Lancaster, "Profits in Gene Splicing Bring the Tangled Issue of Ownership to Fore," *Wall Street Journal*, Dec. 3, 1980.

24. Stanford University Founding Grant, Nov. 11, 1885.

25. Frank Newman to Kenneth M. Cuthbertson, vice president for finance, July 19, 1968. In this letter, Newman promised that "we can operate on a low-risk yet effective basis," and after explaining that there would be "no added or special staff," proposed that $15,000 be set aside for outside attorneys' fees for the pilot program. "The upside chance for gain may be considerable," he notes. He also mentioned "the ⅓ to the university, ⅓ to the department, and ⅓ to the inventor arrangement."

Come with Me, or I'll Go by Myself — Sandra Kurtzig

1. Jon Gertner, *The Idea Factory: Bell Labs and the Great Age of American Innovation* (New York: Penguin Books, 2012): 75–8; Sandra Kurtzig, *CEO: Building a Four Hundred Million Dollar Company from the Ground Up* (New York: W. W. Norton & Co., 1991).

2. William D. Smith, "Computer Time Sharing Grows Up," *New York Times*, Feb. 17, 1969.

3. Ibid.

4. Warner Sinback, CHM interview.

5. Kurtzig, *CEO*: 25. A great account of early women programmers—those working a full generation before Kurtzig—is Jennifer S. Light, "When Computers Were Women," *Technology and Culture* 40, no. 3 (1999): 455–83.

6. The gender breakdown of students is from the program from Stanford's 77th commencement (1968, when Kurtzig graduated). The number and gender of people in the aeronautics and astronautics department is hard to verify, since the department does not have records going back that far. Kurtzig says she was one of two women in a department of 250; the commencement program shows that in 1968, Kurtzig (then Sandra Brody) was the only woman in a group of forty-three students receiving master's degrees in aeronautics/astronautics. Nine doctorates were also awarded in the department that year, all to men. The program is also the source of the statement that her family lived in Beverly Hills.

7. Sandra Kurtzig, interview by author, June 18, 2015.

8. Sinback, CHM interview. Sinback initiated the recruiting program after a friend whose daughter was a math major said that she was having a problem finding a job and Sinback realized that educated women "were potentially great sources of talent."

9. Kurtzig, interview by author, June 24, 2015.

10. Kurtzig, *CEO*: 33.

11. Ibid.

12. Kurtzig, interview by author, June 24, 2015.

13. Kurtzig, *CEO*: 124.

14. Ibid.: 248.

Building: 1972–1975

1. Don C. Hoefler, "Silicon Valley USA," *Electronic News*, Jan. 10, 1971. As late as December 1979, even local papers referred to "the Santa Clara Valley electronics industry; Charles Petit, "Wizard of Silicon Gulch," *Peninsula Times Tribune*, Sept. 21, 1977; Bill Densmore, "The Santa Clara Valley Electronics Industry Comes of Age During the 'Me' Generation Decade," *Peninsula Times Tribune*, Dec. 28, 1979, "The Splintering of the Solid-State Electronics Industry," *Innovation 8*, 1969.

2. The first serious article about the regional economy on the San Francisco Peninsula was Gene Bylinsky, "California's Great Breeding Ground for Industry," *Fortune*, June 1974: 129–224. The name "Silicon Valley" first appeared in the *New York Times* in John H. Allan, "Whither Semiconductor Stocks?," *New York Times*, Nov. 11, 1975.

3. Art Detman (*Forbes* editor) to Regis McKenna, Nov. 8, 1971, RM. McKenna included Detman on a list of editors to receive releases about Silicon Valley startups; Detman wrote to ask to be removed from the list, explaining that he was interested in hearing only about large, publicly held firms.

4. Matt Bowling, "The Massage Parlor Crackdown: Palo Alto's Prostitution Problem," *Palo Alto Daily News*, May 11, 2008.

5. HP-35 manual at http://www.cs.columbia.edu/~sedwards/hp35colr.pdf; "HP-35 Scientific Calculator Awarded IEEE Milestone," HP news release, April 14, 2009, at http://www.ieee.org/documents/hp35_milestone_re lease.pdf.

Have You Seen This Woman? — Sandra Kurtzig

1. Wanted flyer, SK.

2. Sandra Kurtzig, *CEO: Building a Four Hundred Million Dollar Company from the Ground Up* (New York: W. W. Norton & Co., 1991): 17–18. The entrepreneur was Larry Whitaker; the company he launched was Halcyon Communications.

3. *Auerbach Guide to Time Sharing*, January 1973.

4. "Women in Science and Technology: A Report on an MIT Workshop," May 21–23, 1973. Seventy percent of women worked in teaching, sales, or clerical jobs. A later study (Francine D. Blau, "Trends in the Well-Being of American Women," *Journal of Economic Literature*, March 1998: 112–65) estimated that about 45 percent of women comparable to Kurtzig in age, educational level, race, and marriage status (with a working spouse at home) were in the workforce.

5. "Women in the Workforce," 2009 Census presentation at https://www.cen sus.gov/newsroom/pdf/women_workforce_slides.pdf. Ninety-eight percent of women-owned businesses in 1972 were sole proprietorships. *Discussion and Comments on the Major Issues Facing Small Business: A Report of the Select Committee on Small Business, United States Senate to the Delegates of the White House Conference on Small Business*, Dec. 4, 1979: 55.

6. U.S. Department of Labor, "National Survey of Professional, Administrative, Technical, and Clerical Pay," March 1972.

7. "Worldwide IT Spending on Enterprise Software," https://www.statista .com/topics/1823/business-software/.

8. 52.1 percent of venture capital investments in 2015 went to software; "Venture Capital by Industry," Silicon Valley Indicators 2016, http://sil iconvalleyindicators.org/data/economy/innovation-entrepreneurship/venture-capital-by-industry/.

9. Martin Campbell-Kelly, *From Airline Reservations to Sonic the Hedgehog: A History of the Software Industry* (Cambridge, MA: MIT Press, 2003): 4, 6; Jon Levine, "5,000 Entrepreneurs . . . and Counting," *Venture*, January 1982.

10. Martha Reiner, "Top Computer Firm Began as Part-Time Work," *SF Business Journal*, March 2, 1981.

11. Marty Browne, ASK's first employee, estimates that more than 90 percent of ASK's customers in its first decade had never used a computer. "Hewlett Packard Software Workshop—Session 2: Starting HP Software Businesses," session recorded June 5, 2008, Computer History Museum.

12. Kurtzig, *CEO*: 45.

13. "Manufacturing Management," *Tymshare NewsBits*, July 1974; "Manman Adapted to On-line Usage on Tymshare Net," *Computerworld*, Sept. 25, 1974.

14. Kurtzig, *CEO*: 18.

15. Check from Halcyon Communications to ASK, Feb. 25, 1972, SK.

16. Thread at http://www.paloaltoonline.com/square/2006/10/19/the-things-i-remember-about-palo-alto-while-growing-up.

17. "Women in Electronics," *Peninsula Electronic News*, Aug. 27, 1973.

18. Sandra Kurtzig, interview by author, June 24, 2015.

19. Marion M. Woods, "What Does It Take for a Woman to Make It in Management?," *Personnel Journal*, January 1975.

20. Ann Hardy, interview by Janet Abbate, 2002, at http://ethw.org/Oral-History:Ann_Hardy#Interview.

21. Robert Kaestner, Darren Lubotsky, and Javaeria Qureshi, "Mother's Employment by Child Age and Its Implications for Theory and Policy," Sept. 2, 2015, at http://www.sole-jole.org/16113.pdf; "Employment Characteristics of Families Summary," Bureau of Labor Statistics, April 22, 2016.

22. Kurtzig, *CEO*, 55.

23. Shockley papers, SC0222, ARCH-1986-050, Box 2, SUSC.

24. Michael S. Malone, "How Is Kurtzig Doing? You Only Have to ASK," *San Jose Mercury News*, March 6, 1979; Kurtzig, *CEO*: 55.

25. "Women in Electronics." 1973.

26. Kurtzig, interview by author, June 24, 2015.

27. Campbell-Kelly, *From Airline Reservations to Sonic the Hedgehog*: 131. A former Tymshare salesman recalls that the company required contracts with "unusual terms" that charged for CPU seconds, connect time, and storage units; http://corphist.computerhistory.org/corphist/view.php?s=stories&id=136

28. Tymshare Annual Report, 1973.

29. Quist made a private investment in Tymshare and served on the board after helping the company secure an SBIC loan in 1965. Moreover, at nearly the same time that Kurtzig decided to license Tymshare to distribute her program, Tymshare had recently gone back to the SDS 940 (then renamed the XDS 940 after the acquisition by Xerox) after an unsatisfactory effort to use the Sigma 7. The Sigma 7 was the computer that Taylor's group at PARC had refused to work with; Becky McNown, "History of Tymshare."

30. Larry Sonsini, ROHO interview.

31. "Law, Innovation, and Silicon Valley," *Stanford Lawyer* 86 (June 11, 2012); "Wilson Sonsini Goodrich & Rosati: About Us," https://www.wsgr.com/WSGR/Display.aspx?SectionName=about.

32. The exchange with the manager on MAMA is from Kurtzig, *CEO*: 62. Marty Browne, ASK's first programmer, says that "Tymshare required that there be a six-letter reference to a program" and that therefore MAMA was too short (Browne oral history, CHM). Several Tymshare annual reports list programs with names that are not six letters long, but it is possible that six letters were mandatory at the time Kurtzig was choosing a name. It is likewise possible that both stories are true: that Kurtzig had the conversation with the manager *and* that six letters were a Tymshare requirement.

33. Jim A. Thorp, "The Entrepreneur," *The Executive SF*, July 1981.

34. *ASK Computer Systems, Inc.*, Stanford University Graduate School of Business Case S-E-16 (11/1994), Exhibit 3.

35. HP Annual Report, 1975.

36. Kurtzig, *CEO*: 69.

37. "Processor Description: Hewlett-Packard 2100" at http://archive.computerhistory.org/resources/text/HP/HP.2100.1972.102646165.pdf.

38. Kurtzig met Browne at Farinon, the same company whose owner's challenge would change Bill Hambrecht's thinking about small companies going public. Farinon had hired ASK to computerize its book, hundreds of pages long and nicknamed "the bible," that described how to build all the components and subcomponents of the company's antennas, amplifiers, transmitters, oscillators, and receivers used around the world to carry telephone calls over long distances. The early 1970s were a tough time to find a job in Silicon Valley, even for a Stanford math major, so Browne had signed up with a temp agency that had placed him in a $2.75-an-hour job at Farinon, where he did much of the grunt work (and later more sophisticated work, as well) in connection with Kurtzig's program. She was impressed enough with him to offer a job at ASK when the Farinon project ended. Marty Browne, interview by author, July 12, 2015.
39. Liz Seckler, interview by author, July 29, 2015.
40. Kurtzig, interview by author, June 24, 2015.
41. HP, 1975 Annual Report; Jim Leeke, "Career Women Chip Away at the Male Stranglehold at the Top in Silicon Valley," *Electronic Times*, Sept. 25, 1980.
42. Kurtzig, *CEO*: 73–80.
43. Ibid.: 72.

Turn Your Backs on the Origins of Computing! — Bob Taylor

1. Dee F. Andersen (university controller) to Bob Taylor, Dec. 23, 1969; T. C. King to Taylor, March 3, 1970, both RWT. Taylor, "Some Thoughts on Information Processing at the University of Utah," March 10, 1970, RWT.
2. The PARC labs on Porter Drive, and later Coyote Hill, were a mile from the Stanford campus.
3. Stewart Brand, "*Spacewar!*: Fanatic Life and Symbolic Death Among the Computer Bums," *Rolling Stone*, Dec. 7, 1972.
4. Taylor told the author that if he were ever the subject of a biography, he wanted the Leibovitz photo for the cover.
5. Martin Campbell-Kelly and William Aspray, *Computer: A History of the Information Machine* (New York: Basic Books, 1996): 130, 229.
6. Computer History Museum *Revolution* exhibit.
7. Ibid.

8. Chief Scientist Jack Goldman, in Hiltzik, *Dealers of Lightning*: 154. The MIT president was Jerome Wiesner.

9. Taylor, interview by author, March 18, 2013. The first seminar Taylor gave at PARC was called "A Brief History of ARPA-Sponsored Computer Research" (Activity Report for CSL, July 2, 1971, XPA).

10. J. E. Goldman to C. P. McColough, June 23, 1969. In addition to the $5.8 million operating budget, Goldman anticipated a $7 million maximum investment.

11. J. E. Goldman to C. P. McColough, June 23, 1969.

12. George Pake, "Research and Development Management and the Establishment of Xerox Palo Alto Research Center," Remarks for the IEEE Convocation "The Second Century Begins," January 1985, XPA.

13. Pake's role in increased salaries: Michael Hiltzik, *Dealers of Lightning: Xerox PARC and the Dawn of the Computer Age* (New York: Harper-Collins, 2000): 61; Taylor's role: performance reviews, RWT.

14. Taylor's title: Performance Appraisal Notice, March 1, 1971, RWT.

15. In May 1971, Taylor talked about "Xerox's 'Information Company' intent," and he described the emphases for both CSL and SSL as "prototype systems experiments, especially with regard to library systems, office systems, medical systems, and educational systems." At some point in 1972, he began to write about "the PARC prototype office communication system." By May 1975, the "primary mission of the Palo Alto Research Center is to lay the research foundations for Xerox information systems and thus for Xerox business success in the future automated office," according to the "PARC Mission and Relation to Business Goals: Narrative for Long Range Plan, May 1975," XPA. Taylor, performance review, 8/30/70–1/31/71 (completed by Taylor 5/28/71), and performance review for period 9/1/71–9/1/72, RWT.

16. Bob Taylor, interview by author, March 18, 2013.

17. Severo Ornstein, interview by author, Jan. 9, 2014.

18. Taylor's criteria for a good researcher: recognition by peers and superiors; an excellent education, preferably from a top school (he put a great deal of emphasis on that); "singularity of purpose"; and an ability to "communicate about work on the leading edge," even if it meant inventing the vocabulary to describe that work. Taylor, interview by author, Dec. 18, 2013.

19. Bob Metcalfe, interview by author, May 22, 2014. Metcalfe, the coinventor of Ethernet, went on: "Bob was more [about] culture and touchy-feely and charismatic. Jerry was the more technical- and administration-oriented."

20. Paul A. Strassmann, *The Computers Nobody Wanted: My Years with Xerox* (New Canaan, CT: Information Economics Press, 2008): 113. On Larry Tesler (in footnote): Elkind to Pake, "Some Conclusions and Comments on the Tesler Affair" and "Chronology of Our Negotiations with Larry Tesler," both Oct. 1, 1971, RWT.

21. Butler Lampson, interview by Alan Kay, CHM. The research project was Project Genie. For more on Berkeley Computer Company, see "Preliminary Proposal for a Systems Group within the Xerox Palo Alto Research Center," RWT.

22. Taylor, interview by author, April 22, 2013.

23. Taylor, interviews by author, March 18 and April 24, 2013. John Markoff cited several sources who said that Engelbart could not "let go of his creation so the world could use it." An effort to license Engelbart's system to PARC was "stillborn," according to Markoff. John Markoff, *What the Dormouse Said: How the Sixties Counterculture Shaped the Personal Computer Industry* (New York: Penguin Books, 2005): 204.

24. Doug Engelbart, Smithsonian oral history, at http://americanhistory .si.edu/comphist/englebar.htm.

25. Robert Taylor, Forum 79 talk, RWT.

26. Lampson, CHM interview.

27. Metcalfe, interview by author, May 22, 2104.

28. Dealer minutes, Feb. 29, 1972, RWT; Alan Kay, interview by author, May 20, 2014.

29. Taylor, interview by author, March 18, 2013.

30. "These two papers mark the beginning of a controversy which is not going to be settled by any single experiment, but the controversy is healthy and should not be inhibited." Robert W. Taylor, "Man-Computer Input-Output Techniques," *IEEE Transactions on Human Factors in Electronics*, March 1967: 4.

31. Bob Sproull to George Pake, Sept. 28, 1977, RWT.

32. Dealer minutes, Jan. 19, 1972.

33. Dealer minutes, Jan. 9, 1974.

34. Alan Kay recounts this story in his interview with Lampson, CHM.
35. Dealer minutes, Feb. 23, 1972, RWT.
36. Dealer minutes, Jan. 26, 1972, RWT.
37. Severo Ornstein, interview by author, Jan. 9, 2014.
38. Butler Lampson and Chuck Thacker, CHM panel, at http://www.youtube .com/watch?v=2H2BPrgxedY.
39. The price of a PDP-10 system is an estimated cost by the CSL personnel over a five-year lifetime. A Sigma 9 system, they claimed, would cost $5.65 million and a "PARC-built system," $913,000. "MAXC Capital Acquisition Request," June 24, 1971, XPA.
40. A notable exception was chief scientist Jack Goldman, the man behind PARC, who objected to the SDS purchase in the first place.
41. Strassmann, *The Computers Nobody Wanted*: 113.
42. Dealer minutes, March 14, 1972.
43. Janet Abbate, *Inventing the Internet* (Cambridge, MA: MIT Press, 1999): 134–5. See also Howard Frank, Babbage Oral History: 20–1. "ARPA was trying to give away the ARPANET at one time, to get anybody to take it," recalled Frank, who, along with Taylor's successor Larry Roberts, met with AT&T in an effort to convince it to run the network. Frank describes AT&T's reaction to the offer to take ARPANET technology to the public as "complete lack of interest, because they couldn't imagine why anybody would want to send data in the network." AT&T also did not believe packet switching had a future.
44. Notation in Jerry Elkind and Bob Taylor to George Pake, "Activity Report for June 13, 1972, through December 31, 1972," records that on May 22, 1972, Elkind wrote a memo to Jack Goldman ("Xerox Acquisition of the ARPA Network") "discussing the future sale of the ARPANET and the possible implications to and for Xerox," RWT. On June 9, Elkind again wrote to Goldman ("ARPA Network") "suggesting that a group be formed to analyze the opportunity for buying the ARPANET, and recommending the action Xerox should take." In a June 16, 2014, email to the author, Elkind said he had entirely forgotten about the exchange and so could not provide further comment.
45. Email from Janet Abbate to author, June 18, 2014.
46. Dealer minutes, Feb. 23, 1972. Kay says that PARC's creep toward a personal computer began even earlier. He points to the Pendery Papers compiled by the lab in the spring of 1971 in response to a request from Donald

W. Pendery, Xerox's director of product planning. Papers by Jim Mitchell, Alan Kay, and Richard Shoup, among others, make it clear that the lab was already moving toward distributed computing. "Pendery Papers" file, XPA.

47. Dealer minutes, Feb. 29, 1972.

48. Lampson, interview by author, April 18, 2014; Chuck Thacker, interview by author, April 11, 2014.

49. Thacker sent his own memo on a similar theme at the same time. On the network: "We can very easily put in an Aloha-like point-to-point packet network between Alto's, using a coax as the ether (or microwave with a repeater on a hill for home terminals)"; Butler Lampson to CSL, Dec. 19, 1972, RWT.

50. Butler Lampson to CSL, Dec. 19, 1972, RWT.

51. Larry Tesler, Designing Interactions interview: http://www.designingin teractions.com/interviews/LarryTesler.

52. Thacker, interview by author, April 11, 2014.

53. The CSL Archives Notes are filled with these sorts of communications. "Vented Frustrations" is a memo from Metcalfe dated Oct. 16, 1972; "What Am I Doing Here" is from Peter Deutsch, Feb. 12, 1971, RWT.

54. Tesler, interview by author, May 16, 2014.

55. Bob Sproull to George Pake, Sept. 28, 1977, RWT.

56. Taylor, interview by author, July 18, 2013.

Hit in the Ass by Lightning — Al Alcorn

1. Al Alcorn title from résumé in tabbed business plan.

2. The address of Nutting Associates: 500 Logue Avenue.

3. Nolan Bushnell, interview by author, Aug. 1, 2012.

4. Bill Pitts, interview by author, March 21, 2012.

5. Al Alcorn, CHM interview. Alcorn's explanation: "It involved simply making a sync generator, a television sync generator which had, you know, counters to count clock pulses to make a horizontal sync, and then counters to count horizontal sync to make vertical sync, and so you'd get the lines set up. If you had another sync generator and you just had it running at the same time, but not synchronous with it, just the same clock and you decided to take the second sync generator output and make a

spot where horizontal and vertical sync happen the same time, that spot would appear randomly, somewhere on that screen, just by happenstance. You turn the power up, backup would be somewhere else. So now if you had that sitting there and you made the second sync generator a vertical counter, one less than the primary sync generator, that spot would appear to move up, you know, and if you made the count one more than the count, you'd appear to go down and similarly horizontal. So now, you've basically— you're less than 20, 50-cents chips, and you now have a spot, you can put a spot anywhere." For more technical details on *Computer Space*, see Henry Lowood, "Video Games in Computer Space: The Complex History of *Pong*," *IEEE Annals in the History of Computing* 31 (July–September 2009): 5–19.

6. Alcorn, interview by author, Oct. 25, 2012. Bushnell patented the "trick." "Video Image Positioning Control System for Amusement Device," patent 3,793,483, filed by Bushnell Nov. 24, 1972, granted Feb. 19, 1974. The image positioning technique "eventually became a staple of game machines and home computers in the form of sprites," according to the video game historian Henry Lowood.

7. "Syzygy Statement of Owner's Equity, Year Ended December 31, 1971," in Martin Goldberg and Curt Vendel, *Atari Inc.—Business Is Fun* (Carmel, NY: Syzygy Press, 2012): 53.

8. http://retro.ign.com/articles/858/858351p1.html.

9. Alcorn, CHM interview.

10. Bushnell, interview by author, Aug. 1, 2012.

11. Ibid.

12. Alcorn says that he "took half a chip" and "gated out" the tones. http://retro.ign.com/articles/858/858351p1.html.

13. "Atari's Impact on Silicon Valley, 1972–1984," panel organized by the IEEE SV Tech History Committee, Sept. 8, 2016, Santa Clara, California.

14. Joe Decuir, interview by author, Jan. 10, 2011.

15. Alcorn, CHM interview.

16. Alcorn, interview by author, Oct. 25, 2011.

17. Alcorn, CHM interview; Alcorn, interview by author, Jan. 9, 2012.

18. Al Alcorn, speaking at "Atari's Impact on Silicon Valley" panel, Sept. 8, 2016.

19. Nolan Bushnell, quoted in Steven L. Kent, *The Ultimate History of Video*

Games: From Pong to Pokémon and Beyond . . . The Story Behind the Craze that Touched Our Lives and Changed the World (New York: Three Rivers Press, 2010).

20. Alcorn tells this story at http://www.acmi.net.au/talks_gameon_storyof pong.htm. Goldberg and Vendel think it is possible that the players really did know someone who was working on a rival machine.

21. Bushnell, interview by author, Aug. 1, 2012.

22. Alcorn, quoted in Kent, *Ultimate History*, 44.

23. Bushnell, interview by Steve Kent, SUSC.

24. Bushnell, *SV*.

25. In addition to the 3 percent, Atari would receive $4,000 per month. William K. Ford, "Copy Game for High Score: The First Video Game Lawsuit," *Journal of Intellectual Property Law* 20 (2012): fn. 51.

26. Al Alcorn, at http://retro.ign.com/articles/858/858351p1.html.

27. Alcorn, interview by author, Oct. 25, 2011.

28. Atari Business Plan, 1975, Atari Inc. Business Plans Collection, M1641, SSC.

29. Bushnell, interview by author, Aug. 1, 2012.

30. Kent, *The Ultimate History of Video Games*, confirmed by Steve Bristow, interview by author, Nov. 9, 2011.

31. Kurt Wallace, interview by author, April 18, 2012.

32. Eddie Adlum, quoted in Kent, *The Ultimate History of Video Games*.

33. "Atari: Beginning to End," panel with Al Alcorn, Steve Mayer, Bill Reebok, and John Skruch, CGE 2k4 keynote (2004), at http://www.digitpress.com/cge/2k4_mp3/2k4_mp3.htm.

34. It would be years before the trade press that followed jukeboxes and pinball machines would begin to cover video games.

35. At high interest rates, and using essentially all the company's assets as security, Atari secured some $2.25 million in loans from General Electric Credit Corporation and Bank of America. Atari, Notes to Consolidated Financial Statements, June 2, 1973, and June 1, 1974.

36. Alcorn, interview by author, Jan. 12, 2011.

37. Promotional materials for *Pong* advertised the game's "Low Key Cabinet, Suitable for All Sophisticated Locations." Raiford Guins, "Beyond the Bezel: Coin-op Arcade Video Game Cabinets as Design History," *Journal of Design History*, Oct. 7, 2015: 7.

38. "The Duke" brochure, SB; *The Gospel According to St. Pong*. http://www
.digitpress.com/library/newsletters/stpong/st_pong_v1n6.pdf.

39. Bushnell, interview by author, Aug. 1, 2012.

40. Ibid.

41. Atari agreed to buy Dabney's ownership share for $246,418, paid in ten
equal installments beginning one year later, in March 1974. To support
himself before the payments kicked in, Dabney bought the pinball route
that Atari owned. Because Dabney, like the company itself, had little
cash, he paid for the route with a $100,000 IOU. Details from Atari S-1
mockup, Al Alcorn, papers relating to the history of video games, 1973–
1974, M1758, SUSC.

42. Bushnell, interview by Peter Sellers, Aug. 8, 2003, cited in John Sellers,
Arcade Fever: The Fan's Guide to the Golden Age of Video Games (Philadelphia: Running Press, 2001).

43. Bushnell, SV.

44. *The Gospel According to St. Pong*, Aug. 8, 1973.

45. Bushnell, SV; Stanley T. Kaufman (Actuarial Systems, Inc.) to Atari, Inc.,
March 29, 1974, courtesy Greg Kovacs.

46. Atari Business Plan, 1974, Section IV, Atari, Inc., Business Plans, 1974–
1975, M1641, SUSC.

47. Atari's 1973 list of competitors selling "computer video games" numbered
a dozen: Allied Leisure, Midway (a Bally subsidiary), Ramtek, Williams
(a Seeburg subsidiary), For-Play, Chicago Coin (a division of Chicago Dynamics), Amutronics, Alca, Sega Enterprises (a Gulf+Western
subsidiary), Taito, Nutting Associates, and U.S. Billiards. Of these, three
(along with Atari) were in the Bay Area; three were in Chicago, the traditional home of gaming companies; two were in Japan; and one each
was in Florida, New York, New Jersey, and the United Kingdom. See
"Exhibit 3, Major Coin-Operated Amusement Machine Manufacturers,"
tabbed business plan, Atari Business Plans Collection, SUSC.

48. From a business plan written in Spring 1974: "Atari sold almost 7,000
Pongs. . . . The total volume for both Pong and Pong-type copies hit 45,000
units in 1973." There was one legitimate competitor: Atari had sold Bally a
license to manufacture *Pong*.

49. Bound business plan, Atari Business Plans collection, SUSC.

50. Bushnell, interview by author, Aug. 1, 2012.

51. Alcorn's experience as head of R&D after outside management was brought in: Alcorn, interview by author; Alcorn, CHM interview; description of "Moose" in Wieder, "Fist Full of Quarters."

52. "Cumulative Unit Trade Sales Year-to-Date, March 74," tabbed business plan, Atari Business Plans collection, SUSC.

53. Bristow, interview by author, Nov. 9, 2011.

54. Alcorn, interview by author, Oct. 25, 2011.

55. Steve Mayer, interview by author, Feb. 3, 2012.

56. David Owen, "The Second Coming of Nolan Bushnell," *Playboy*, June 1983.

57. "Atari Sells Itself to Survive Success," *BusinessWeek*, Nov. 15, 1976.

58. Bristow, interview by author, Nov. 9, 2011.

59. Bushnell to Engineering, Aug. 3, 1973; Alcorn's response, same date, AA.

60. *Historical Statistics of the United States*, Table Dg117-130, "Radio and television—stations, sets produced, and households with sets: 1921–2000."

61. Alcorn, CHM interview. Steve Mayer confirmed the Intel visit in his interview with the author, Feb. 3, 2012.

62. Valentine received 12,500 shares of Atari stock in exchange for "advice on organization, manufacturing, and marketing, as well as development of a business plan" that called for him to help the company raise $2 million.

63. Don Valentine, interview by author, Nov. 7, 2012. Valentine worked at Fairchild for eight years before leaving in 1967 with a number of other Fairchild employees, including Charlie Sporck and Floyd Kvamme, to revivify the moribund National Semiconductor.

64. Valentine, interview by author, Nov. 7, 2012. "They liked the purity of the mutual fund business, the fact that it was regulated [and] we know how to behave, and we're good at it."

65. Valentine, interview by author, Nov. 7, 2012.

66. Ibid.; Jeff Moad, "When Your Investors Are Entrepreneurs," *Venture*, October 1980.

67. Ibid.; "jester": Valentine, SV.

68. Hot-tub board meetings were mentioned by Alcorn, Bushnell, and Valentine in interviews with the author and also by Jac Holzman, who became a director after the acquisition by Warner; see Connie Bruck, *Master of the Game: Steve Ross and the Creation of Time Warner* (Simon & Schuster, 1994): 172.

69. Valentine, *SV*.

70. Daniel Raff and Peter Temin, "Sears Roebuck in the Twentieth Century: Competition, Complementarities, and the Problem of Wasting Assets," *NBER Working Paper Series on Historical Factors in Long-Run Growth*, June 1997: 30, 25.

71. Valentine, *SV*.

72. Atari, Notes to Consolidated Financial Statements, May 31 and Nov. 29, 1975 (the interest rate was prime + 1.25 percent); purchase contract from Sears, signed by Thomas F. Quinn, March 17, 1975, Al Alcorn collection; Sears loan, executed Aug. 1, 1975, per Sept. 18, 1975, board minutes, Atari Business Plan Collection, SUSC.

73. Noyce told Joe Keenan this. Joe Keenan, interview by author, Dec. 3, 2013.

74. The small company that built the cases was called Crafts West. Alcorn, interview by author, Jan. 9, 2012.

75. Sears sold *Pong* under the name *Tele-Game*. An example is at http://www .atarimuseum.com/videogames/dedicated/homepong/homepong-pt2. htm. An AC power adapter sold for an additional $7.99.

76. Raiford Guins to author, Nov. 14, 2016.

77. "Space-Age Pinball," *Time*, April 1, 1974, tabbed business plan, SUSC.

78. Atari S-1 mockup, proof of June 24, 1976 , Alcorn papers, SUSC.

Make It Happen — Niels Reimers

1. Ferns and memorabilia: Debby Fife, "The Marketing of Genius," *Stanford Magazine*. "MAKE IT HAPPEN": Randy Block, "Wanted: Ideas to Manage," *Stanford Daily*, March 29, 1976.

2. Most of the money came from companies paying for licenses, not from royalties earned by sales of marketed products. During the pilot period, Reimers and Newman implemented a flat 15 percent administrative charge, in addition to out-of-pocket expenses, against gross royalty income to cover costs. At the end of 1969, Newman "bowed out" of the licensing program to begin work on the so-called Newman Report on educational policy. See "Technology Licensing Program Now Has Earned over $200,000," *Campus Report*, Nov. 15, 1972; Frank Newman to Kenneth Cuthbertson, Nov. 26, 1969, SUOTL; Niels Reimers to Patterson, Frank Newman, and Earl Cilley, Nov. 5, 1969, SUOTL; Al Miller to Deans, Department Heads, and

Principal Investigators, June 10, 1979, Lederberg Papers SC186: 22(B), SUSC; and Richard Lyman to Department Chairmen and Principal Investigators, May 16, 1969, Paul Berg Papers, SC0358: 17, SUSC.

3. "Stanford OTL Revenue/Expense/Breakeven Analysis (10/1/01)" and "OTL Income and Other Figures (2013)," SUOTL.

4. 1971 Annual Report—Technology Licensing, SUOTL. The sound system inventor was John M. Chowning.

5. Ibid.

6. "OTL Income and Other Figures (2013)," SUOTL. In 1971, Reimers hired an associate, John Poitras.

7. Siddhartha Mukherjee, *The Gene: An Intimate History* (New York: Scribner, 2016): 237.

8. Victor K. McElheny, "Animal Gene Shifted to Bacteria; Aid Seen to Medicine and Farm," *New York Times*, May 20, 1974, SUOTL. Since March 2015, the article has appeared online under the title "Gene Transplants Seen Helping Farmers and Doctors."

9. Niels Reimers, interview by author, Nov. 5, 2014.

10. All Cohen description from Stanley Cohen, ROHO interview, including the Falkow introduction.

11. Stanford Medical History Flickr photo stream and *Stanford Daily*, May 1974.

12. Cohen explains in his ROHO interview, "Our discoveries were dependent partly on the earlier discovery of DNA ligase and on ten years of basic research with plasmids." He specifically cites work by Paul Berg, Dale Kaiser, H. Gobind Khorana, D. A. Jackson, Paul Lobban, Janet Mertz, Vittorio Sgaramella, R. H. Symons, and J. H. van de Sande.

13. Reimers, ROHO interview.

14. Stanford had an institutional patent agreement with the National Institutes of Health that fell under the purview of the Department of Health, Education, and Welfare.

15. Reimers, ROHO interview.

16. In the Office of Technology Licensing's first eighteen months, when only five inventions had been licensed, Reimers reported that "three petitions for title in inventions to sponsors (HEW twice and NSF) were first denied but were eventually granted after much effort including meetings in Washington, D.C. with agency officials"; Licensing Program Progress Report, July 20, 1970, SUOTL.

17. Cohen in "Campus to Commerce: Trailblazers of Technology Transfer," https://www.youtube.com/watch?v=HA6SYaQ6ZZw.

18. Reimers, ROHO interview.

19. Double Helix Medals Dinner, https://www.youtube.com/watch?v=wTuy4_e9O08.

20. Genentech GenenLab notebook at http://blog.zymergi.com/2013/01/origins-biotech-genentech.html.

21. Boyer's lab learned that EcoR1 could be used in recombinant DNA after reading Janet E. Mertz and Ronald W. Davis, "Cleavage of DNA by RI Restriction Endonuclease Generates Cohesive Ends," *Proceedings of the National Academy of Sciences*, November 1972: 3370–4. Many thanks to Doogab Yi for his technical review of this chapter.

22. Stanley Cohen, interview by author, June 3, 2015.

23. Cohen, in "Stanley Cohen and Herb Boyer, Co-Recipients of 1996 Lemelson-MIT Prize," https://www.youtube.com/watch?v=G3H-Uzts108. Cohen says something very similar in Stanley Cohen, "DNA Cloning: A Personal View After 40 Years," *Proceedings of the National Academy of Sciences USA* 110, no. 39 (Sept. 24, 2013): 15521–9.

24. Stanley Cohen, in "Campus to Commerce" video.

25. Niels Reimers to File, Notes on a conversation with Stan Cohen, July 29, 1974, SUOTL: "Herb Boyer is willing to cooperate, also does not want personal gain."

26. See Stanley Cohen to Niels Reimers, June 9, 1975 and Niels Reimers to Josephine Opalka, June 26, 1974; Cohen, "DNA Cloning."

27. Stanley Cohen to Bertram Rowland, Jan. 22, 1975, SUOTL.

28. Keith Yamamoto, in "Campus to Commerce" video.

29. Stanley Cohen to Bertram Rowland, Jan. 22, 1975, SUOTL.

30. S. N. Cohen, A.C.Y. Chang, H. W. Boyer, and R. B. Helling, "Construction of Biologically Functional Bacterial Plasmids *in Vitro*," *Proceedings of the National Academy of Sciences* 70, no. (1973): 3240–4. Helling was a researcher working in Boyer's lab.

31. Josephine Opalka (UC patent administrator) to Niels Reimers, July 30, 1974, SUOTL.

32. Niels Reimers to Josephine Opalka, Aug. 2, 1974, SUOTL.

33. Reimers, ROHO interview.

34. Niels Reimers to Cassius L. Kirk, Aug. 22, 1974, SUOTL.

35. Niels Reimers to Gerald Lieberman (dean of research and vice provost), Dec. 11, 1979, SUOTL.

36. On the UC patent office, see Mark Owens (assistant vice president at UC and former patent administrator), "Internal Administration of Technology Transfer: Organization of a University Patent Office," in *Technology Transfer University Opportunities and Responsibilities. A Report on the Proceedings of a National Conference on the Management of University Technology Resources* (conference held at Case Western Reserve University, October 1974): 58–65.

37. Reimers speech ("Mechanisms for Technology Transfer: Marketing University Technology") and Owens speech at the Case Western Reserve conference, October 1974.

38. Reimers, ROHO interview. One historian has pointed out that "until the last two decades of the century, the focus of UC patent policy was on making faculty inventions available to the private sector; marketing and making money from them were secondary concerns. It was a far cry from the situation at Stanford." Sally Smith Hughes, "Making Dollars Out of DNA," *Isis* 92 (2001): 541–75.

39. Agreement Concerning Rights In Invention, Aug. 29, 1975, SUOTL.

40. The $107 million to Stanford would be evenly split among inventor, department, and the university. Precise amounts: total royalties: $254,763,248; total admin fee: $38,214,63; total expenses: $1,801,761; total share for each school: $107,373,658. "Cohen-Boyer Royalties and Distribution FY 1979–80 Through FY 1998–1999," SUOTL.

41. Reimers, ROHO interview.

42. Sally Smith Hughes pointed out that Genentech and other companies began using recombinant DNA technology even before the Cohen-Boyer patent was issued in 1980. Nonetheless, she added, the broad claims in the patent that required anyone using recombinant DNA to buy a license, coupled with the landmark *Chakrabarty* decision that allowed the patenting of living organisms, meant that "the issuance of the Cohen-Boyer patent served to reinforce confidence that commercial biotechnology had a future and was a sound investment opportunity." Hughes, "Making Dollars Out of DNA": 572.

43. See Bill Carpenter to Stanley N. Cohen, Sept. 4, 1974; Bill Carpenter to File, Sept. 18, 1974, Re: Dr. Herbert Boyer [handwritten on yellow legal pad]; Bill Carpenter to Niels Reimers, Oct. 18, 1974, all SUOTL.

Notes to Pages 141–143   423

44. "Bertram Rowland and the Cohen/Boyer Cloning Patent," GWU Law posting, https://web.archive.org/web/20160105155929/http://www .law.gwu.edu/Academics/FocusAreas/IP/Pages/Cloning.aspx. Rowland worked for the firm Flehr, Hohbach, Test, Albritton & Herbert.

45. In 1959, scientists discovered that bacteria contain plasmids, in addition to chromosomes. "Recombinant DNA in the Lab," Smithsonian, http:// americanhistory.si.edu/collections/object-groups/birth-of-biotech?ogmt _page=recombinant-dna-in-the-lab.

46. Stanley N. Cohen, interview by author, June 3, 2015.

47. Rowland says that "Boyer was not available for discussion," and Cohen, in his ROHO interview, says that Boyer did not speak with Rowland before the patent application was drafted.

48. Rowland: "Cohen asked me why I was limiting the claims to bacteria, as plasmids are also available in eukaryotes or one could use viruses for cloning in mammalian cells. So far as eukaryotic plasmids, there was one yeast plasmid known and viruses had not been previously manipulated and shown to be capable of introducing foreign DNA into a mammalian host and the foreign DNA replicated." "Bertram Rowland and the Cohen/ Boyer Cloning Patent."

49. Niels Reimers to Norman Latker, Aug. 20, 1974, SUOTL.

50. P. Berg and J. E. Mertz, "Personal Reflections on the Origins and Emergence of Recombinant DNA Technology," *Genetics* 184 (2010): 9–17.

51. Cohen, ROHO interview.

52. Told Reimers did not want: Reimers, ROHO interview: 22. Cohen using the Arpanet: Cohen oral history and Josephine Opalka to Sally Hines, Sept. 8, 1976, which references a note dated Nov. 3, 1975, that was sent to Cohen in England.

53. Maxine Singer and Dieter Soll to Philip Handler, July 17, 1973, at http:// profiles.nlm.nih.gov/ps/retrieve/ResourceMetadata/CDBBCG.

54. "Potential Biohazards of Recombinant DNA Molecules," *Science*, 26 July 1974, letter reprinted in James D. Watson and John Tooze, *The DNA Story: A Documentary History of Gene Cloning* (San Francisco: H. W. Freeman and Co., 1981): 11.

55. Mukherjee, *The Gene*: 230.

56. Michael Rogers, "The Pandora's Box Congress," *Rolling Stone*, June 19, 1975. The conference was the second on recombinant DNA risks that was

held at Asilomar—the first was in January 1973—but it was so extraordinary that it has come to be known as *the* Asilomar conference. Only six of the 150 Asilomar scientists "now fiddling with the basic mechanics of reproduction," as one journalist put it, were female, but one woman played a pivotal role: the molecular biologist Maxine Singer was an organizer of the conference and among the very first to call attention to the potential risks.

57. Robert Pollack of the Cold Spring Harbor Laboratory, quoted in "Microbiology: Hazardous Profession Faces New Uncertainties," *Science*, Nov. 9, 1973, reprinted in Watson and Tooze, *The DNA Story*.

58. Berg, quoted in "Secret of Life: Playing God," https://www.youtube.com/watch?v=M3wg-W3Slow.

59. Mukherjee, *The Gene*: 232.

60. Berg and Mertz, "Personal Reflections."

61. Anonymous review included in Josephine Opalka to Niels Reimers, July 11, 1975, SUOTL.

62. Niels Reimers to William F. Massy, Feb. 10, 1975, SUOTL.

63. Doogab Yi, *The Recombinant University: Genetic Engineering and the Emergence of Biotechnology at Stanford, 1959–1980*, Princeton University doctoral dissertation, 2008: 216.

64. Reimers, "Mechanisms for Technology Transfer."

65. Niels Reimers to Bruce Hinchliffe, April 18, 1975, SUOTL.

66. Niels Reimers to William F. Massy, March 12, 1976, SUOTL.

67. Norm Latker, quoted in Debby Fife, "The Marketing of Genius," *Stanford Magazine*.

68. Reimers, quoted in Barbie Fields, "Special Office Aids in Securing Patents for Campus Inventors," *Stanford Daily*, Nov. 11, 1977. The historian Doogab Yi has described the position Reimers adopted thus: "They claimed that the 'public' ownership of inventions was fundamentally at odds with the promotion of the public interest through development and licensing." Yi, *The Recombinant University*: 177.

That's What I Did on Mondays — Mike Markkula

1. Notes to Consolidated Financial Statements, Dec. 31, 1971, and Dec. 31, 1972, ACM; *Moody's OTC Industrial Manual 1973*: 391.

2. Mike Markkula, interview by author, Dec. 4, 2015.

3. Markkula, interview by author, Feb. 24, 2016.

4. Robert N. Noyce and Marcian E. Hoff, Jr., "A History of Microprocessor Development at Intel," *IEEE Micro*, February 1981: 13.

5. Federico Faggin, "The Birth of the Microprocessor," *Byte*, March 1992; Bill Davidow, quoted in Davidow, Flath, and Noyce oral history, courtesy Intel.

6. Tom R. Sawyer to author, Feb. 7, 2016.

7. Mimi Real, *A Revolution in Progress: A History of Intel to Date* (Palo Alto, CA: Intel, 1983): 9.

8. Ibid.: 26.

9. Andy Grove, interview by author, Aug. 19, 2003; Grove, quoted in Sporck, *Spinoff*: 199–200.

10. Markkula, interview by author, Feb. 24, 2016.

11. Arthur Rock, interview by author, April 1, 2016.

12. Grove, conversation with author, Nov. 11, 2015.

13. Grove, quoted in Charles E. Sporck and Richard Molay, *Spinoff: A Personal History of the Industry That Changed the World* (Saranac Lake, NY: Saranac Lake Publishing, 2001): 199–200.

14. Ann Bowers, interview by author, Nov. 7, 2015.

15. Markkula, SUSG interview.

16. Richard Immel, "After Bad 2nd Half, Semiconductor Firms See More of the Same for Much of This Year," *New York Times*, Jan. 9, 1975; Ramon C. Sevilla, "Employment Practices and Industrial Restructuring: A Case Study of the Semiconductor Industry in Silicon Valley, 1955–1991," PhD dissertation, UCLA, 1992, Table 3.10.

17. Paul Plansky, "Protests Mark Wema Meeting," *Electronic News,* Dec. 2, 1974.

18. Although Carsten was named vice president and director of marketing within the components division, his was the only marketing title in the company, aside from someone in the digital watch organization.

19. "I did not feel comfortable with Jack Carsten at all" is Markkula's only comment. Markkula, interview by author, Dec. 4, 2015. Carsten was supposed to begin work in January 1975, but illness delayed his start until April. Carsten, CHM interview.

20. Even without taking into account any additional options he might have been granted while at Intel, the two stock splits since Markkula had joined

in January 1971 meant that by 1975 he owned 2.25 times more shares than his original generous grant. The strike price (what he had paid for each share) was $6.22; meanwhile, Intel stock was trading at about $45/share.

21. Markkula, quoted in "The Milliard Dollar Armas Clifford Markkula Realized His Dream," translation of an unidentified Finnish article published in 1981, ACM.

22. Markkula, interview by author, Dec. 4, 2015.

23. Allan Tommervik, "Exec Apple: Mike Markkula," *Softalk*, June 1981.

24. Markkula began serving on the Cupertino Planning Commission in January 1977 and resigned in April of the following year. Beth Ebben, Cupertino planning office, to author, Dec. 23, 2015.

25. Markkula, CHM interview.

26. Markkula, SUSG interview.

27. Markkula, interview by author, Feb. 24, 2016; Markkula, CHM interview.

28. Markkula, *SV*.

Challenges: 1976–1977

1. Scott Herhold, "The Story Behind Joe Colla's Famous 1976 Highway Stunt," *San Jose Mercury News*, Oct. 16, 2013. The overpass was at the intersection of Highways 101, 280, and 680 in San Jose.

2. *Surveillance Technology. Joint Hearing Before the Subcommittee on Constitutional Rights of the Committee on the Judiciary and the Special Subcommittee on Science, Technology, and Commerce of the Committee on Commerce, United States Senate 94th. Congress, First Session on Surveillance Technology*, June 23 and Sept. 9–10, 1975. An excellent summary of the Data Center debate is *Federal Data Banks and Constitutional Rights: A Study of Data Systems on Individuals Maintained by Agencies of the United States Government*, prepared by the staff of the Subcommittee on Constitutional Rights of the Committee on the Judiciary, United States Senate, as part III of the subcommittee's study of federal data banks, computers, and the Bill of Rights (1974), pp. xv–xviii. For more on the privacy debate in the 1970s, see Frederick S. Lane, *American Privacy: The 400-Year History of Our Most Contested Right* (Boston: Beacon Press, 2011).

I Needed to Land Behind a Desk — Fawn Alvarez

1. History of California Minimum Wage, State of California Department of Industrial Relations, http://www.dir.ca.gov/iwc/minimumwagehistory.htm.
2. Alexis Madrigal, "Not Even Silicon Valley Escapes History," *The Atlantic*, July 23, 2013.
3. ROLM Annual Report, 1977.
4. ROLM S-1, Sept. 15, 1976.
5. Ibid.
6. Half of employees: ROLM's S-1, filed Sept. 15, 1976, stated that ROLM had 397 employees, 200 of whom were in production.
7. Fawn Alvarez Talbott, interview by author, June 19, 2015.
8. Ibid. Jeff Smith, an engineer who worked in manufacturing, agrees that "there really wasn't a lot of discipline on the line; everybody was kind of free to do it their own way." Jeff Smith, interview by author, Feb. 10, 2017.
9. The relevant clip of the "Job Changing" episode that aired Sept. 15, 1952, is hilarious and can be found at https://www.youtube.com/watch?v=8NPzLBSBzPI.
10. Bob Maxfield, interview by author, Jan. 30, 2017.
11. Ibid.
12. Talbott, interview by author, Jan. 26, 2017.
13. Marcie Axelrad, "Profile of the Electronics Industry Workforce in the Santa Clara Valley: A Preliminary Report from the Project on Health and Safety in Electronics (PHASE)," July 1979, PSC.
14. Michael W. Miller, "Unions Curtain Organizing in High Tech," *Wall Street Journal*, Nov. 13, 1984; Ramon C. Sevilla, "Employment Practices and Industrial Restructuring: A Case Study of the Semiconductor Industry in Silicon Valley, 1955–1991," PhD dissertation, UCLA, 1992: 172, 292, 299.
15. "Non-Union Seminar Slated," *Palo Alto Times*, Dec. 26, 1973, PSC.
16. Many thanks to Martin Manley, who worked as a machinist and organizer in Silicon Valley at that time, for sharing the perspective of unions toward organizing in the Valley. Manley left his work as a machinist to study at the Harvard Business School. He later served as assistant secretary of labor for President Bill Clinton and became an Internet entrepreneur, founding Alibris, among the world's largest online rare- and used-book sellers. Martin Manley, interview by author, Feb. 21, 2013.

17. Talbott, interview by author, Jan. 26, 2017.

18. Smith, interview by author, Feb. 9, 2017.

19. Ibid.

20. Talbott, interview by author, July 24, 2013.

21. Vineta Alvarez Eubank, interview by author.

22. Talbott, interview by author, July 24, 2013.

This Is a Big Fucking Deal — Al Alcorn

1. Manny Gerard, interview by author, Jan. 23, 2012.

2. Ibid.

3. Warner holdings from Warner Communications Annual Report, 1976.

4. Atari S-1 mockup, proof of June 24, 1976, Alcorn Papers, SUSC.

5. Doubling the price: Atari board minutes, Sept. 2, 1975, Atari Business Plan collection, SUSC; author's interviews with Al Alcorn and Nolan Bushnell.

6. Alcorn, interview by author, Oct. 31, 2013.

7. Stella was the name for the chip that talked to the video display and sound in the four-chip system. Tekla Perry and Paul Wallich, "Design Case History: The Atari Video Computer System," *IEEE Spectrum*, March 1983: 45–51. Key players on the Stella team from Grass Valley included Joseph Decuir, Larry Emmons, Steve Mayer, Ron Milner, and Jay Miner.

8. Arthur Young & Company to William White [vice president for finance, Atari], June 18, 1975, Alcorn Papers, SUSC.

9. Alcorn, interview by author, Jan. 9, 2012.

10. "Pong Inventor Rich in Year," *Bakersfield Californian*, July 18, 1976.

11. All Gerard quotes: Manny Gerard, interview by author, Jan. 23, 2012.

12. Alcorn, speaking at "Atari's Impact on Silicon Valley" panel, Sept. 8, 2016.

13. Ralph Blumenthal, "'Death Race' Game Gains Favor, but Not with the Safety Council," *New York Times*, Dec. 28, 1976.

14. Ibid.; Carly A. Kocurek, "The Agony and the Exidy: A History of Video Game Violence and the Legacy of Death Race," *Game Studies*, September 2012, http://gamestudies.org/1201/articles/carly_kocurek; Bushnell quote: Kent, *Ultimate History*: 92.

15. Gerard wrote the copy for the 1976 Warner annual report announcing the acquisition of Atari: "Toys and games of skill go back to the early history

of human life. Stones, bones, and wood were early materials for games, and many of those are still highly salable products today . . . virtually every game that was ever enjoyed by a lot of people is still made and sold. Electronic games are a logical step in this historic process."

16. *BusinessWeek* estimated that Atari had earned $3.5 million on sales of $39 million in fiscal year 1976. "Atari Sells Itself to Survive Success," *Business-Week*, Nov. 15, 1976.

17. Warner Communications, 1976 Annual Report.

18. At Atari, 5 percent of pretax profit was split among twenty-four employees, the top seven of whom were compensated based on salary; the rest received bonuses based on performance. At Warner, the bonus system was even richer. Manny Gerard was not surprised to hear that the bonus had gotten better; Ross, he said, had offered wonderful incentive compensation without being asked for it. Ross once told *BusinessWeek* that "the way a division is run is up to the CEO." "How Steve Ross's Hands-off Approach Is Backfiring at Warner," *BusinessWeek*, Aug. 8, 1983.

19. In their interviews with the author and elsewhere, Don Valentine and Nolan Bushnell both recalled the flight with Eastwood and Locke. Bruck, *Master of the Game*, confirms the story and adds the detail about the hotel. Valentine claims that Dirty Harry even made Bushnell a sandwich, but Joe Keenan says that the movie stars had no interaction with the men from Atari.

20. Bushnell's ownership percentage: "Games, Inc." [summary of Atari ownership structure at the time of the Warner acquisition]; owner requested anonymity.

21. Bushnell, in *Stella at 20* video.

22. Bushnell, interview by author, Aug. 1, 2012.

23. Alcorn, interview by author, Oct. 31, 2013.

24. Gerard, interview by author, Jan. 23, 2012; Alcorn, interview by author, Jan. 9, 2012.

25. Gerard, interview by author, Jan. 23, 2012. Bushnell's ex-wife allegedly threatened to challenge their divorce agreement after reading an article about Bushnell in the July 18, 1976, issue of the *Bakersfield Californian*. Goldberg and Vendel say that Bushnell's ex-wife received $300,000 in exchange for dropping all claims (*Atari: Business Is Fun*: 205).

26. Gerard, interview by author, Jan. 23, 2012.

27. Bushnell, interview by author, Aug. 1, 2012.

28. Valentine and the other venture capitalists received cash for their stake in the company.

29. Gerard, interview by author, Jan. 23, 2012.

One More Year or Bust — Sandra Kurtzig

1. In ASK's 1974 balance sheet, Kurtzig valued the to-be-released minicomputer version of MANMAN at $150,000, twice the Tymshare version and six times more than the batch-processing version. ASK Computer Services, Inc., Balance Sheet, July 1, 1974, SK; ASK Computer Services, Business Review, Dec. 1, 1976, SK.

2. Sandra Kurtzig, *CEO: Building a Four Hundred Million Dollar Company from the Ground Up* (New York: W. W. Norton & Co., 1991): 88.

3. Marty Browne, interview by author, July 12, 2015. "There is not a program that Sandy wrote that we didn't have to rewrite," he says.

4. Browne, interview by author, July 12, 2015.

5. Kurtzig, *CEO*: 111.

6. ASK, Income Statement, June 30, 1975, SK.

7. Sandra Kurtzig, interview by author, June 24, 2015.

8. Campbell-Kelly, *From Airline Reservations*: 131.

9. Jack McNamee, quoted in Esther Surden, "Manufacturing Mini Gives Hughes 'More Control,'" *Computerworld*, Aug. 30, 1976.

10. "A Mini Is Not Just a Small Big Computer," HP advertisement, probably 1977, SK.

11. Kurtzig, *CEO*: 114.

12. Ibid.: 116; *APICS News*, November 1976, courtesy Marty Browne.

13. Kurtzig, quoted in "The Woman President," *Computer Manufacturing Opportunities*, September 1979: 17.

14. HP charged ASK a discounted $67,000 ($29,000 down, balance due later) for the computer, which retailed for about $80,000. Kurtzig, *CEO*: 113.

15. ASK 1976 business plan, "Capitalization," SK.

16. Liz Seckler, interview by author, July 29, 2015.

17. Stanford Computer Science timeline, http://www-cs.stanford.edu/timeline.

18. Seckler, interview by author, July 29, 2015.

19. Howard Klein, interview by author, July 14, 2015.
20. Intelligent Factory: 1976 business plan, SK.
21. Kurtzig, *CEO*: 122.
22. Ibid.: 124.
23. ASK Statement of Operations for years ending June 30, 1975, and June 30, 1976, SK. Kurtzig's salary increased from $12,000 in 1975 to $24,000 in 1976.
24. Joint Venture Silicon Valley, "Venture Capital by Industry," Silicon Valley Index 2015; Jon Levine, "5,000 Entrepreneurs . . . and Counting," *Venture*, January 1982.
25. Larry Ellison, quoted in Silicon Valley Historical Association, "Billionaires: Silicon Valley Entrepreneurs Who Made Their Fortune Without Venture Capital," https://www.youtube.com/watch?v=5vXs6JqMu7U.
26. ASK S-1, Aug. 6, 1981.
27. Kurtzig, *CEO*: 152.

No Idea How You Start a Company — Niels Reimers and Bob Swanson

1. Liebe F. Cavalieri, "New Strains of Life—or Death," *New York Times Magazine*, Aug. 22, 1976.
2. "Dr. Jekyll": *Friends of the Earth*, May 1976. "Doomsday": *Time* (cover story), April 18, 1977; "Pulling Back from the Apocalypse," *Not Man Apart*, January 1977.
3. Charles McCabe, "On Playing God," *San Francisco Chronicle*, April 4, 1977, reprinted in Watson and Tooze, *DNA Controversy*: 165.
4. "Gene-Splicing: At Grass-Roots Level a Hundred Flowers Bloom," *Science*, Feb. 11, 1977.
5. Alfred E. Vellucci to Phillip Handler, May 16, 1977, reprinted in James D. Watson and John Tooze, *The DNA Story: A Documentary History of Gene Cloning* (San Francisco: H. W. Freeman and Co., 1981): 206. Vellucci quoted in "Test Tube Laboratory Fine, but Build It Somewhere Else," *Palo Alto Times*, June 16, 1976. On Harvard Yard: "Recombinant DNA: Cambridge City Council Votes Moratorium," *Science,* July 1976.
6. Gene Bylinsky, "DNA Can Build Companies, Too," *Fortune*, June 16, 1980.
7. "Gene Splicing Sheds Its Mad-Scientist Image," *BusinessWeek*, May 16, 1983.

8. Kennedy's opening statement, Watson and Tooze, *DNA Story*: 144. Norman J. Latker, Patent Branch Chief at DHEW, read out loud to Reimers over the phone the proposed testimony that the director of the NIH, Donald Fredrickson, planned to give during the hearings. Niels Reimers to File, Sept. 14, 1976, SUOTL.

9. "Gene Splicing: Cambridge Citizens OK Research but Want More Safety," *Science*, January 1977; "Gene-Splicing: At Grass-Roots Level a Hundred Flowers Bloom," *Science*, Feb. 11, 1977.

10. "Who Should Control Recombinant DNA?," *Chronicle of Higher Education*, March 21, 1977. Jeremy Rifkin led the protesters, who called themselves the People's Business Commission. Chant detail is from Jeffrey Fox, "Genetic Engineering Industry Emerges," *Chemical and Engineering News*, March 17, 1980.

11. Donald Fredrickson to Robert Rosenzweig, May 2, 1978, SUOTL. Technically, it was a letter written by Stanford's vice president for public affairs, requesting "new consideration of the propriety of our proceeding as planned and authorized," that prompted the review. Robert Rosenzweig to Distribution, June 30, 1976; Rosenzweig to Those Interested in Recombinant DNA, June 4, 1976; Rosenzweig to Donald Frederickson, June 18, 1976; Rosenzweig to Joseph Califano, Jr. Feb. 15, 1977, SUOTL.

12. Doogab Yi, *The Recombinant University: Genetic Engineering and the Emergence of Biotechnology at Stanford, 1959–1980*, Princeton University doctoral dissertation, 2008: 218.

13. Paul Berg, at a May 21, 1976, meeting with Reimers, Cohen, Joshua Lederberg, and several Stanford administrators, quoted in Yi, *The Recombinant University*: 209.

14. Niels Reimers, interview by author, May 15, 2015.

15. See also Cohen, "Recombinant DNA: Fact and Fiction," statement prepared for a meeting of the Committee on Environmental Health, California Medical Association, Nov. 18, 1976, and "The Nobel Letters," Watson and Tooze, *The DNA Story*: 112.

16. Niels Reimers to Rodney Adams [Stanford treasurer], April 8, 1976, SUOTL.

17. "Licensing Plan," Niels Reimers to File, July 13, 1976, SUOTL.

18. "So, Bob": Manny Levinson, introduction to Swanson, ROHO interview.

19. Peter Meyer, *Eugene Kleiner: Engineer, Venture Capitalist, Founding*

Father of Silicon Valley (Brooklyn, NY: Polytechnic Institute, 2006), EKF.

20. Robert Lenzner, "General Doriot: Time for Credit," *Boston Sunday Globe*, March 25, 1979; Doriot, *Manufacturing Class Notes* (Boston: Board of Trustees, The French Library in Boston, 1993): 7.

21. The company, called Antex, made light-emitting diodes.

22. Tom Perkins, *Valley Boy: The Education of Tom Perkins* (New York: Nicholas Brealey Publishing, 2007): 115.

23. Cetus's original product was going to be a labor-saving machine that screened microorganisms (identifying ones with a resistance to antibiotics, for example).

24. Perkins, quoted in Sally Smith Hughes, *Genentech: The Beginnings of Biotech* (Chicago: University of Chicago Press, 2011): 32.

25. David Arscott, interview by author, May 13, 2015.

26. Bob Swanson, ROHO interview.

27. Perkins, quoted in an undated manuscript, "The High Rollers," with handwritten notes by Kleiner, EKF.

28. Swanson, ROHO interview. In his ROHO interview, Perkins says that it was Kleiner, in particular, who wanted to fire Swanson. This is impossible to verify, given the deaths of all three men.

29. Swanson, ROHO interview.

30. Ibid.

31. Brook Byers, interview by author, Oct. 8, 2015. Byers joined Kleiner & Perkins after Swanson left and remains a partner at the firm. Byers is one of the world's premier investors in biotechnology, a passion he says was ignited, in part, by his conversations with Swanson.

32. Bob Swanson, speech Sept. 17, 1978, EKF. The speech was given in New York, but the audience is not indicated.

33. Herb Boyer, ROHO interview.

34. Ibid.

35. Stanley Falkow, "I'll Have the Chopped Liver Please, or How I Learned to Love the Clone," *ASM News* 67, no. 11 (2001).

36. Manny Levinson, introduction to Bob Swanson, ROHO interview.

37. Brook Byers, ROHO interview.

38. Hughes, *Genentech*: 35.

39. Swanson, quoted in Hughes, *Genentech*: 37; Genentech S-1, Oct. 14, 1980.

40. Randall Rothenberg, "Robert A. Swanson, Chief Genetic Officer," *Esquire*, December 1984.

41. Swanson, ROHO interview. This is also the interview referenced in the footnote.

42. "Introduction," undated document [but clearly 1982], EKF.

43. *Testimony by Thomas J. Perkins Before the U.S. Department of Commerce Hearing on the Future for New Technology Based Ventures*, April 3, 1976.

44. Swanson, ROHO interview.

45. Ibid.

46. Ibid.

47. *Report of Committee on Rules and Jurisdiction*, report from the Divisional Senate Committee on Rules and Jurisdiction, Jan. 11, 1979, SSH.

48. Boyer, ROHO interview. "We played a cruel trick" (in footnote): Sally Smith Hughes, *Genentech: The Beginnings of Biotech* (Chicago: University of Chicago Press, 2011): 62.

49. "Genentech Outline for Discussion Kleiner & Perkins, 1 April 1976," SSH.

50. See Cohen, ROHO interview, and Boyer, ROHO interview.

51. Swanson pitched Crocker Capital in March 1976.

52. Fred Middleton says of Perkins in late 1975–early 1976, "I don't think that at the time he had a tremendous amount of confidence in Bob" (Middleton, ROHO interview). Regis McKenna, who worked with Genentech, has a "vague memory" that Swanson had approached him for advice on asking Perkins for an investment. Swanson, McKenna says, was concerned because he knew that "Tom wasn't high on Bob as an associate." McKenna, email to author, April 18, 2015.

53. Eugene Kleiner to Nathaniel Weiner [Corporations Counsel], May 7, 1976, SSH.

54. Tom Perkins, ROHO interview.

55. Ibid.

56. "Genentech Outline for Discussion, Kleiner & Perkins," April 1, 1976, SSH. Genentech called itself "a development stage company" until the hormone was cloned.

57. Swanson talks about Perkins's insistence on exclusivity in his ROHO interview; David Arscott, who had worked with Swanson at Citibank and was in touch with him through that period, distinctly remembers Swanson telling him about Perkins's stipulation. (Arscott mentioned this

without any prompting.) Typical amounts invested in companies in the first K&P fund ranged from roughly $120,000 to $500,000. Kleiner & Perkins, Summary of Venture Investments (1), EKF.

58. Perkins, quoted on p. 17 of an undated manuscript, "The High Rollers," with handwritten notes from Kleiner, EKF.

59. At its founding Genentech acquired interest in the Boyer and Swanson partnership valued at $24,000 (48,000 shares of stock). Genentech Inc., A Development Stage Company, Financial Statements for Year Ended December 31, 1977. Courtesy Genentech.

60. Boyer, ROHO interview.

61. Genentech, December 1976 Business Plan, quoted in Hughes, *Genentech*: 47.

62. Swanson, ROHO interview.

63. Swanson to Stanford University, University of California, April 19, 1976, SSH. There is no mention of receipt of a letter in the Office of Technology Licensing files until June 1976, and the letter itself is not included.

64. See, e.g., Niels Reimers to Distribution, Nov. 15, 1976, SUOTL; Niels Reimers, "Mechanisms for Technology Transfer," Case Western Reserve conference, October 1974.

65. Niels Reimers to Robert Rosenzweig, June 11, 1976, SUOTL. Stan Cohen, in his interview with the author, says that he urged Reimers to license nonexclusively.

66. See, e.g., John Poitras to File, Telecon with Bob Swanson, March 17, 1977, SUOTL.

67. Niels Reimers to File, July 15, July 22, Aug. 11, Aug. 28, Sept. 13, Sept. 14, and Nov. 2, 1976. For a meticulous accounting of the steps to patenting the Cohen-Boyer process, see Sally Smith Hughes, "Making Dollars Out of DNA," *Isis* 92 (2001).

68. The patent filed Nov. 4, 1974 (ser. no. 520,591), the so-called parent application, was abandoned in favor of a continuation-in-part, ser. no. 687,430, filed May 17, 1976. Ser. no. 687,430 was then abandoned for two continuations-in-part: ser. no. 959,288 (filed Nov. 9, 1978) and ser. no. 1,000,021 (filed on Jan. 4, 1979). Sally Hines to Leroy B. Randall [acting patent chief, Health and Human Services], July 15, 1980, SUOTL.

69. Kathleen J. Sullivan, "Richard W. Lyman, Stanford's Seventh President,

Dead at 88," *Stanford Report*, May 27, 2012; Gene I. Maeroff, "'Harvard of the West' Climbing in the Ratings," *New York Times*, Oct. 10, 1977.

70. Robert Rosenzweig to Those Interested in Recombinant DNA, June 4, 1976, SUOTL.

71. Stanley Cohen to Niels Reimers, June 14, 1976.

72. Ibid. In his ROHO interview, Cohen explained, "It became clear to me that whatever position the university took, it was not realistic for me to simply say, 'Well, I have nothing to do with it.'"

73. Niels Reimers to Bertram Rowland, March 17, 1978, SUOTL.

74. Handwritten note to Josephine Opalka, June 2, 1975, reporting on a call from Reimers. "He seemed worried that nothing would happen until your return on 7/7," the note says. In response, Opalka wrote at the bottom, "Let him worry a little bit." SSH.

75. Sally Hines, interview by author, May 11, 2015.

76. Interviewee requested anonymity.

77. Reimers, interview by author, May 15, 2015.

78. "Cohen-Boyer Royalties and Distribution FY 1979–80 Through FY 1998–1999," SUOTL.

79. "Stanford OTL Revenue/Expense/Breakeven Analysis (10/1/01" and "OTL Income and Other Figures (2013)," SUOTL.

80. Niels Reimers to Arnold, Feb. 17, 1978, SUOTL.

81. Niels Reimers to File, "Various Conversations Regarding Recombinant DNA," July 15, 1976, SUOTL.

82. "Flashback to 1970," recollection by Sally Hines, in *40 Years of Discovery*, Office of Technology Licensing anniversary publication. Genentech held Friday-afternoon celebrations that they called Ho-hos. Tandem and ROLM were also renowned for their Friday-afternoon beer bashes.

That Flips My Switch — Mike Markkula

1. Mike Rose to "Bob," June 23, 1976, Apple Computer Original Advertising File, Misc 551, SUSC.

2. Gordon Moore, quoted in Gene Bylinsky, "How Intel Won Its Bet on Memory Chips," *Fortune*, November 1973: 143.

3. Mike Markkula, interview by author, Feb. 24, 2016.

4. Ibid.

5. Trip Hawkins, interview by author, May 20, 2016; he was describing a different early hobby store, but the clientele was the same.

6. *b* indicated a blank space and (RET) the return key. *Apple-1 Operation Manual*: 2.

7. Paul Freiberger and Michael Swaine, *Fire in the Valley: The Making of the Personal Computer* (New York: Osborne/McGraw-Hill, 1984): 209; at p. 203, Wozniak calls computers "the love of my life."

8. Campbell-Kelly says the typical cost of a minicomputer was $20,000 for complete installation. Martin Campbell-Kelly and William Aspray, *Computer: A History of the Information Machine* (New York: Basic Books, 1996): 238–9. Farrah Fawcett swimsuit poster: Kurt Lassen, "The Farrah Fawcett-Majors Phenomenon: A Case of Too Much Fame Too Soon?," *Nashua Telegraph*, April 30, 1977. Smithsonian curator comment: Jessica Gresko, "Farrah Fawcett's Red Swimsuit Goes to Smithsonian," *Washington Post*, Feb. 2, 2011.

9. Steve Wozniak and Gina Smith, *iWoz: Computer Geek to Cult Icon: How I Invented the Personal Computer, Co-Founded Apple, and Had Fun Doing It* (New York: W. W. Norton & Co, 2006): 196.

10. Having the BASIC interpreter written directly into ROM meant that (1) it was immediately available when the machine turned on and the user hit a few keys, and (2) users would not have the then-common frustration of hitting the wrong key and losing the program language itself, in addition to the user's own set of instructions.

11. Wozniak's description of his demo to Peddle; Veit's description of the demo he saw in the garage is at http://www.pc-history.org/apple.htm.

12. Markkula, SUSG interview.

13. Ibid.

14. Markkula, interview by author, Feb. 24, 2016.

15. Wozniak and Smith, *iWoz*.

16. Size and location of Homebrew meetings: Freiberger and Swaine, *Fire in the Valley*: 106. *Witch Mountain*: Christopher Pollock, *Reel San Francisco Stories: An Annotated Filmography of the Bay Area* (San Francisco: Castor & Pollux P., 2013): 75–6.

17. Campbell-Kelly, *Computer*: 240.

18. The music played by "picking up buss harmonics and running a clever timing loop." Bob Lash, "Memoir of a Homebrew Computer Club Member," www.bambi.net/bob/homebrew.html.

19. Michael Moritz, *Return to the Little Kingdom: How Apple and Steve Jobs Changed the World* (London: Overlook Press, 2010): 157.

20. Wozniak, "Homebrew and How the Apple Came to Be," http://www.at ariarchives.org/deli/homebrew_and_how_the_apple.php. Felsenstein is listed as the "Girl Friday" in the *Berkeley Barb* issue of April 11–17, 1969.

21. Keith Britton, quoted in Freiberger and Swaine, *Fire in the Valley*: 104.

22. Wozniak and Smith, *iWoz*: 176.

23. The engineer was Steve Landesberg. Phil Roybal, who joined Apple very early on, also worked with Carter on the business plan. Carter, CHM interview.

24. Victor McElheny, "Revolution in Silicon Valley," *New York Times*, June 20, 1976.

I've Never Seen a Man Type That Fast — Bob Taylor

1. "XEROX World Conference 1977" (brochure), XPA.

2. "Shape of Tomorrow": film shown at Boca Raton World Conference, XPA.

3. David T. Kearns and David A. Nadler, *Prophets in the Dark: How Xerox Reinvented Itself and Beat Back the Japanese* (New York: HarperBusiness, 1992): 97.

4. 1977 share price from Yahoo Finance.

5. C. Peter McColough, World Conference 1977 speech; Kearns, "About Xerox People," Nov. 9, 1977, XPA.

6. N. M. Beyer to John Ellenby, "Boca Raton—Cost Estimate," Oct. 25, 1977, XPA. Beyer writes that the total estimate is "$218.7 (in 000s)," which would be $2.2 million, but that number seems too high, unless it factors in the value of the equipment ($1.6 million, according to a handwritten note titled "Boca Raton Video" in XPA).

7. Dealer minutes, July 30, 1974, RWT.

8. Jerry Elkind to Jack Goldman, "A Computer Business Opportunity," Feb. 11, 1974, RWT. Though little immediately followed from this letter, it appears that Goldman, the head of research of Xerox and the driving force behind PARC, was sympathetic; he met with Elkind and Taylor several times.

9. "Let me apologize in advance for that phrase—I was probably conditioned

unconsciously by 9 years of Washington exposure to such phrases as 'missile gap' and 'credibility gap.'" Bob Taylor to George Pake, LRP Working Paper no. 2, May 2, 1975, XPA.

10. "Intro to Office System Research: Robert Taylor CSL, August 1976," video V07, XPA.

11. Chuck Geschke, interview by author, April 21, 2014.

12. Chuck Geschke to Distribution, "Final Boca Raton Materials List," Oct. 21, 1977, XPA; Geschke, interview by author, Apr. 21, 2014; Timothy Mott to Distribution, "Schedule," Nov. 7, 1977, XPA.

13. W. Graham Price [Corporate Security] to John Ellenby, "Boca Raton and Worldstage," Sept. 29, 1977, XPA.

14. Taylor, interview by author, March 18, 2013.

15. Ibid.

16. 80 percent: ibid.; 18 percent: Bob Taylor to Bob Spinrad, "Is the message one of appreciation?," Aug. 18, 1982.

17. Pake's handwritten notes titled "1977 Xerox World Conference—What It Was," XPA. They appear to be notes for a talk he planned to give upon returning to PARC after Boca Raton.

18. Taylor, interview by author, July 18, 2014.

19. Performance reviews, RWT. "Because Bob is good" (in footnote): Sproull to Sutherland (acting director of PARC), May 4, 1977, RWT.

20. Michael Hiltzik, *Dealers of Lightning: Xerox PARC and the Dawn of the Computer Age* (New York: HarperCollins, 2000): 379. Beyond the startling comparison (and Hiltzik points it out as such), Pake rarely spoke a word against Taylor. However, when Pake made a list of "pioneering managers" at PARC, he never elevated Taylor above "associate manager," and he listed that qualified title before Taylor's name, while every other manager was named before his title was given. George Pake, "R&D Management and the Establishment of the Xerox Palo Alto Research Center, January 1985," remarks for the IEEE convocation "The Second Century Begins," XPA.

21. On sneaking into Futures Day: Taylor, interview by author, March 18, 2013. In his interview with the author, Geschke said, "I think I knew he was down there and was going to try to get in."

22. Description of video and presentation from "The Shape of Tomorrow," video shown at Futures Day, XPA.

23. Elizabeth Peer, "How to Work the Thing," *Newsweek,* Feb. 22, 1982.

24. Fawn Alvarez Talbott, interview by author, July 24, 2014.

25. That was the intention, but the machine had a long way to go, as a quick look at the 127-page *Alto User's Handbook* (1976), with its lists of instructions, makes clear.

26. Kearns and Nadler, *Prophets in The Dark*: 100. When George Pake got back to Palo Alto, he called Futures Day "a phenomenal success" with "tremendous impact on senior management, managers and officers. Spouses [were] wide-eyed!" Pake, "1977 Xerox World Conference—What it Was," handwritten note, XPA.

27. Paul A. Strassmann, *The Computers Nobody Wanted: My Years with Xerox* (New Canaan, CT: Information Economics Press, 2008): 122.

28. Phil Roybal, quoted in "Now the Office of Tomorrow," *Time*, Nov. 17, 1980.

29. The spreadsheets VisiCalc and then Lotus 1-2-3 were among the bestselling business software titles of the early personal computer era. Datamation article (in footnote): Robert B. White, "A Prototype for the Automated Office," *Datamation*, April 1977: 83–90.

30. In 1977, Pake wrote, "PARC is increasingly an object of envy and jealousy in Xerox. Morale of those who are fighting daily in the C/D [copier/duplicator] trenches has, at least in some cases suffered [due to] inadequate sensitivity in how to tell them the corporation's future lies elsewhere." George Pake, handwritten notes on 1977 Xerox World Conference, XPA.

31. Kearns and Nadler, *Prophets in the Dark*: 104.

32. Strassmann, *The Computers Nobody Wanted*: 113.

33. Dealer minutes, March 21, 1972, RWT.

34. Bob Taylor, performance review, 9/72–9/73, RWT; Stewart Brand, *II Cybernetic Frontiers* (New York: Random House, 1974): 88. Taylor's boss, Jerry Elkind, was gone on a long-term assignment, and Taylor was acting as head of the lab when the magazine crew came by. Taylor says he assumed the *Rolling Stone* people had gotten permission to come to the lab from Elkind or someone else at the lab.

35. Bob Metcalfe, interview by author, May 22, 2014.

36. The CSL Activity Report for March 15–June 12, 1972, references Peter Deutsch's volunteer work with Resource One. Kay's library order: Fred Turner, *From Counterculture to Cyberculture: Stewart Brand, the Whole*

Earth Network, and the Rise of Digital Utopianism (Chicago: University of Chicago Press, 2006): 124.

37. Chuck Thacker, interview by author, April 11, 2014.
38. Paul A. Strassmann, CBI interview. *Computer Lib* is also the title of a book published by Ted Nelson in 1974.
39. Kearns and Nadler, *Prophets in the Dark*.

There Are No Standards Yet — Mike Markkula

1. Steve Wozniak and Gina Smith, *iWoz: Computer Geek to Cult Icon: How I Invented the Personal Computer, Co-Founded Apple, and Had Fun Doing It* (New York: W. W. Norton & Co, 2006): 184.
2. Ibid.: 186.
3. Ibid.: 180.
4. Ibid.: 196.
5. Outline for Apple Computer Buisness [*sic*] Plan, Nov. 18, 1976, ACM.
6. "I also designed the Apple I because I wanted to give it away for free to other people," Wozniak wrote, explaining why he passed out copies of the machine's schematics at Homebrew meetings. Wozniak and Smith, *iWoz*: 157.
7. Jobs worked at Atari from September 1974 through December 1975.
8. Mike Markkula, interview by author, Feb. 24, 2016.
9. Markkula, interview by author, May 3, 2016.
10. Outline for Apple Buisness [*sic*] Plan, Nov. 18, 1976, ACM.
11. Stan Veit, "PC—History: Apple II" at http://www.pc-history.org/apple.htm.
12. Hambrecht profile in Robert Levering, Michael Katz, and Milton Moskowitz, *The Computer Entrepreneurs: Who's Making It Big and How in America's Upstart Industry* (New York: New American Library, 1984): 429.
13. Markkula, interview by author, May 3, 2016.
14. US Senator John Tunney, in "First Session of Surveillance Technology," June 23 and Sept. 9 and 10, 1975, *Joint Hearings Before the Subcommittee on Constitutional Rights of the Committee on the Judiciary and the Special Subcommittee on Science, Technology, and Commerce of the Committee on Commerce, U.S. Senate*: 62.
15. Markkula, interview by author, Feb. 21, 2012.
16. Markkula, interview by author, Feb. 24, 2016.

17. Ibid. Markkula said, "I was surprised that it had got me excited." "Wozniak recalls that" (in footnote): Michael Moritz, *Return to the Little Kingdom: How Apple and Steve Jobs Changed the World* (New York: Overlook Press, 2010): 186.

18. The remaining shares were set aside for yet-to-be-hired employees. Confidential Private Placement memorandum, Nov. 18, 1977: 9, ACM; Apple IPO Prospectus: 25. "Big time" (in footnote): Markkula, interview by author, Feb. 24, 2016. On Ron Wayne (in footnote): Ronald G. Wayne, *Adventures of an Apple Founder* (2010: 512k Entertainment): 64, 105-6; *Atari Standards Drafting Manual*, SB.

19. "Apple Computer (A)," Graduate School of Business, Stanford University, S-BP-229(A): 4.

20. Wendy Quiones, "Pioneering a Revolution: Apple's Steve Jobs and Steve Wozniak," *Boston Computer Update*, July–August 1981.

21. Norman Sklarewitz, "A Used Volkswagen Van and a $500 Commission Were the Starting Capital for Apple Computer," *SF Executive*, December 1979.

22. Scott, quoted in Michael Moritz, *Return to the Little Kingdom: How Apple and Steve Jobs Changed the World* (London: Overlook Press, 2010): 187.

23. The Sporck-Noyce comparison is Scott's and one that Markkula, in interviews with the author, agreed was accurate. See Bruce Entin, "Can Apple Keep Its Piece of the Pie?," *San Jose Mercury News*, Aug. 17, 1981.

24. Markkula, interview by author, Feb. 24, 2016. "Scott once said of Jobs" (in footnote): Walter Isaacson, *Steve Jobs* (New York: Simon & Schuster, 2011): 83.

25. John Hall, interview by author, April 26, 2016. Markkula and Hall had met at a party, after which they occasionally played mixed-doubles tennis with their wives. The startup was a company called Educational Products that developed training materials (slides, workbooks, etc.). Hall says the company was "not very successful."

26. Hall, interview by author, April 26, 2015, and emails from John Hall to author, April 28 and 29, 2016. Trip Hawkins quote (in footnote): interview by author, May 20, 2016.

27. Confidential Private Placement Memorandum, Nov. 18, 1977, Section II, ACM.

28. Confidential Private Placement Memorandum, June 30, 1979: 21, 27; "Apple Education Foundation Advances Learning Methods Through Microcomputers," press release, Aug. 13, 1979; Apple Computer, Inc., Records, M1007, ser. 9, no. 6: 4, SUSC.

29. Markkula, interview by author, Feb. 24, 2016. Confidential document referenced (in footnote): Slides for Executive Briefing Tour, Sept. 1983, Apple Computer Inc. Records M1007, Series Administrative/organization (I) 6:9, SUSC.

30. The national legislation was known as the "Computer Education Contribution Act of 1982." Rob Price, *So Far: The First Ten Years of a Vision* (Apple anniversary book, 1987); Apple Education Foundation Awards Microcomputer Systems to Northampton Educator" [press release], March 11, 1982; Apple Computer Inc. Records M1007, ser. 9, no. 6: 4, SUSC; Harry McCracken, "The Apple Story Is an Education Story: A Steve Jobs Triumph Missing from the Movie," *The Seventy Four*, Oct. 26, 2015.

31. "He works for me and I work for him. How about that, folks? Do we care about who's the CEO? All we care about is who's going to do what," Markkula, interview by author, Feb. 24, 2016. See also Markkula, SUSG interview.

32. Markkula, *SV*; Gene Carter, *Wow! What a Ride: A Quick Trip Through Early Semiconductor and Personal Computer Development* (Raleigh, NC: Lulu Press, 2016): 67.

33. Markkula, *SV*.

34. Markkula, quoted in Norman Sklarewitz, "Born to Grow," *Inc.*, April 1979.

35. Don [Kobrin] to Frank, Hal, Regis, Subject: Apple Computer Company, June 22, 1976, RM

36. The account executive was William Kelley. "Birth of an Industry, 1976–1977," http://www.kelleyad.com/histry.htm.

37. "Who's Apple," Dec. 4, 1976, entry in Regis McKenna's notebook, RM.

38. The designer was Rob Janoff. Holden Firth, "Unraveling the Tale Behind the Apple Logo," CNN, Oct. 7, 2011, http://www.cnn.com/2011/10/06/opinion/apple-logo/.

39. William Kelley (account executive) to Steve Jobs, March 14, 1977.

40. West Coast Computer Faire brochure, http://www.digibarn.com/collec tions.

41. "Birth of an Industry, 1976–1977," http: //www.kelleyad.com/histry.htm.

42. Wozniak and Smith, *iWoz*: 201, 205. "The dress code is" (in footnote): Carter to Roybal et al., San Jose Computer Faire Booth Duty—March 2, 3, 4, 5 [1978], Apple Computer Collection, M1007, ser. 5, no. 2: 9, SUSC.

43. Martin Campbell-Kelly and William Aspray, *Computer: A History of the Information Machine* (New York: Basic Books, 1996): 247.

44. David H. Ahl, "The First West Coast Computer Faire," *Creative Computing* (1977) at http://www.atariarchives.org/bcc3/showpage.php?page=100.

45. http://www.sfgate.com/technology/businessinsider/article/These-Pic tures-Of-Apple-s-First-Employees-Are-5096350.php.

46. Image of Markkula's office at https://www.flickr.com/photos/munnecket /143884812/in/album-72057594130786050/. See also photos and office description at URL in note 45.

47. Gene Carter, interview by author, Jan. 7, 2016; Carter to author, March 9, 2016; Carter, *Wow! What a Ride!*

48. Markkula, SUSG interview.

49. Carter, interview by author, Jan. 7, 2016.

50. Price: *So Far*.

51. Carter, interview by author, Jan. 7, 2016. Though Carter's title was orig-inally director of dealer marketing, he and Markkula agree that the plan was always to have Carter run sales.

52. Confidential private offering memo, probably September 1977, ACM.

53. George Sollman, a former Shugart Associates executive, recalls demon-strating a floppy drive at a Homebrew meeting before Apple introduced its drive, but Wozniak says, "I had never been around or even used a floppy disk in my life." Wozniak and Smith, *iWoz*: 211. Sollman story: see "The Floppy Disk," http://www-03.ibm.com/ibm/history/ibm100/us/en/ icons/floppy/.

54. Gene Carter, *Wow! What a Ride: A Quick Trip Through Early Semiconductor and Personal Computer Development* (Raleigh, NC: Lulu Press, 2016): 63.

55. Wozniak and Smith, *iWoz*: 153; Kamradt, quoted in Moritz, *Return to the Little Kingdom*: 125. "One day he asked me" (in footnote): Gregg Williams and Rob Moore, "Guide to the Apple," *Byte*, December 1984.

56. Williams and Moore, "Guide to the Apple."

57. Paul Laughton, an engineer at Shepardson Microsystems who wrote the disk operating system, says, "When Woz showed me the designs of the disk controller hardware and software driver. I was truly amazed. At that time, all disk drive controllers were big cards with dozens of large- and small-scale integrated circuits. The design Woz created required only seven small-scale integrated circuits. What was even more amazing was that Woz's design had significantly better performance (data density, reliability, cost) than existing controllers." Paul Laughton, "Apple Computer: The Early Days, a Personal Perspective," http://www.laughton .com/Apple/Apple.html. See also http://www.digibarn.com/collections /business-docs/apple-II-DOS/index.html and http://www.computerhis tory.org/atchm/apple-ii-dos-source-code/.

58. http://www.computerhistory.org/atchm/apple-ii-dos-source-code/.

59. "The Floppy Disk," http://www-03.ibm.com/ibm/history/ibm100/us /en/icons/floppy/. Apple comparison (in footnote): Gene Carter to All Authorized Apple Dealers, July 29, 1978, Gene Carter Collection, M1009.

60. The TRS-80 disk drive system (two drives and a controller), for example, cost $1,297, whereas a similar Apple system cost $1,090. The Apple system provided 30 percent more capacity, 25 percent faster transfer rate, and a 250 percent improvement in access time.

61. Markkula, interview by author, Feb. 24, 2016.

62. Carter, *Wow! What a Ride*: 64.

63. "Digitizing" (Talk of the Town), *The New Yorker*, Nov. 14, 1977.

64. Apple Computer Preliminary Confidential Offering Memorandum (probably September 1977): 20. The company was still considering, and would soon reject, an alternative approach to sell through department stores such as Sears, Roebuck and Montgomery Ward.

65. Gene Carter to Mark Radcliffe, Nov. 7, 1988, GC.

66. Gene Carter to Authorized Apple Dealer, March 1, 1978; "AppleSOURCE: A Semi-Regular Newsletter for Apple Computer Dealers," both in navy binder in Gene Carter Collection M1059, Box 2, SUSC.

67. William Bates, "Home Computers—So Near and Yet . . . ," *New York Times*, Feb. 26, 1978.

68. V. I. Vyssotsky and Ed Feigenbaum, quoted in J. F. Traub, "Quo Vadimus:

Computer Science in a Decade," *Communications of the ACM* 24, no. 6 (June 1981).

69. Norman Sklarewitz, "Born to Grow," *Inc.*, April 1979. Before Apple was even a year old, it launched a study to estimate its order-processing needs over the next five years, hired consultants to identify the best hardware and software suppliers, and bought a system that would be able to meet the company's needs up to the $500 million level. Almost from the beginning, Apple stored orders, credit files, and shipping records in databases.

70. "A module for the Apple II fills 30,000 square feet, requires a crew of 70, and produces between 450 and 500 units per day." Apple Annual Report, 1981: 10. On Chris Espinosa and red book (in footnote): https://archive.org/details/applerefjan78.

71. "Blast from the Past—Floor Plan of Apple's Original Bandley 1 Headquarters," press release, http://www.cultofmac.com/128374/blast-from-the-past-floor-map-of-apples-original-bandley-1-headquarters/; "Apple Computer Announces New International Headquarters," Dec. 16, 1977.

72. Trip Hawkins, interview by author, May 20, 2016.

73. Apple Computer Preliminary Confidential Offering Memorandum (probably September 1977): 1, ACM. Andre Sousan, former vice president of engineering and board member at Commodore, ran the company's European operations, called Eurapple.

74. Confidential Private Placement Memorandum, Nov. 18, 1977, ACM.

75. "Home and Non-Home Market for Personal Computers—Systems Selling at Less than $15,000," Creative Strategies International, n.d., but clearly 1982, ACM.

76. Carole Kolker, "Venture Capital Greats: A Conversation with Peter O. Crisp," Oct. 21, 2008, NVCA interview.

77. Between 1977 and 1980, Rock served on the board of another computer firm, Qantel. He waited until that company sold before joining the Apple board. Arthur Rock, interview by author, April 1, 2016; Rock to Markkula, March 7, 1978, AR.

78. Markkula, *SV*; Don Valentine, interview by author, Nov. 7, 2012.

79. Apple Computer Preliminary Confidential Offering Memorandum (probably September 1977): 2, 6, ACM. This was clearly a draft of the official memorandum issued later in the fall.

80. Ibid.: 22–4.
81. Ibid.

Triumph: 1979–1981

1. "The Electronics Outlook: Next Quarter, Next Year, Next Decade," transcript of a talk Ben Rosen gave to the Wescon Annual Marketing Conference on Sept. 17, 1979, reprinted in Rosen's *Morgan Stanley Electronics Letter*, Sept. 28, 1979. On Ben Rosen (in footnote): Philip Elmer-DeWitt, "Steve Jobs Through Rosen-Colored Glasses," *Fortune*, 23 Oct. 2011.
2. Susan Benner, "Storm Clouds over Silicon Valley," *Inc.*, September 1982: 84.
3. Pamela G. Hollie, "Companies Compete to Lure Employees: California Computer Suppliers Trying Novel Incentive Plans," *New York Times*, April 30, 1980.
4. Association of Bay Area Governments, "Silicon Valley and Beyond: High Technology Growth for the San Francisco Bay Area" (Working Papers on the Region's Economy, no. 2): 1; "The Silicon Valley Economy," *FRBSF [Federal Reserve Bank of San Francisco] Weekly Newsletter*, no. 92-22 (May 29, 1992). (Jobs increased from 380,000 to 665,000.) Wells Fargo Bank, "Economic Forecast," 17.
5. Marilyn Chase, "Electronics Firms in 'Silicon Valley' Start Exodus Due to Lack of 'Roofs and Folks,'" *Wall Street Journal*, March 13, 1980; Susan Benner, "Storm Clouds over Silicon Valley," *Inc.*, September 1982. The classified page count is from 1979.
6. Marilyn Chase, "Venture Capitalists Rush In to Back Emerging High-Technology Firms," *Wall Street Journal*, March 18, 1981.
7. *Discussion and Comments on the Major Issues Facing Small Business, A Report of the Select Committee on Small Business, United States Senate to the Delegates of the White House Conference on Small Business*, Dec. 4, 1979: 9.
8. "Big Is Powerless," *Wall Street Journal*, Sept. 8, 1981.
9. The commission had only seven members from private industry, so the high-technology industry was exceptionally well represented. Speaking at a venture capital luncheon, the governor proudly listed a dozen bills and resolutions designed to promote computer literacy, reduce state taxes on employee stock ownership plans, provide tax credits for donations of computers to schools, and improve job training in the "expanded industries."

Jerry Brown to Eugene Kleiner, Oct. 13, 1982, and "Status Report on the Recommendations of the California Commission on Industrial Innovation," Oct. 12, 1982, both EKF.

10. A former spokesman for President Clinton who later worked at Uber and now runs a boutique public relations firm in Silicon Valley calls the Valley "an assisted living facility for political vets." Former White House officials now work at Uber, Amazon, Apple, Facebook, Square, and SpaceX. Matt McKenna, quoted in Michael Shear and Natasha Singer, "Next Job for Obama? Silicon Valley Is Hiring," *New York Times*, Oct. 24, 2016.

11. Joan Chatfield-Taylor, "California Casual—Plus Hard Work," *San Francisco Chronicle*, Sept. 23, 1980.

Looks Like $100 Million to Me! — Niels Reimers and Bob Swanson

1. "'Astonishing' Report on Gene Research," *San Francisco Chronicle*, Oct. 28, 1976; "Geneticists Spur Chemical Action in Living Cells," *New York Times*, Oct. 28, 1976; "A 'Triumph' in Genetic Engineering," *San Francisco Chronicle*, Dec. 2, 1977; "The Bold Entrepreneurs of Gene Engineering," *San Francisco Chronicle*, Dec. 2, 1977; "A Commercial Debut for DNA Technology," *BusinessWeek*, Dec. 12, 1977; "DNA: Industrial Research Grows," *New York Times*, May 29, 1977; "One for the Gene Engineers," *Time*, June 6, 1977; Jerry E. Bishop, "Gene-Splicing Field Is Swiftly Approaching the Commercial Stage," *Wall Street Journal*, June 24, 1980. (In footnote) Information about large companies acquiring stakes in biotech firms: Marcella Rosene, "Why Corporations Back Entrepreneurs," *Venture*, May 1980.

2. In 1999, Genentech paid an additional $200 million to the university, both parties agreeing that "this settlement is not an admission that Genentech infringed UC's patent or used the genetic material in question." http://www.gene.com/media/press-releases/4887/1999-11-19/university-of-california-and-genentech-s.

3. Bob Swanson, ROHO interview.

4. Fred Middleton, ROHO interview.

5. Brook Byers, interview by author, Oct. 8, 2015.

6. "Kleiner & Perkins, Venture Capital, and the Chairmanship of Genentech, 1976–1995," oral history of Thomas J. Perkins, ROHO interview. Perkins

had started a laser company that was acquired by Spectra-Physics, which H&Q took public.

7. Bill Hambrecht, *ROHO* interview. Hambrecht, profile in Robert Levering, Michael Katz, and Milton Moskowitz, *The Computer Entrepreneurs: Who's Making It Big and How in America's Upstart Industry* (New York: New American Library, 1984): 424.

8. Hambrecht, ROHO interview.

9. Levering, Katz, and Moskowitz, *Computer Entrepreneurs*: 424.

10. Hambrecht, ROHO interview.

11. Kathryn Christensen, "Hambrecht and Quist's Underwriting Success Stems from Focus on High-Tech Firms," *Wall Street Journal*, Dec. 2, 1980. The four senior managing partners (Hambrecht, Quist, William Timken, and Roy L. Rogers), each owned 16 percent of H&Q; twelve general partners and thirteen limited partners owned the rest. In the mid-1970s, the firm had been in such trouble that both Hambrecht and Quist had mortgaged their homes to pay investors who wanted out. Hambrecht profile in Levering, Katz, and Moskowitz, *The Computer Entrepreneurs*.

12. Fred Middleton, ROHO interview.

13. Middleton, ROHO interview; Swanson, ROHO interview.

14. Middleton, ROHO interview.

15. Swanson, ROHO interview.

16. Byers, interview by author, Oct. 8, 2015.

17. Kathy Christensen, "Gene Splicers Develop a Product: New Breed of Scientist-Tycoons," *Wall Street Journal*, Nov. 24, 1980.

18. Sally Smith Hughes, *Genentech: The Beginnings of Biotech* (Chicago: University of Chicago Press, 2011): 161.

19. Siddhartha Mukherjee, *The Gene: An Intimate History* (New York: Scribner, 2016): 234.

20. Tom Perkins, ROHO interview.

21. http://www.nobelprize.org/nobel_prizes/chemistry/laureates/1980/. Stanford Medicine's "Legacy of Innovation" site describes Berg's work thus: "1972—Biochemist Paul Berg successfully combines the DNA of two different organisms." The description of Cohen's work: "1973—First Expression of a Foreign Gene Implanted in Bacteria by Recombinant DNA Methods," http://med.stanford.edu/about/highlights.html.

22. Christensen, "Gene Splicers Develop a Product."

23. Stan Cohen, ROHO interview.

24. Herb Boyer, quoted in Jane Gitschier, "Wonderful Life: An Interview with Herb Boyer," *PLOS Genetics*, September 2009.

25. Cohen, interview by author, June 3, 2015.

26. Niels Reimers to Stan Cohen, Sept. 12, 1986, SUOTL.

27. Cohen, interview by author, June 3, 2015.

28. Niels Reimers to File, Dec. 2, 1980, SUOTL. "In March, Rowland" (in footnote): Reimers Memo to file, March 14, 1980; Reimers to Beyers, June 5, 1980, OTL

29. Robert Rosenzweig, vice president for public affairs at Stanford, quoted in "Genetic Engineering Could Pay Off for Stanford," *San Jose Mercury News*, Dec. 1, 1980.

30. "Where Genetic Engineering Will Change Industry," *BusinessWeek*, Oct. 22, 1979. The first recombinant DNA product, DNA ligase (an enzyme) from *E. coli* produced from a cloned gene, reached the market in 1975. Nicholas Wade, "Cloning Gold Rush Turns Basic Biology into Big Business," *Science*, May 16, 1980.

31. Niels Reimers to William F. Massy, May 15, 1979, SUOTL.

32. Licensees were charged $10,000 per year until they had products, at which point they would switch to paying royalties that looked low (1 percent and as low as .5 percent for sales over $10 million) but would translate to high numbers for Stanford and the University of California if the licensees had profitable product sales. Royalties were calculated as a percentage of end products, which were the most valuable. Some companies paid royalty rates on other types of products (basic genetic products such as plasmids, process improvement products such as enzymes used for chemical manufacturing, and bulk products, such as a dipeptide sold as a sweetener to beverage companies), but those rates were higher than the rates for end products (10 percent, 10 percent, and 1–3 percent respectively).

33. Niels Reimers to William F. Massy, May 15, 1979; Niels Reimers to Donald Kennedy, Jerry Lieberman, and William F. Massy, Dec. 17, 1981; Niels Reimers to Stanley N. Cohen, Mike Hudnall, and Bertram Rowland, Sept. 14, 1978, all SUOTL.

34. Reimers to Roger Ditzel (UC patent administrator), Jan. 12, 1981; form letter dated June 25, 1981, SUOTL.

35. Andy Barnes, interview by author, March 18, 2015.

36. Advertisement text, June 25, 1981, SUOTL. Advertisements ran on July 31 in *Science*, August 3 in the *Wall Street Journal*, and August 6 in *Nature*.

37. Barnes, interview by author, March 18, 2015. Jeffrey Fox, "Genetic Engineering Industry Emerges," *Chemical and Engineering News*, March 17, 1980.

38. Andy Barnes to File, Chicago/New York Trip—October 1–6, 1981, SUOTL.

39. Andy Barnes to File, European Trip—October 24 through November 7, 1981, SUOTL.

40. Barnes, interview by author, March 18, 2015.

41. Reimers, interview by author, May 15, 2015.

42. Niels Reimers to Donald Kennedy, Sept. 11, 1981, SUOTL.

43. Donald Kennedy to Kent Peterson, "End-Game on Cohen-Boyer," Dec. 2, 1981, email printout, SUOTL. Peterson noted in another email in the chain, "Don [Kennedy] pounced on Niels during the Admin Council briefing on Monday on these points, so we knew he was concerned. . . . I think we will find out that the end game has been fairly well thought out, but I was surprised as well as Don by the amount of the litigation reserve that Niels thinks may be required."

44. Even as Reimers hoped that he would not have to fight, he privately used a good bruiser nickname for the law firm he retained: he called the firm of Finnegan, Henderson, Farabow & Garrett "Bruno, Boris, and Brunhilde." The joking name was suggested by Charlie Lipsey, an attorney at the firm.

45. Kent Peterson to Niels Reimers and William F. Massy, Dec. 5, 1981, SUOTL.

46. Andy Barnes to Dorothy Bender, Sept. 28, 1981: "We need to have the file in the computer and ready for data input by December 1, 1981."

47. Sally Hines, interview by author, May 13, 2015.

48. In their interviews with the author, both Reimers and Barnes recalled that, as Barnes put it, "At the eleventh hour, we were still sweating getting Genentech."

49. Niels Reimers to Donald Kennedy, Jerry Lieberman, and William F. Massy, Dec. 17, 1981, SUOTL.

Sitting in a Kiddie Seat — Al Alcorn

1. "Computers for People" brochure; "Atari Inc., Division of Warner Communications Inc., Enters Personal-Home Computer Industry," WCI Press Release, Nov. 28, 1978, both SB.

2. Al Alcorn, interview by author, Oct. 31, 2013.

3. "History of Atari," http://www.youtube.com/watch?v=tOUxetHEn6Y&feature=fvwrel.

4. Mark Simon, "Folger Estate Up for Sale in Woodside," *San Francisco Chronicle*, April 25, 1996.

5. Joe Keenan, quoted in Peter W. Bernstein, "Atari and the Video-Game Explosion," *Fortune*, July 27, 1981.

6. Tekla E. Perry and Paul Wallich, "Design Case History: The Atari Video Computer System," *IEEE Spectrum*, March 1983: 45–51.

7. Bernstein, "Atari and the Video-Game Explosion."

8. Manny Gerard, interview by author, Jan. 23, 2012.

9. Angela Taylor, "A Busy Executive Who Didn't Want to 'Sit Around' on Weekends," *New York Times*, June 25, 1972. "Historical Information," collection description in "Burlington Industries, Inc. Records, 1844–2001," University of North Carolina Libraries, http://www2.lib.unc.edu/mss/inv/b/Burlington_Industries,Inc.html.

10. At the same time that he began working with Atari, Kassar launched a textile import business.

11. Ray Kassar, quoted in Tristan Donovan, "The Replay Interviews: Ray Kassar," April 29, 2011," http://www.gamasutra.com/view/feature/134733/the_replay_interviews_ray_kassar.php?page=3.

12. Nolan Bushnell, interview by author, Aug. 1, 2012.

13. Bushnell, quoted in John Hubner and William F. Kistner, "What Went Wrong at Atari," *InfoWorld* 5, no. 49 (Dec. 5, 1983).

14. It is hard to track Atari's spending on research and development after the acquisition, but the company spent 7.6 percent of revenues in 1974, 5.4 percent in 1975, and 4.2 percent in 1976. Atari mockup S-1.

15. Bushnell, quoted in Steve Fulton, "Atari: The Golden Years—A History, 1978–1981," at http://www.gamasutra.com/view/feature/3766/atari_the_golden_years__a_.php?print=1.

16. Kassar, quoted in Donovan, "The Replay Interviews: Ray Kassar."

17. "History of Atari" video at http://www.youtube.com/watch?v=tOUx etHEn6Y&feature=fvwrel.

18. "Ray Kassar: Former CEO, Atari," in Lucinda Watson, *How They Achieved: Stories of Personal Achievement and Business Success* (New York: John Wiley & Sons, 2001).

19. Bushnell says he wanted to lower the price for the unit that attaches to the television and increase the price of the cartridges. Bushnell, interview by author, Aug. 1, 2012. In the video "Stella at 20," made for the twentieth anniversary of the VCS, Bushnell said that at the Warner budget meeting, he "was so convinced that the real way to take control of the business was to drastically drop the price of the unit" that he told the board that "the best and smartest thing we could do would be to take a 747 and airdrop the 2600 over greater Los Angeles with little parachutes, and we will be positive cash flow by that evening, and we will have absolute market dominance."

20. "The pricing strategy [and] advertising and marketing plans, including the decision regarding the video pinball release, have all hurt the consumer division. I believe that all those Warner decisions have materially and wrongfully jeopardized my investment in my Atari debentures and my compensation under the bonus pool arrangement." Nolan Bushnell to Manny Gerard, Jan. 26, 1979, MG.

21. Nolan Bushnell to Manny Gerard, Jan. 26, 1979.

22. "Commerce and Industry," *Wall Street Journal*, Jan. 4, 1979.

23. Connie Bruck, *Master of the Game: Steve Ross and the Creation of Time Warner* (Simon & Schuster, 1994): 168.

24. Gerard's estimate may be high. A Harvard Business School case on Atari claims 67 percent gross margins. Activision's gross margin in 1983 was 73 percent. Peter J. Coughlan and Debbie Freier, "Competitive Dynamics in Home Video Games (A): The Age of Atari"(HBS case); Activision S-1: 9.

25. Bruce Entin, "How a New York Textile Executive Rebuilt an Exciting, Red-Ink Stained Company," *San Jose Mercury News*, March 15, 1981.

26. "Atari had more than an 80 percent share of the home game hardware and software market. Despite George Plimpton's earnest commercials, Mattel's Intellivision, the company's nearest competitor, had only a 15 percent market share." Hubner and Kistner, "What Went Wrong at Atari": 157. 4 million game systems: Bruce Entin, "How a New York Textile Executive

Rebuilt an Exciting, Red-Ink Stained Company," *San Jose Mercury News*, March 15, 1981.

27. Steve Bloom, "Atari: From Cutoffs to Pinstripes," http://www.atarimu seum.com/articles/10yranniversary/cut2pin.html.

28. *Once upon Atari* video, Episode 4.

29. Kent, *Ultimate History*: 186–7.

30. Howard Scott Warshaw, DigitPress interview, http://www.digitpress.com /library/interviews/interview_howard_scott_warshaw.html.

31. The VCS could access only 4 kilobytes of external ROM. A technique called bank switching, developed by Larry Wagner and first used by Tod Frye in the *Asteroids* cartridge, enabled the VCS to process 8-kilobyte games. Herman, *Phoenix: The Rise and Fall of Video Games* (Rolenta Press, 1997): 56.

32. Carla Meninsky, interview by author, Nov. 3, 2011.

33. *Once upon Atari* video, episode 1.

34. Bruce Entin, "Ray Kassar and Atari: It's All in the Game," *San Jose Mercury News*, Sept. 27, 1982.

35. Howard Anderson, managing director of the Yankee Group management consulting firm, quoted in Sally O'Neil, "Researcher Sees Trouble for Atari: Competition, Management Could Make Firm Vulnerable," *Peninsula Times Tribune*, Oct. 20, 1981.

36. Rob Fulop, quoted in Steve Fulton, "Atari: The Golden Years—A History, 1978–1981," http://www.gamasutra.com/view/feature/3766/atari_the_gold en_years__a_.php?print=1.

37. Rich Moran, interview by author, May 28, 2015.

38. "Atari's Struggle to Stay Ahead," *BusinessWeek*, Sept. 13, 1982.

39. Stuart Hung, "In the Chair: David Crane," *Retro Gamer*, 2010: 86–93. William H. Draper, III, *The Startup Game: Inside the Partnership Between Venture Capitalists and Entrepreneurs* (New York: St. Martin's Press, 2011): 86.

40. Draper, *The Startup Game*: 87; Carol Shaw, interview at http://www.vin tagecomputing.com/index.php/archives/800.

41. A third company to come directly out of Atari was the Learning Company, which was purchased by SoftKey International in 1995 for $600 million and then by Mattel in 1998 for $4.2 billion before being sold again for far less. Andrew Cave, "Mattel Sale Ends $3.6bn Fiasco," *Telegraph*, Sept. 30, 2000.

42. Meninsky, interview by author, Nov. 3, 2011.

43. Atari released thirteen cartridges in 1978, twelve in 1979, and six in 1980, after the four key programmers left. "Atari 2600—Cartridges Released," 1983, MG.

44. Bruce Entin, "How a New York Textile Executive Rebuilt an Exciting, Red-Ink Stained Company," *San Jose Mercury News*, March 15, 1981.

45. Alcorn, interview by author, Oct. 31, 2013, Al Alcorn, email to author, Dec. 18, 2013.

46. Alcorn, interview by author, Oct. 31, 2013.

47. Roger Van Och, "Idea Killers," 1979, photocopy in Alcorn Papers, SUSC.

48. The characters are C3-PO and Chewbacca. Victoria Woollaston, "Could Star Wars' Holographic Chess Become a Reality?," *Daily Mail*, Oct. 15, 2013.

49. Alcorn, CHM interview.

50. Alcorn, interview by author, Oct. 31, 2013.

51. "Consumer Electronics Show Focuses on Videodisc Marketing War," *Ad-Week*, Jan. 12, 1981.

52. Ibid.; "Atari Extends Gaming Offers," *CES Trade News Daily* (newspaper of the Consumer Electronics Show), Jan. 9, 1981; *Video News*, Washington, DC, Jan. 21, 1981.

53. Al Alcorn, at http://retro.ign.com/articles/858/858351p1.html.

54. The Atari-Tel business plan is at http://www.atarimuseum.com/archives/pdf/misc/ataritel.pdf.

55. Al Alcorn, quoted in http://retro.ign.com/articles/858/858351p1.html

56. HP president John Young, quoted in "View from Silicon Valley: A Long Climb in the Quest for Special Tax Treatment," *Wall Street Journal*, Feb. 27, 1981.

57. Al Alcorn, memo to self, April 3, 1981, Alcorn Collection, memos folder, SUSC.

Can You Imagine Your Grandmother Using One? — Bob Taylor

1. A mention of this meeting, held April 16, 1975, at the Peninsula School in Menlo Park, is at http://www.bambi.net/homebrew/gordon_notes.jpg.

2. Larry Tesler, interview by author, May 16, 2014.

3. Alan Kay, interview by author, May 20, 2014; Bill Pitts, interview by author, March 21, 2012.

4. Mark Seiden to author, Aug. 25, 2016.

5. "For those universities which are not well enough endowed with money to afford a variety of expensive machines, I think that the local computer store offers some very interesting options." He suggested that underfunded academic departments "turn [graduate students] loose on available, inexpensive computing." Taylor, quoted in J. F. Traub, "Quo Vadimus: Computer Science in a Decade," *Communications of the ACM* 24, no. 6 (June 1981).

6. Apple Confidential Private Placement Memo, June 30, 1979: 25.

7. Trip Hawkins, interview by author, May 20, 2016.

8. Linda F. Runyan, "A Trial Balloon in D.C.," *Datamation*, October 1979.

9. Alan Kay would have given the demo, had he not been on medical leave. The demos, which have taken on mythical status in the history of Silicon Valley, have been detailed and rehashed so many times that only the broadest outlines are necessary here. The best, most detailed account of the PARC/Apple demo is in Michael Hiltzik, *Dealers of Lightning: Xerox PARC and the Dawn of the Computer Age* (New York: HarperCollins, 2000).

10. Roughly fifty Xerox stores carried Apple computers, along with the Xerox 820 and machines from Hewlett-Packard and Osborne, until July 1982, when the agreement between the parent companies expired. Scott Mace and Paul Freiberger, "Xerox Stores Take Aim at Retail Computer Market," *Info World*, March 29, 1982; Jeff Brown, "Apples Picked off Shelves of Xerox Corp's Retail Stores," *Info World*, July 26, 1982; Bertil Nordin to Arthur Rock, 12 July 1982, AR.

11. The hiring of Tesler and others who joined him from PARC, combined with the knowledge that the graphical user interface and other features of the Alto were in fact doable, helped the Apple team figure out how to implement and go beyond those features in the Lisa and Macintosh.

12. Steve Lohr, "A Microsoft Pioneer Leaves to Strike Out on His Own," *New York Times*, Sept. 17, 2002; "The World's Billionaires: Charles Simonyi," at http://www.forbes.com/profile/charles-simonyi/.

13. Chuck Geschke, interview by author, April 21, 2014.

14. Douglas K. Smith, *Fumbling the Future: How Xerox Invented, Then Ignored, the First Personal Computer* (New York: iUniverse, 1999).

15. Though Gary Starkweather began working on the laser printer at Xerox's Webster Research Center in Rochester, the work came to fruition only

once he moved to PARC and began working with Butler Lampson and others on a digital control system and character generator.

16. Donald E. Knuth to David T. Kearns, Sept. 23, 1983, RWT.
17. Butler Lampson, interview by author, April 18, 2014.
18. Geschke, interview by author, April 21, 2014.

Young Maniacs — Mike Markkula

1. Arthur Rock, interview by author, April 4, 2016. It could not have hurt matters that Richard B. Fisher, the leading senior executive at Morgan Stanley (he would be named president in 1984) had been one of Bill Hambrecht's closest friends since college (Hambrecht ROHO interview). In 1980, Markkula said that Apple had received "at least 20, maybe 30 calls" from investment banks offering to service the public offering. Philip Shenon, "Investment Climate Is Ripe for Offering by Apple Computer," *Wall Street Journal*, Aug. 20, 1980.
2. Don Valentine, interview by author, Nov. 7, 2012.
3. Apple sold directly to retailers in the United States and internationally via twenty-one distributors. International sales accounted for a quarter of Apple's sales (by dollar value). Apple Computer Prospectus, Dec. 12, 1980: 11, 15, 20.
4. In 1981, Apple's analysis indicated that games/entertainment software was most popular (with 86 percent of buyers having bought or planning to buy the software), followed by word processing (65 percent) and finance (63 percent). "Software for Apple II," Gene Carter Collection, M1059, SUSC.
5. John Markoff, "Radio Shack: Set Apart from the Rest of the Field," *InfoWorld* 5 (July 1982): 43.
6. Gene Carter, interview by author, Jan. 7, 2016.
7. John Hall, the consultant who helped write Apple's business plan, was a controller at Intel when VisiCalc was in beta testing; his division was one of several beta sites within Intel, all of them running VisiCalc on Apple II machines. (One of the founders of VisiCalc's parent company, Personal Software, had previously worked at Intel.)
8. Dan Flystra, "The Creation and Destruction of VisiCalc," edesber.com /visicorp-history; Apple Computer Prospectus, Dec. 12, 1980: 13; Dan

Bricklin of VisiCalc, quoted in Daniel Terdiman, "The Untold Story Behind Apple's $13,000 Operating System," http://www.cnet.com/news/the-untold-story-behind-apples-13000-operating-system/.

9. Anthony Hilton, "Drop-Out Duo Cash in $460m Chip," *Sunday Times*, date unknown.

10. State regulators required a company's book value to be at least 20 percent of its market value.

11. Robert J. Cole, "An 'Orderly' Debut for Apple," *New York Times*, Dec. 13, 1980. Another 400,000 shares were sold by investors not in management at Apple.

12. Carter, interview by author, Jan. 7, 2016.

13. Joe Shelpela (Personnel) to Distribution, Sept. 25, 1980. Apple Computer M1007, ser. 2, no. 8: 21, SUSC. The three invited firms were Arthur Young & Co., The Portola Group, and Robert P. Martin Accountancy.

14. This is Jobs's description of himself and other young tech entrepreneurs in 1985. See Jobs's interview by David Sheff in *Playboy*, February 1985: "*Playboy*: It's interesting that the computer field has made millionaires of— Jobs: Young maniacs, I know."

15. "Apple Exec in Guarded Condition with Injuries from Plane Crash," *San Jose Mercury News*, Feb. 9, 1981.

16. Gregg Williams and Rob Moore, "Guide to the Apple," *Byte*, December 1984.

17. Harriet Stix, "A UC Berkeley Degree Is Now the Apple of Steve Wozniak's Eye," *Los Angeles Times*, May 14, 1986.

18. Marilyn Chase, "Technical Flaws Plague Apple's New Computer," *Wall Street Journal*, April 15, 1981. Apple III prices ranged from $4,300 to nearly $8,000, compared to the Apple II systems at about half the cost.

19. Apple fixed the problems and brought the Apple III back in late 1981— "Let me re-introduce myself," one advertisement began—but not much software was written for the machine, and it was not anywhere near as popular as the Apple II. (Sales were around 1,000 per month versus the Apple II's 15,000.) By one estimate (Brent Schlender and Rick Tetzeli, *Becoming Steve Jobs: The Evolution of a Reckless Upstart into a Visionary Leader* [New York: Crown Business, 2015]: 72), before the Apple III was discontinued in 1984, only 120,000 had been sold. In the same period between the Apple III's introduction and its demise, the company sold

nearly 2 million Apple IIs. Andrew Pollack, "Next, a Computer on Every Desk," *New York Times*, Aug. 23, 1981; Bruce Entin, "Can Apple Keep Its Piece of the Pie?," *San Jose Mercury News*, Aug. 17, 1981.

20. The drop was from $285,000 in December 1980 to $170,000 in August 1981, though the annualized rate for the year was at $209,000. Ann Bowers to Executive Staff, Aug. 3, 1981, Apple Computer, M2007, ser. 2, no. 8: 32, SUSC.

21. Jean Richardson (advertising director), quoted in Michael Moritz, *Return to the Little Kingdom: How Apple and Steve Jobs Changed the World* (London: Overlook Press, 2010): 248.

22. "Apple Controls Animated Dragon at Studio," *Industry Week*, July 6, 1981; John Schneidawind, "Oldest Profession Jumps into Newest Technology," *MIS Week*, Aug. 12, 1981; Cinde Chorness, "Silicon Valley Clergyman Computes Their Wedding Vows," *Valley Journal*, July 29, 1981.

23. Kathleen K. Wiegner, "Apple Loses Its Polish," *Forbes*, April 13, 1981.

24. Marilyn Chase, "Apple Computer Takes a Bruising as Analysts Lower Earnings Estimates for Fourth Quarter," *Wall Street Journal*, Sept. 28, 1981.

25. *So Far: The First Ten Years of a Vision* (Apple 10th Anniversary Document, 1987): n.p.; "40 Workers Laid Off at Apple," *Peninsula Times Tribune*, Feb. 26, 1981. The date of the firing was February 25.

26. Scott's managerial style had always been forceful and confrontational, but according to Markkula, something changed at the end of 1980. "He was not himself," Markkula says of Scott. Bowers will say only that after fielding complaints about Scott, she told Markkula that he "had to go."

27. The computer became the employee's after one year. Ann Bowers to All Apple Employees, Dec. 23, 1980, Apple Computer M2007, ser. 2, no. 8: 21, SUSC.

28. Dana Scott to [Exec team] Subject: Travel Policy, Jan. 28, 1981, Apple Computer M2007, ser. 2, no. 8: 32, SUSC.

29. Dana Scott to Ken Zerbe, Subject: Style, July 17, 1981, M1007, ser. 1, no. 5: 14, SUSC. This photocopy of Scott's letter includes a note from Scott's assistant JoAnn Burton Glock, indicating that she had typed it. (Scott did not respond to my requests for an interview.) See Bruce Entin, "Can Apple Keep Its Piece of the Pie?," *San Jose Mercury News*, Aug. 17, 1981; Bruce Entin, "Why Three Top Electronics Execs Are Out of Work," *San Jose Mercury News*, July 23, 1981; Bruce Entin, "For Mike Scott, Future Is a $60 Million Question," *San Jose Mercury News*, April 24, 1981; "Ex-Chief

Exits Apple," *Electronic News*, July 27, 1981; "Apple's First President Turns In Resignation," *San Francisco Chronicle*, July 22, 1981; Jay Yarow, "Interview with Apple's First CEO, Mike Scott," *Business Insider*, May 24, 2011; Scott to Zerbe, Subject: Style, July 17, 1981, M1007, ser. 1, no 5: 14, SUSC.

30. Bruce Entin, "For Mike Scott, Future Is a $60 Million Question," *San Jose Mercury News*, April 24, 1981.
31. Yarow, "Interview with Apple's First CEO."
32. Trip Hawkins, interview by author, May 20, 2016.
33. Comment from stevewoz at http://www.cultofmac.com/96939/apples-first-ceo-says-young-steve-jobs-could-be-trusted-with-detail-but-not-with-a-staff/.
34. Ann Bowers, interview by author, Nov. 7, 2015.
35. "The overriding consideration was pride—Mike just wasn't going to let anything stand in the way of making Apple a huge success and fulfilling the implied commitment to the Apple employees." Introductory remarks given by Arthur Rock at the Harvard Business School Entrepreneurial Award Dinner Honoring Steve Jobs and Mike Markkula, San Francisco Olympic Club, Feb. 3, 1983, AR.
36. Mike Markkula, interview by author, Feb. 24, 2016.
37. Steve Jobs, quoted in "Apple Shuffling Reflects Growth," *Computer Systems News*, April 13, 1981.
38. Entin, "Can Apple Keep Its Piece of the Pie?"
39. Markkula, interview by author, May 3, 2016.
40. Trip Hawkins, quoted in Moritz, *Return to the Little Kingdom*: 217.
41. Ken Zerbe to Board of Directors, Aug. 21, 1979. Apple had two private offerings, one in 1977 and another in 1979.
42. Markkula called the process by which a customer develops a sense of a company's character "imputing." See, e.g., Jobs quoting Markkula when explaining why he wanted the Apple stores to be large, elegant, and centrally located. Walter Isaacson, *Steve Jobs* (New York: Simon & Schuster, 2011): 370.
43. Jobs, quoted in Isaacson, *Steve Jobs*: 78.
44. Don Valentine, interview by author, Nov. 7, 2012.
45. Apple Annual Report, 1982.
46. Will Houde replaced Tom Whitney to run personal computers, Del Yocam was named corporate vice president of manufacturing, John Vennard was

named vice president, peripherals, and Thomas Lawrence was named vice president of European operations. "Corporate V-P Chosen at Apple," *Electronic News*, Sept. 7, 1981.

47. Markkula, interview by author, May 3, 2016.
48. Hawkins, interview by author, May 20, 2016.
49. Markkula, interview by author, May 3, 2016.
50. Markkula, CHM interview.
51. Larry Tesler and Chris Espinosa, "Origins of the Apple User Interface," talk given Oct. 28, 1997, http://web.archive.org/web/20040511051426/http://computerhistory.org/events/lectures/appleint_10281997/appleint_xscript.shtml. Andy Hertzfeld recalls that Bill Atkinson, a member of the Lisa group, attended Mac group meetings. Hertzfeld, "Credit Where Due," http://www.folklore.org/StoryView.py?project=Macintosh&story=Credit_Where_Due.txt.
52. Regis McKenna, interview by author, April 11, 2000.
53. Hawkins, interview by author, May 20, 2016.
54. The full interview appears in *Computers and People*, July–August 1981: 8. The magazine was edited by the computer scientist and social activist Edmund C. Berkeley.
55. "When we invented the personal computer, we created a man-machine partnership" (advertisement), *Wall Street Journal*, Feb. 25, 1981.
56. Wendy Quiones, "Pioneering a Revolution: Apple's Steve Jobs and Steve Wozniak," *Boston Computer Update*, July–August 1981.
57. Jobs, quoted in Kay Mills, "The Third Wave: Whiz-Kids Make a Revolution in Computers," *Los Angeles Times*, July 5, 1981. Ben Rosen's comments referenced in the footnote appeared in the Sept. 28, 1979, edition of his widely read *Electronics Newsletter*.
58. Jobs, quoted in Mills, "The Third Wave."
59. Ibid.
60. Sally O'Neil, "Apple Uses Dick Cavett for Its 'Intelligent Pitch,'" *Peninsula Times Tribune*, July 22, 1981. This number seems reasonable, given that Apple's Schedule X Supplementary Income Statement shows $6.4 million spent on advertising for the six months ended March 7, 1981—halfway through the fiscal year, which had begun in September 1980. Apple Computer M1007, ser. 5, no. 2: 18.
61. Apple Annual Report, 1981: 3.

62. "To Each His Own Computer," *Newsweek*, Feb. 22, 1982.
63. "Potential Competition," Confidential Private Placement Memorandum, Nov. 18, 1977.
64. Entin, "Can Apple Keep Its Piece of the Pie?"
65. George Anders, "IBM's New Line Likely to Shake the Market for Personal Computers," *Wall Street Journal*, Aug. 13, 1981.
66. Entin, "Can Apple Keep Its Piece of the Pie?"
67. Advertisement in *Byte* magazine, January 1982.
68. Martin Campbell-Kelly and William Aspray, *Computer: A History of the Information Machine* (New York: Basic Books, 1996): 257.
69. Alice Rawsthorn, "The Clunky PC That Started It All," *New York Times*, July 31, 2011; "Personal Computers: And the Winner Is IBM," *BusinessWeek*, Oct. 3, 1983: 76.
70. David E. Sanger, "Philip Estridge Dies in Jet Crash; Guided IBM Personal Computer," *New York Times*, Aug. 5, 1985.
71. "Macintosh vs. IBM PC at One Year," *InfoWorld*, Jan. 14, 1985.
72. Markkula, CHM interview.
73. Otto Friedrich, "Machine of the Year: The Computer Moves In," *Time*, Jan. 3, 1983.
74. Marilyn Chase, "Apple Computer Takes a Bruising as Analysts Lower Earnings Estimates for Fourth Quarter," *Wall Street Journal*, Sept. 28, 1981. The price dropped by 2⅛ to 14¼.
75. "Home and Non-Home Market for Personal Computers—Systems Selling at Less than $15,000," Creative Strategies International, n.d., but clearly 1982, ACM.
76. 98 percent of the 82.37 million American households had televisions in 1981, according to the Statistical Abstract of the United States.
77. Cary Pepper, "Putting a Finger on the Future: Personal Computers," *Town & Country*, January 1982.
78. Entin, "Can Apple Keep Its Piece of the Pie?"

What in the Hell Are You Trying to Say? — Fawn Alvarez

1. Jeff Smith, interview by author, May 19, 2015.
2. Fawn Alvarez Talbott, interview by author, July 24, 2013. Smith did not recall the particulars but agrees that there was a transition during which

writing change orders moved from the engineers' responsibility to a task performed by women in the department.

3. After an exhaustive search, Sarah Reis, Reference Librarian at Stanford University's Robert Crown Law Library, was unable to find any statutory provision in the 1970s that forbade women from owning or acquiring shares in a risky company. Reis did find a number of peer-reviewed articles confirming that women were "viewed as inept or irrational investors" at the time. Sarah Reis to author, April 18, 2017.

4. Bob Maxfield, interview by author, Jan. 30, 2017.

5. John Young, quoted in Mark Blackburn, "Allure of Silicon Valley Fades," *New York Times*, April 14, 1980.

6. Ibid.

We Don't Need Any Money — Sandra Kurtzig

1. Molly Upton, "HP 3000 Sales Exceeding Quota," *Computerworld*, July 3, 1974; Upton, "Sales of 3000s Twice Those of Last Year, HP Reports," *Computerworld*, Sept. 17, 1975; Hewlett-Packard, 1980 Annual Report: 9.

2. Shipping the source code also benefited ASK in later years. If the software had problems, ASK could dial into a customer's computer and recompile the code immediately, logging the problem and bringing it back to R&D for a fix before the next release. Computer-savvy customers also appreciated that since MANMAN was written in FORTRAN, it would be easy to find coders to help maintain the product if ASK suddenly disappeared, a common fate of small software businesses in the late 1970s.

3. Sandra Kurtzig, *CEO: Building a Four Hundred Million Dollar Company from the Ground Up* (New York: W. W. Norton & Co., 1991): 158.

4. Burt McMurtry, interview by author, Nov. 26, 2012; McMurtry, CHM interview. McMurtry credits investments in ROLM, NBI, Triad Systems, and KLA for his venture fund's success.

5. The company was informally called by its original name, Cullinet. John Cullinane, CBI interview.

6. "Missing Computer Software," *BusinessWeek*, Sept. 1, 1980, quoted in Martin Campbell-Kelly, *From Airline Reservations to Sonic the Hedgehog: A History of the Software Industry* (Cambridge, MA.: MIT Press, 2003): 2.

7. Gary Slutsker, "The New 'Publishers' in Computer Software," *Venture*, September 1980: 79.

8. "Many of us had no idea what it meant to 'go public.'" David V. Goeddel and Arthur D. Levinson, "Obituary: Robert A. Swanson (1947–99)," *Nature* 403 (Jan. 20, 2000): 264.

9. Burt McMurtry to author, Sept. 20, 2015.

10. Kurtzig, *CEO*: 187–8; McMurtry to author, Sept. 18, 2015.

11. Kurtzig, *CEO*: 188.

12. Betty Lehan Harragan, *Games Mother Never Taught You: Corporate Gamesmanship for Women* (Warner Books, 1978): 299, 310.

13. In *CEO*, Kurtzig says that the original IPO date was October 1980. But according to ASK's S-1 filed with the SEC, the new board members Ken Oshman, Ron Braniff, and Tommy Unterberg, who joined in July 1980, did not receive stock grants until October, which makes it unlikely that October was the original IPO date, particularly given that a letter from L. F. Rothschild, Unterberg, Towbin to Kurtzig dated May 7, 1980, proposed a March 1981 date (letter courtesy SK).

14. Kurtzig, *CEO*: 192. Although H&Q's hybrid structure was unusual, it was not unique in 1981. Allen and Co. had a similar structure, and a few investment banks (Smith Barney Harris Upham & Co. and Donaldson, Lufkin & Jenrette, for example) had venture capital arms. Hambrecht notes that "there were a lot of people that thought that might be a conflict of interest," but he and Quist believed that investing in a company before taking it public meant "you knew what you were dealing with" (Hambrecht, ROHO interview). Among those who worried the structure might lead to a conflict of interest was the National Association of Securities Dealers (NASD). As the vice president of NASD put it in 1981, "It can be hard to tell if something is a legitimate venture capital deal or a way of increasing underwriting fees." Dennis Hensley, quoted in Dave Lindorff, "Investment Bankers Take the Venture Plunge," *Venture*, January 1981.

15. Nancy Anderson (general manager, Computer Systems Division), quoted in Charles H. House and Raymond L. Price, *The HP Phenomenon: Innovation and Business Transformation* (Stanford, CA: Stanford Business Books, 2009): 212; *HP 1980 Annual Report*, Marty Browne communication, Sept.14, 2015.

16. Ken Oshman, quoted in Kurtzig, *CEO*: 200.

17. Sandra Kurtzig, interview by author, June 24, 2015.

18. *Going Public—The IPO Reporter 1979, 1980.* By year-end, 237 companies had gone public.

19. "High Technology: Wave of the Future or a Market Flash in the Pan?," *BusinessWeek*, Nov. 10, 1980.

20. "Investment Advisers Give Clues on 'Going Public,'" *Peninsula Times Tribune*, Dec. 5, 1980.

21. Tom Lavey, interview by author, Aug. 4, 2015.

22. ASK, 1984 Annual Report.

23. Campbell-Kelly, *From Airline Reservations to Sonic the Hedgehog*: 154.

24. Michael S. Malone, "Dangerous Woman Breaks Barriers," *San Jose Mercury News*, March 6, 1979; Martha Reiner, "Top Computer Firm Began as Part-Time Work," *SF Business Journal*, March 2, 1981; Joan A. Tharp, "The Entrepreneur," *The Executive SF*, July 1981; "Raising Output Levels," *Informatics*, Jan. 19, 1981; *Computer Systems News*, March 2, 1981; *The Executive Woman*, April–May 1980; "Women Rise as Entrepreneurs," *BusinessWeek*, Feb. 25, 1980.

25. Tharp, "The Entrepreneur."

26. Kurtzig, interview by author, June 24, 2015.

27. ASK S-1, Aug. 6, 1981.

28. Office expansion: *Newsman* [ASK newsletter], March 1981, MB. Kurtzig changed the name from ASK Computer Services to ASK Computer Systems.

29. Kurtzig, interview by author, June 23, 2015.

30. Marty Browne, CHM interview.

31. Browne, interview by author, July 12, 2015.

32. Kurtzig, *CEO*: 226.

33. The S&P closed the first week of January 1981 at 136.34 and the last week of September at 112.77. The NASDAQ saw a 14 percent drop over the same time.

34. Kurtzig had rejected private investment most recently in the spring of 1980, after her bankers suggested she do a private placement before the IPO. L. F. Rothschild, Unterberg, Towbin to Kurtzig, May 7, 1980: Table 3, SK.

35. Vector Graphic went public on the NASDAQ, underwritten by Hambrecht & Quist, along with Shearson American Express, on Oct. 14, 1981.

Benj Edwards to author, Sept. 30, 2015. For an excellent history of Vector Graphic, see Benj Edwards, "How Two Bored 1970s Housewives Helped Create the PC Industry," *Fast Company*, http://www.fastcompany.com /3047428/how-two-bored-1970s-housewives-helped-create-the-pc -industry.

36. Kurtzig, *CEO*: 236–7. Kurtzig gave Liz Seckler, who had recently bought herself a watch, a bracelet.
37. Kurtzig, *CEO*: 113.
38. ASK S-1; quote in footnote: Sandy Kurtzig to author, August 11, 2017.
39. The attorney was Craig W. Johnson; the banker was Tommy Unterberg.
40. The equity structure had been with ASK since its earliest days. In 1976, for example, Kurtzig owned 2.5 million shares. Options on a total of 50,000 shares had been granted to the company's other six employees. 1976 business plan: 23, SK.
41. Fox, quoted in House and Price, *HP Phenomenon*: 479. Details on HP's stock option plans described in the footnote are from a spreadsheet, "Hewlett-Packard Company Stock Timeline," provided by the company.
42. Chuck House to author, Oct. 5, 2015. Thanks also to John Minck, Jim Hall, Al Steiner, and Curt Gowan of the HP Alumni Association for their memories.
43. Sandra Kurtzig showed the author the photo.
44. Rankings of market value of entrepreneurs' holdings at the time of their IPOs: "100 Who Made Millions in 1981," *Venture*, April 1982.
45. Kurtzig, *CEO*: 251; Timothy C. Gartner, "Divorce Prompts Sale of ASK Computer Shares," *San Francisco Chronicle*, Dec. 7, 1982.

Transition: 1983–1984

1. David Djean, "Westward Ha! A Visitor's Guide to Silicon Valley," *PC-Computing*, June 1989: 99.
2. Morgan Stanley Dean Witter, "U.S. Business Capital Equipment Spending—Information Technology (IT) Spending vs. Non-IT Spending, 1960–2000," Exhibit 7, *Technology and Internet*, April 5, 2001. The *Wall Street Journal* introduced its "Small Business" column and *Forbes* its "Up-and-Comers" section around that time. *BusinessWeek* began an "Information

Processing" section and *Fortune* began featuring an "Entrepreneurs" slug. Andrew Feinberg, "Why Entrepreneurs Make Good Copy," *Venture*, July 1982: 44.

3. Ronald Reagan, "Address Before a Joint Session of the Congress on the State of the Union: January 25, 1983," *Papers of the Presidents: Administration of Ronald Reagan*, 107. Reagan also established an Innovation and Entrepreneurship Task Force whose members included Bob Noyce, Spectra-Physics founder Herb Dwight, and Larry Sonsini's law partner Mario Rosati.

4. This figure is for Santa Clara County only. Lenny Siegel, *Testimony Prepared for the Subcommittee on Science, Research, and Technology of the House Committee on Science and Technology and the Task Force on Education and Employment of the House Budget Committee*, June 16, 1983, 1100–1, PSC.

The Rabbits Hopped Away — Bob Taylor

1. Hamilton [Bruce Hamilton, based in El Segundo, California], ES to StarArchInterest, StarOnTajo, "LISA in the Flesh," May 3, 1983, RWT.

2. Taylor's handwritten notes, probably from 1976, appended to a copy of T. George Harris, "The Religious War over Truths and Tools," *Psychology Today* (January 1976), RWT.

3. Alan Kay, quoted in M. Mitchell Waldrop, *The Dream Machine: J.C.R. Licklider and the Revolution That Made Computing Personal* (New York: Viking, 2001): 437.

4. Harlan M. Averitt [PARC Personnel] to Bob Taylor, Aug. 24, 1982, RWT.

5. Graphs, "Source: E Steffensen 6/12/80," courtesy Taylor.

6. Quote is from Taylor performance review, 2/1/80–2/1/81, RWT. Hiltzik mentions the five-year plan in *Dealers*, 354–5.

7. Bill Spencer, interview by author, May 1, 2014.

8. Michael Hiltzik, *Dealers of Lightning: Xerox PARC and the Dawn of the Computer Age* (New York: HarperCollins, 2000): 377–8.

9. Spencer, interview by author, May 1, 2014.

10. Bob Taylor, interview by author, March 18, 2013.

11. Spencer, interview by author, May 1, 2014.

12. Ibid.

13. "PARC Spin-Outs," reflections by Bill Spencer, sent to the author May 1, 2014; Ralph R. Jacobs and Donald R. Scifres, "Recollections on the Founding of Spectra Diode Labs, Inc. (SDL, Inc.)," *IEEE Journal of Selected Topics in Quantum Electronics*, 6, no. 6 (November–December 2000): 1228–30.

14. Spencer, interview by author, May 1, 2014.

15. Bill Spencer to Bob Taylor, "Management Counselling [*sic*] Session," Aug. 22, 1983, RWT.

16. Jerry Elkind, interview by author, June 2, 2014. Elkind says that by the time he came to PARC, Taylor had already set the culture for the lab.

17. Bob Taylor to Bill Spencer, "Your memo of 22 August 83," Aug. 28, 1983.

18. Spencer, interview by author, May 1, 2014.

19. Hiltzik, *Dealers of Lightning*: 381–2.

20. Handwritten note, clipped to handwritten note from Spencer dated Sept. 19, 1983, that reads, "You are correct time is short today. Therefore, to reduce any overlap in what you and I tell CSL, I plan to attend your 10:30 meeting," RWT.

21. Chuck Thacker to Bill Spencer, Sept. 19, 1983, RWT.

22. Spencer, interview by author, May 1, 2014

23. Spencer, interview by author, June 13, 2014.

24. The group included Mike Schroeder, "Mark, Bill, Severo, Paul, and Ed." Mike Schroeder to CSL Only, Sept. 23, 1983, RWT.

25. Ibid.

26. Donald E. Knuth to David Kearns, Sept. 23, 1983, RWT.

27. Dana Scott to David Kearns, Oct. 4, 1983, RWT.

28. Richard Karp to David Kearns, Sept. 23, 1983, RWT.

29. Brian K. Reid to David Kearns, Sept. 26, 1983, RWT. Roger Needham at Cambridge University also wrote a letter of support.

30. J.C.R. Licklider to David Kearns, Sept. 28, 1983, RWT.

31. Spencer, interview by author, May 1, 2014.

32. Sam Fuller to Bob Taylor, Nov. 18, 1983, RWT. Fuller would go on to serve as chief scientist at DEC, CTO and vice president of research and development at Analog Devices, and a visiting scientist at MIT.

33. "A Parable of Rabbits," n.d., unsigned, RWT.

34. By the end of 1984, former PARC employees at Digital's Systems Research Center (DEC SRC) included Andrew Birrell, Marc Brown, Leo Guibas,

Jim Horning, Steve Jeske, Butler Lampson, Roy Levin, Carol Peters, Phil Petit, Lyle Ramshaw, Paul Rovner, Mike Schroeder, Larry Stewart, Chuck Thacker, Mary-Claire van Leunen, and John Wick.

35. Citation as Fellow of the Royal Society; Crystal Lu, "The Genius: Mike Burrows' Self-Effacing Journey Through Silicon Valley" at http://web.ar chive.org/web/20080217003150/http://www.stanford.edu/group/gpj/cgi -bin/drupal/?q=node/60.

Video Nation — Al Alcorn

1. Charles P. Alexander, "Video Games Go Crunch!," *Time*, Oct. 17, 1983.
2. Steve Fulton, "Atari: The Golden Years—A History, 1978–1981," http:// www.gamasutra.com/view/feature/3766/atari_the_golden_years_a_.php ?print=1.
3. *City of Mesquite v. Aladdin's Castle, Inc.*, 455 U.S. 283 (1982), No. 80-1577, argued November 10, 1981, decided February 23, 1982, 455 U.S. 283.
4. Peter W. Bernstein, "Atari and the Video-Game Explosion," *Fortune*, July 27, 1981; Lynn Langway et al., "Invasion of the Video Creatures," *Newsweek*, Nov. 16, 1981.
5. Jonathan Greer, "Atari's $310 Million Loss Breaks Record," *San Jose Mercury News*, July 22, 1983.
6. Bruck, *Master of the Game*: 180.
7. "Atari 2600—Cartridges Released," 1983, MG.
8. "Chase the Chuckwagon," Steven L. Kent, *The Ultimate History of Video Games: From Pong to Pokémon and Beyond . . . The Story Behind the Craze That Touched Our Lives and Changed the World* (New York: Three Rivers Press, 2010): 235; *Custer's Revenge*, Leonard Herman, *Phoenix: The Fall and Rise of Home Video Games* (Union, NJ : Rolenta Press, 1994): 75; Deborah Wise, "Video-Pornography Games Cause Protest," *InfoWorld*, Nov. 8, 1982. Oklahoma City outlawed *Custer's Revenge*.
9. Connie Bruck, *Master of the Game: Steve Ross and the Creation of Time Warner* (Simon & Schuster, 1994): 181–3.
10. Laura Landro, "Warner's Atari Is Trying to Regain Top Spot in Consumer Electronics," *Wall Street Journal*, July 6, 1983; "Employees at Warner's Atari Unit Turn Down Bid to Unionize Them," *Wall Street Journal*, Dec. 2, 1983.

11. Alexander, "Video Games Go Crunch."

12. Trip Hawkins, quoted in Jeffrey Fleming, "We See Farther—A History of Electronic Arts," www.gamasutra.com/view/feature/130129/we_see_far ther_a_history_of_.php?print=1.

13. Hawkins, quoted in Fleming, "We See Farther"; John Gaudiosi, "Madden: The $4 Billion Video Game Franchise," *CNN Money*, Sept. 5, 2013, money.cnn.com/2013/09/05/technology/innovation/madden-25.

14. "Atari's Struggle to Stay Ahead," *BusinessWeek*, Sept. 13, 1982.

15. "Ted Hoff: Engineer's Engineer," *Peninsula Times Tribune*, March 29, 1983.

16. David E. Sanger, "Warner Sells Atari to Tramiel," *New York Times*, July 3, 1984.

17. "Top 25 Videogame Consoles of All Time," http://www.ign.com/top-25 -consoles/1.html.

Knew It Before They Did — Niels Reimers

1. Kimberly Brown, "Proposal Would Simplify Bio-Tech Patent License," *Stanford Daily*, March 30, 1983.

2. The $270 billion figure is as of 2013. Grand View Research, "Biotechnology Market Analysis by Technology," September 2015, http://www.grand viewresearch.com/industry-analysis/biotechnology-market.

3. Hans U. D. Wiesendanger, "History and Operation of the Office of Technology Licensing, Stanford University, 1996": n.p., Stanford University Office of Technology Licensing Records 1996–2007, SC729 ACCN 2016-037, SUSC.

4. Kennedy said that $11 million of the $60 million raised by Stanford in the previous year had come from corporations. He was also aware of the criticism that Harvard had faced in 1981 when it considered investing in a company founded by Professor Mark Ptashne. Harvard president Derek Bok ultimately decided not to make the investment. Donald Kennedy, *Academic Duty* (Cambridge, MA: Harvard University Press, 1997): 256. Don Kennedy, "The University and Industry," speech given to the Colorado Council of Deans Association, American Medical Colleges, March 30–31, 1981: 10, Donald Kennedy Personal Papers, SC708, Accn 2009-139, 2: 20, SUSC.

5. Kennedy, "The University and Industry": 9.

6. Seth H. Lubove, "College Funds Taking Steps to Raise Yields," *New York Times*, Aug. 10, 1983.

7. Kristen Christopher, "Stanford Tops Ranking in Technology Revenues," *Stanford Daily*, Jan. 11, 1983.

8. "Taking Stock of Equity," *Stanford Technology Brainstorm*, http://otl.stanford.edu/about/brainstorm/1203_equity.html.

9. https://web.archive.org/web/20080917185106/http://otl.stanford.edu/about/resources/equity.html. This page, from Sept. 17, 2008, is the last archiving of the Office of Technology Licensing's "Partial List of Licensees Whose License Involved Equity."

10. Lisa M. Krieger, "Stanford Earns $336 Million off Google Stock," *San Jose Mercury News*, Dec. 1, 2005.

11. "Stanford held equity in 121 companies as a result of license agreements. For institutional conflict-of-interest reasons and insider trading concerns, the Stanford Management Company sells our public equities as soon as Stanford is allowed to liquidate rather than holding equity to maximize return," *Stanford Office of Technology Licensing Annual Report 2014–2015*.

12. Ken Auletta, "Get Rich U.," *The New Yorker*, April 30, 2012.

13. "Niels Reimers, Long-Time Director of Stanford's Patent Office Resigns; OTL to Gross $24 Million This Year," Stanford News Release, March 20, 1991, Stanford News Service.

14. Robert E. Alvarez and Albert Macovski, "X-Ray Spectral Decomposition Imaging System," patent no. 4029963.

15. "OTL Financial Data 1970–2016," KK.

No One Thought They Would Sell — Fawn Alvarez

1. Ann Magney Kieffaber, interview by author, Feb. 15, 2017.

2. Ibid.

3. Fawn Alvarez Talbott, interview by author, Feb. 17, 2017.

4. "IBM and ROLM Cope with Prenuptial Jitters," *BusinessWeek*, Nov. 19, 1984.

5. Ron Raffensperger, interview by author, Feb. 4, 2017.

6. ROLM, 1984 Annual Report.

7. Thomas C. Hayes, "At ROLM, an Independent Style," *New York Times*, Sept. 27, 1984; Joan A. Tharp, "M. Kenneth Oshman Tackles Telecommunications Goliaths," *The Executive SF*, March 1982.

8. Raffensperger, interview by author, Feb. 4, 2017; Jeff Smith used almost identical language to describe his reaction to the acquisition.

9. Ann Magney Kieffaber, interview by author, Feb. 15, 2017.

10. Vineta Alvarez Eubank, interview by author, Feb. 14, 2017.

11. "IBM and ROLM Cope with Prenuptial Jitters," *BusinessWeek*, Nov. 19, 1984.

12. Fawn Alvarez Talbott, interview by author, June 9, 2015.

13. Talbott, interview by author, July 24, 2013.

The Entire World Will Never Be the Same — Mike Markkula

1. Friedrich, "Machine of the Year."

2. Party description and quotes: Christopher Menkin, "Computer Reconnaissance," *Daily Commercial News*, Feb. 23, 1983.

3. Sculley called his hiring as CEO a "big mistake" in an interview by Leander Kahney, Oct. 14, 2010, at http://www.cultofmac.com/63295/john -sculley-on-steve-jobs-the-full-interview-transcript/.

4. "Home Computers in a Slump," graphic based on data from Software Access (Mountain View), n.d., but probably 1985. The graphic appears in Regis McKenna, "The Wisdom and Inspiration of Steve Jobs," slide show shared with the author, RM.

5. Elizabeth Peer, "How to Work the Thing," *Newsweek*, Feb. 22, 1982. The article appeared in an issue with a cover featuring the iconic *Whistler's Mother* sitting at a personal computer.

6. "Personal Computers: And the Winner is IBM," *BusinessWeek*, Oct. 3, 1983: 78.

7. "Apple will be a Fortune 500 company by 1985," Markkula says in Susan A. Thomas, "Early Entry in Personal Computer Market Makes Sales Easy as Pie for Apple," *San Francisco Business Journal*, Jan. 7, 1980.

8. Mike was "like a father, and I always cared for him." Steve Jobs, quoted in Walter Isaacson, *Steve Jobs* (New York: Simon & Schuster, 2011): 319.

She Works Hard for the Money — Sandra Kurtzig

1. ASK annual report, 1984. Ranking of fastest-growing companies (footnote) is from the *Inc.* 100 of May 1983.

2. Sandra Kurtzig, *CEO: Building a Four Hundred Million Dollar Company from the Ground Up* (New York: W. W. Norton & Co., 1991): 244.

3. Tom Lavey, interview by author, Aug. 4, 2015.

4. Ibid. Compaq was funded by Ben Rosen, the technology-analyst-turned-venture-capitalist who also funded Electronic Arts.

5. Sandra Kurtzig, interview by author, June 24, 2015.

6. Ibid.

7. "30 Minutes" video and accompanying note, HK.

Conclusion: Wave After Wave

1. On the Facebook campus: "Facebook Campus Project," City of Menlo Park, http://www.menlopark.org/643/Facebook-Campus-Project; Julie Bort, "A Tour of Facebook's Disneyland-Inspired Campus," *Business Insider*, Oct. 13, 2013; Alyson Shontell, "Why Sun's Logo Is on Sign Outside Facebook's Campus," *Business Insider*, Dec. 7, 2014; Facebook farmers' market: advertisement, *Palo Alto Weekly*, June 3, 2016.

2. "Sun Microsystems Inc. History," http://www.fundinguniverse.com/company-histories/sun-microsystems-inc-history/.

3. Burt McMurtry, interview by author, Nov. 26, 2012.

4. "The real precursor for the Sun was the Xerox Alto," says Andy Bechtolsheim, who worked as a "no-fee consultant" at PARC and who as a graduate student was assigned to build a computer based on the Alto but with off-the-shelf parts; quoted in John Markoff, "Even Sun Microsystems Had Its Roots at Xerox PARC," *New York Times*, May 28, 2014.

5. The story of the Sun invention disclosure is a bit murky. Because Bechtolsheim was a Stanford student at the time he developed the Sun workstation, he filed a patent disclosure with the Office of Technology Licensing (Sun originally stood for Stanford University Network). He was granted all rights to the design after a review determined that it was only "marginally patentable" (whether that determination came from the Office of Technology Licensing or the Sponsored Projects Office is unclear). Sun did receive a patent on one aspect of Bechtolsheim's design (high-speed memory and memory management system, filed July 2, 1982, granted July 2, 1985, patent no. 4527232). The Office of Technology Licensing, which learned about the patent only after it had been granted, negotiated

an agreement with Sun "whereby Stanford would be provided sufficient rights only to be able to grant this license to the government." Brenda Whitmarsh [Office of Technology Licensing] to Mr. Robin Simpson, Office of Naval Research. Oct. 4, 1988, KK. Bill Osborn, interview by author, May 14, 2015 (source of the phrase "marginally patentable"); Kathy Ku, interview by author; Niels Reimers, interview by author. The original disclosure is in Forest Baskett, Andreas Bechtolsheim, Bill Nowicki, and John Seamons, "The Sun Workstation: A Terminal System for the Stanford University Network," March 30, 1980, KK.

6. Lavey, interview by author, Aug. 4, 2015.
7. Markoff, "Even Sun Microsystems Had Its Roots at Xerox PARC."
8. "Von Bechtolsheim: I Invested in Google to Solve My Own Problem," *Deutsche Welle*, http://www.dw.com/en/von-bechtolsheim-i-invested-in-google-to-solve-my-own-problem/a-4557608; "Andy Bechtolsheim: An Engineering Hero Talks Innovation, Success, and Engineering," http://www.fundinguniverse.com/company-histories/sun-microsystems-inc-history/.
9. Silicon Valley Indicators, Joint Venture Silicon Valley, http://www.joint venture.org/images/stories/pdf/index2013.pdf.
10. See, e.g., David Dayen, "The Android Administration: Google's Remarkably Close Relationship with the Obama White House," *The Intercept*, April 22, 2016.
11. Statistics are for Santa Clara County. Data for 1980: 13.5 percent of SCC population was foreign-born, vs. 6.2 percent of US population as a whole (Nativity and Language for Counties [Table 172], 1980 Census). Today 12.9 percent of the US population is foreign-born. http://www.bay areacensus.ca.gov/counties/SantaClaraCounty50.htm.
12. "Components of Population Change," Silicon Valley Indicators, Joint Venture Silicon Valley, http://siliconvalleyindicators.org/snapshot/. *Joint Venture Index 2016* reports that nearly 74 percent of "computer and mathematical workers ages 25–44" are foreign-born.
13. Stuart Anderson, "Immigrants and Billion Dollar Startups," NFAP Policy Brief, March 2016, National Foundation for American Policy, http://nfap.com/wp-content/uploads/2016/03/Immigrants-and-Billion-Dollar-Startups.NFAP-Policy-Brief.March-2016.pdf.

Postscript: The Troublemakers Today

1. Al Alcorn, CHM interview.

2. Zowie Intertainment was acquired by Lego in April 2000.

3. Martin Campbell-Kelly, *From Airline Reservations to Sonic the Hedgehog: A History of the Software Industry* (Cambridge, MA: MIT Press, 2003): 223.

4. Ibid.: 156.

5. ASK's tools and methodologies, developed for powerful machines with large memories, were ill suited to the personal computer market.

6. Steve Lohr, "Computer Associates to Buy ASK," *New York Times*, May 20, 1994.

7. For more on Echelon's start, see David Lane and Robert Maxfield, "Building a New Market System: Effective Action, Redirection and Generative Relationships," in *Complexity Perspectives in Innovation and Social Change*, ed. D. Lane et al. (Berlin: Springer Science + Business Media, 2009): 263–88.

8. "History of the Markkula Center for Applied Ethics," https://www.scu.edu/ethics/about-the-center/history/.

9. Mike Markkula, CHM interview.

10. Betrayed: John Markoff, "An 'Unknown' Co-Founder Leaves After 20 Years of Glory and Turmoil," *New York Times*, Sept. 1, 1997. Unethical: Markkula, interview by author, May 3, 2016.

11. "In some ways, he's the John McEnroe of American capitalism: arrogant, self-centered, and too wealthy for his own good." Robert J. Samuelson, "Steve Jobs and Apple Pie," *Newsweek*, Oct. 7, 1985.

12. Markkula's description of his internal reaction to Jobs's request as "Whoopee! No problem at all!" Markkula, interview by author, May 3, 2016.

13. John Sculley, quoted in Regis McKenna's notes on January 11, 1985, board meeting, reproduced in Regis McKenna, "'The Journey Is the Reward': My Thoughts on Steve Jobs and 35 Years of Being Inside/Outside Apple," unpublished manuscript, revision of Jan. 29, 2012, RM.

14. Regis McKenna, notebook entry, n.d., but likely April 10 or 11, 1985, RM.

15. Steve Jobs, quoted in Regis McKenna's notes on January 11, 1985, board meeting, reproduced in Regis McKenna, "'The Journey Is the Reward': My Thoughts on Steve Jobs and 35 Years of Being Inside/Outside Apple," unpublished manuscript, revision of Jan. 29, 2012, RM.

16. Regis McKenna, "The Wisdom and Inspiration of Steve Jobs," slide show shared with author, RM.

17. "OTL, Financial Data 1970–2016," KK.

18. Dave Merrill, Blacki Migliozzi, and Susan Decker, "Billions at Stake in University Patent Fights," Bloomberg.com, May 24, 2016.

19. Susan Decker, "Apple Told to Pay University $234 Million over Processor Patent," Bloomberg Technology, Oct. 16, 2015, https://www.bloomberg .com/news/articles/2015-10-16/apple-told-to-pay-university-234-mil lion-over-processor-patent.

20. Andrew Pollack, "Robert A. Swanson Dies at 52; Early Leader in Biotechnology," *New York Times*, Dec. 8, 1999.

21. Ibid.

22. David V. Goeddel and Arthur D. Levinson, "Obituary: Robert A. Swanson (1947–99)," *Nature* 403 (Jan. 20, 2000): 264.

23. Andrew Pollack, "Roche Agrees to Buy Genentech," *New York Times*, March 12, 2009.

24. Genentech had a 3-for-2 stock split in March 1983 and then 2-for-1 splits in 1986, 1987, 1999, 2000, and 2004. Roche paid $95 per share at the acquisition.

25. Bob Taylor to author, May 11, 2013. He pointed out that not only most awards but also the reward structure at universities is geared strictly to individuals. An individual gets tenure, for example, not a group or lab.

Index